Assimilation of Remote Sensing Data into Earth System Models

Assimilation of Remote Sensing Data into Earth System Models

Special Issue Editors

Jean-Christophe Calvet
Patricia De Rosnay
Stephen G. Penny

MDPI • Basel • Beijing • Wuhan • Barcelona • Belgrade

Special Issue Editors
Jean-Christophe Calvet
CNRM
France

Patricia De Rosnay
European Centre for Medium-Range Weather Forecasts UK

Stephen G. Penny
University of Colorado
National Oceanographic and Atmospheric Administration (NOAA)
USA

Editorial Office
MDPI
St. Alban-Anlage 66
4052 Basel, Switzerland
This is a reprint of articles from the Special Issue published online in the open access journal *Remote Sensing* (ISSN 2072-4292) from 2018 to 2019 (available at: https://www.mdpi.com/journal/remotesensing/special_issues/dataassimilation_rs)

For citation purposes, cite each article independently as indicated on the article page online and as indicated below:

LastName, A.A.; LastName, B.B.; LastName, C.C. Article Title. *Journal Name* **Year**, *Article Number*, Page Range.

ISBN 978-3-03921-640-6 (Pbk)
ISBN 978-3-03921-641-3 (PDF)

Contents

About the Special Issue Editors

Jean-Christophe Calvet joined Centre National de Recherches Météorologiques, Toulouse, in 1994, where he has been Head of a Land Surface. His background is in land surface modeling, microwave remote sensing, and data assimilation. His research interests include land–atmosphere exchange modeling and the use of remote sensing over land surfaces for meteorology. His most recent works concern the joint analysis of soil moisture and vegetation biomass using data assimilation techniques.

Patricia De Rosnay is leading the Coupled Assimilation Team of the European Centre for Medium-Range Weather Forecasts (ECMWF). Her team is in charge of the development of coupled Earth system assimilation, land surface assimilation, and ocean assimilation in ECMWF NWP systems. She received her Ph.D. degree in climate modeling from University Pierre et Marie Curie (France) in 1999. She worked on land surface and climate modeling at LMD/IPSL (Laboratoire de Météorologie Dynamique/Institut Pierre Simon Laplace) from 1994 to 2002, and on land surface remote sensing at the Centre d'Etudes Spatiales de la Biosphère (CESBIO) until 2007. She has been at ECMWF since 2007. She is member of the SMOS Quality Working Group, the H-SAF project Team, the SRNWP (Short-Range Numerical Weather Prediction) surface expert team. She is also member of the Steering Group of Global Cryosphere Watch, a program of the World Meteorological Organization (WMO) and co-chair of its Snow Watch Team.

Stephen G. Penny currently works in the Physical Sciences Division (PSD) at the Earth System Research Laboratory (ESRL) of the National Oceanographic and Atmospheric Administration (NOAA), via the Cooperative Institute for Research in Environmental Sciences (CIRES) at the University of Colorado. His current research focuses on coupled data assimilation (CDA) and machine learning for improving subseasonal-to-seasonal (S2S) prediction. Dr. Penny was previously a research professor at the University of Maryland, where he developed the Hybrid Global Ocean Data Assimilation System (Hybrid-GODAS) that is currently in use by the Climate Prediction Center (CPC) at the National Centers for Environmental Prediction (NCEP).

remote sensing

Editorial

Editorial for the Special Issue "Assimilation of Remote Sensing Data into Earth System Models"

Jean-Christophe Calvet [1,*], Patricia de Rosnay [2,*] and Stephen G. Penny [3,4,*]

1 CNRM, Université de Toulouse, Météo-France, CNRS, 31057 Toulouse, France
2 European Centre for Medium-Range Weather Forecasts (ECMWF), Reading RG2 9AX, UK
3 Cooperative Institute for Research in Environmental Sciences (CIRES), University of Colorado, Boulder, CO 80309, USA
4 Physical Sciences Division (PSD), Earth System Research Laboratory (ESRL), National Oceanic and Atmospheric Administration (NOAA), Boulder, CO 80305-3328, USA
* Correspondence: jean-christophe.calvet@meteo.fr (J.-C.C.); patricia.rosnay@ecmwf.int (P.d.R.); steve.penny@noaa.gov (S.G.P.)

Received: 17 September 2019; Accepted: 18 September 2019; Published: 19 September 2019

Abstract: This Special Issue is a collection of papers reporting research on various aspects of coupled data assimilation in Earth system models. It includes contributions presenting recent progress in ocean–atmosphere, land–atmosphere, and soil–vegetation data assimilation.

Keywords: data assimilation; Earth system models; atmospheric models; ocean models; land surface models

1. Introduction

A transition is currently occurring in multiple fields in the Earth sciences towards an integrated Earth system approach, with applications including numerical weather prediction, hydrological forecasting, climate impact studies, ocean dynamics estimation and monitoring, carbon cycle monitoring. These approaches rely on coupled modeling techniques, using Earth system models (ESMs) that account for an increased level of complexity of (coupled) processes and interactions between atmosphere, ocean, sea ice, and terrestrial surfaces [1]. A crucial component of Earth system approaches is the development of coupled data assimilation (CDA) of satellite observations to ensure consistent initialization at the interface between the different subsystems [2]. For example, a coupled ocean–atmosphere data assimilation system ensures consistent sea surface temperature and near-surface atmospheric conditions [3], and coupled land–atmosphere assimilation produces consistent soil moisture and air temperature analyses [4].

There is a large range of CDA approaches, from weakly coupled (coupled forecast model but separate analyses) to strongly coupled assimilation (single cost function and control vector). Intermediate levels of coupling (quasi-CDA) allow observations in one subsystem to provide increments in other subsystems [5]. CDA development in ESMs will open possibilities to further exploit satellite observations that are sensitive to both the lowest levels of the atmosphere and the underlying system (land, urban surfaces, ocean, or sea ice).

The integration of satellite-derived observations into ESMs or into ESM modules can also help minimize modeling uncertainties [6,7]. The assimilation of new remote sensing products is expected to benefit a wide range of applications, including weather, subseasonal to seasonal (S2S), seasonal and interannual climate prediction, and climate reanalysis [8,9]. Satellite-derived climate data records of essential climate variables are now available for the different components of the Earth system, including terrestrial and ocean surfaces.

2. Overview of Contributions

The contributions reported in this Special Issue include key aspects of CDA involving several components of ESMs: ocean–atmosphere interactions, land–atmosphere interactions including hydrological processes, and interactions within the soil–plant system.

2.1. Ocean–Atmosphere Data Assimilation

In this Special Issue, state-of-the-art developments in operational coupled ocean–atmosphere developments are presented by Browne et al. [10]. Recent advances in Earth system components assimilation are presented, including (1) assimilation of Global Positioning System (GPS) radio occultation (RO) in atmospheric models (Banos et al. [11]) as well as (2) sea level interpolation using an analog data assimilation approach to improve high resolution current representation in ocean general circulation models (Lguensat et al. [12]). Assimilation of observations from the Tropical Rainfall Measuring Mission (TRMM) and the Integrated Multi-satellitE Retrievals for Global Precipitation Measurement (GPM IMERG) is also investigated by Yi et al. [13]. A consistent benefit of atmospheric assimilation is shown on both atmosphere and hydrological components.

2.2. Land–Atmosphere Data Assimilation

This Special Issue also addresses the relationship between soil moisture and different land surface–atmosphere fields (precipitation, surface air temperature, total cloud cover, and total water storage) and Pangaluru et al. [14] show that assimilation of Advanced Microwave Scanning Radiometer for Earth Observing System (AMSR-E) over India improves their consistency. Furthermore, Yi et al. [15] compare the use of different precipitation products in the hydrological modeling of the Wangjiaba (WJB) watershed in China with the Soil Moisture Active and Passive (SMAP) data used to validate soil moisture. They show that although in situ precipitation reports provide the most reliable local information, precipitation from numerical weather prediction models provides the gridded and future information necessary for flood forecasting. With the paper of Massari et al. [16], this Special Issue further studies the strong physical connection between soil moisture dynamics and rainfall. This work shows that in Mediterranean areas, correction of precipitation is most relevant for high flow representation, whereas soil moisture assimilation brings slightly more benefit in low flow conditions.

2.3. Soil–Vegetation Data Assimilation

Drought propagation from soil moisture to vegetation dynamics is investigated by Sawada [17] using a newly developed eco-hydrological land reanalysis. Results from Leroux et al. [18] show a positive impact of the joint assimilation of leaf area index (LAI) and surface soil moisture in a global Land Data Assimilation System (LDAS-Monde) over the Euro-Mediterranean area. Vegetation sun-induced fluorescence (SIF) is used as an independent observational system to validate the added value of the assimilation. The work of Albergel et al. [19] published in this Special Issue confirms the positive impact of soil moisture and LAI joint assimilation over the contiguous United States. They point out that soil moisture and LAI satellite observations assimilated in LDAS-Monde for reanalysis purposes have the potential to be used to monitor extreme events such as agricultural droughts.

3. Conclusions

Going towards strongly coupled data assimilation involving all Earth system components is a subject of active research. This Special Issue shows that a lot of progress is being made in the ocean–atmosphere domain, but also over land. As atmospheric models now tend to address subkilometric scales, assimilating high spatial resolution satellite data into the land surface models used in atmospheric models is critical. This evolution is also challenging for hydrological modeling.

Author Contributions: The three authors contributed equally to all aspects of this editorial.

Acknowledgments: The Guest Editors would like to thank the authors who contributed to this Special Issue and the reviewers who dedicated their time and provided the authors with valuable and constructive recommendations. They would also like to thank the editorial team of Remote Sensing for their support.

Conflicts of Interest: The authors declare no conflict of interest.

References

1. Balsamo, G.; Agustì-Panareda, A.; Albergel, C.; Arduini, G.; Beljaars, A.; Bidlot, J.; Bousserez, N.; Boussetta, S.; Brown, A.; Buizza, R.; et al. Satellite and in situ observations for advancing global Earth surface modelling: A review. *Remote Sens.* **2018**, *10*, 2038. [CrossRef]
2. Mulholland, D.P.; Laloyaux, P.; Haines, K.; Alonso Balmaseda, M. Origin and impact of initialization shocks in coupled atmosphere-ocean forecasts. *Mon. Weather Rev.* **2015**, *143*, 4631–4644. [CrossRef]
3. Laloyaux, P.; Balmaseda, M.; Dee, D.; Mogensen, K.; Janssen, P. A coupled data assimilation system for climate reanalysis. *Q. J. R. Meteorol. Soc.* **2016**, *142*, 65–78. [CrossRef]
4. De Rosnay, P.; Balsamo, G.; Albergel, C.; Muñoz-Sabater, J.; Isaksen, L. Initialisation of land surface variables for numerical weather prediction. *Surv. Geophys.* **2014**, *35*, 607–621. [CrossRef]
5. Penny, S.G.; Akella, S.; Alves, O.; Bishop, C.; Buehner, M.; Chevallier, M.; Counillon, F.; Draper, C.; Frolov, S.; Fujii, Y.; et al. *Coupled Data Assimilation for Integrated Earth System Analysis and Prediction: Goals, Challenges and Recommendations*; Technical Report; World Meteorological Organisation: Geneva, Switzerland, 2017.
6. Reichle, R.H.; Koster, R.D.; Liu, P.; Mahanama, S.P.P.; Njoku, E.G.; Owe, M. Comparison and assimilation of global soil moisture retrievals from the Advanced Microwave Scanning Radiometer for the Earth Observing System (AMSR-E) and the Scanning Multichannel Microwave Radiometer (SMMR). *J. Geophys. Res.* **2007**, 112. [CrossRef]
7. Barbu, A.L.; Calvet, J.-C.; Mahfouf, J.-F.; Albergel, C.; Lafont, S. Assimilation of Soil Wetness Index and Leaf Area Index into the ISBA-A-gs land surface model: Grassland case study. *Biogeosciences* **2011**, *8*, 1971–1986. [CrossRef]
8. Dee, D.P. The ERA-Interim reanalysis: Configuration and performance of the data assimilation system. *Q. J. R. Meteorol. Soc.* **2011**, *137*, 553–597. [CrossRef]
9. Hersbach, H.; de Rosnay, P.; Bell, B.; Schepers, D.; Simmons, A.; Soci, C.; Abdalla, S.; Alonso-Balmaseda, M.; Balsamo, G.; Bechtold, P.; et al. Operational global reanalysis: Progress, future directions and synergies with NWP. *ERA Report Series* **2018**, *27*, 65.
10. Browne, P.A.; de Rosnay, P.; Zuo, H.; Bennett, A.; Dawson, A. Weakly Coupled Ocean–Atmosphere Data Assimilation in the ECMWF NWP System. *Remote Sens.* **2019**, *11*, 234. [CrossRef]
11. Banos, I.H.; Sapucci, L.F.; Cucurull, L.; Bastarz, C.F.; Silveira, B.B. Assimilation of GPSRO Bending Angle Profiles into the Brazilian Global Atmospheric Model. *Remote Sens.* **2019**, *11*, 256. [CrossRef]
12. Lguensat, R.; Viet, P.H.; Sun, M.; Chen, G.; Fenglin, T.; Chapron, B.; Fablet, R. Data-Driven Interpolation of Sea Level Anomalies Using Analog Data Assimilation. *Remote Sens.* **2019**, *11*, 858. [CrossRef]
13. Yi, L.; Zhang, W.; Wang, K. Evaluation of Heavy Precipitation Simulated by the WRF Model Using 4D-Var Data Assimilation with TRMM 3B42 and GPM IMERG over the Huaihe River Basin, China. *Remote Sens.* **2018**, *10*, 646. [CrossRef]
14. Pangaluru, K.; Velicogna, I.; Mohajerani, Y.; Ciracì, E.; Charakola, S.; Basha, G.; Rao, S.V.B. Soil Moisture Variability in India: Relationship of Land Surface–Atmosphere Fields Using Maximum Covariance Analysis. *Remote Sens.* **2019**, *11*, 335. [CrossRef]
15. Yi, L.; Zhang, W.; Li, X. Assessing Hydrological Modelling Driven by Different Precipitation Datasets via the SMAP Soil Moisture Product and Gauged Streamflow Data. *Remote Sens.* **2018**, *10*, 1872. [CrossRef]
16. Massari, C.; Camici, S.; Ciabatta, L.; Brocca, L. Exploiting Satellite-Based Surface Soil Moisture for Flood Forecasting in the Mediterranean Area: State Update Versus Rainfall Correction. *Remote Sens.* **2018**, *10*, 292. [CrossRef]
17. Sawada, Y. Quantifying Drought Propagation from Soil Moisture to Vegetation Dynamics Using a Newly Developed Ecohydrological Land Reanalysis. *Remote Sens.* **2018**, *10*, 1197. [CrossRef]

18. Leroux, D.J.; Calvet, J.-C.; Munier, S.; Albergel, C. Using satellite-derived vegetation products to evaluate LDAS-Monde over the Euro-Mediterranean area. *Remote Sens.* **2018**, *10*, 1199. [CrossRef]
19. Albergel, C.; Munier, S.; Bocher, A.; Bonan, B.; Zheng, Y.; Draper, C.; Leroux, D.J.; Calvet, J.-C. LDAS-Monde Sequential Assimilation of Satellite Derived Observations Applied to the Contiguous US: An ERA-5 Driven Reanalysis of the Land Surface Variables. *Remote Sens.* **2018**, *10*, 1627. [CrossRef]

remote sensing

MDPI

Article

Weakly Coupled Ocean–Atmosphere Data Assimilation in the ECMWF NWP System

Philip A. Browne *, Patricia de Rosnay, Hao Zuo, Andrew Bennett and Andrew Dawson

European Centre for Medium-Range Weather Forecasts (ECMWF), Shinfield Road, Reading RG2 9AX, UK; patricia.rosnay@ecmwf.int (P.d.R.); hao.zuo@ecmwf.int (H.Z.); andrew.bennett@ecmwf.int (A.B.); andrew.dawson@ecmwf.int (A.D.)

* Correspondence: p.browne@ecmwf.int

Received: 19 December 2018; Accepted: 19 January 2019; Published: 23 January 2019

Abstract: Numerical weather prediction models are including an increasing number of components of the Earth system. In particular, every forecast now issued by the European Centre for Medium-Range Weather Forecasts (ECMWF) runs with a 3D ocean model and a sea ice model below the atmosphere. Initialisation of different components using different methods and on different timescales can lead to inconsistencies when they are combined in the full system. Historically, the methods for initialising the ocean and the atmosphere have been typically developed separately. This paper describes an approach for combining the existing ocean and atmospheric analyses into what we categorise as a weakly coupled assimilation scheme. Here, we show the performance improvements achieved for the atmosphere by having a weakly coupled ocean–atmosphere assimilation system compared with an uncoupled system. Using numerical weather prediction diagnostics, we show that forecast errors are decreased compared with forecasts initialised from an uncoupled analysis. Further, a detailed investigation into spatial coverage of sea ice concentration in the Baltic Sea shows that a much more realistic structure is obtained by the weakly coupled analysis. By introducing the weakly coupled ocean–atmosphere analysis, the ocean analysis becomes a critical part of the numerical weather prediction system and provides a platform from which to build ever stronger forms of analysis coupling.

Keywords: ocean–atmosphere assimilation; weakly coupled data assimilation; numerical weather prediction

---

1. Introduction

As of June 2018, the European Centre for Medium-Range Weather Forecasts (ECMWF) has a coupled forecasting system for all timescales; every forecast from the high-resolution 10-day deterministic forecasts, the ensemble forecasts and the monthly forecasts to the seasonal prediction system runs an Earth system model. Specifically, this means that there is a three-dimensional ocean model and a sea ice model that runs coupled to the atmosphere, wave and land surface components. Such a multi-component Earth system model needs initialisation of each of its components, and this manuscript is concerned with how the ocean and atmosphere are initialised together for the purposes of numerical weather prediction (NWP).

In order to advance numerical weather prediction, ECMWF is developing its modelling and its data assimilation toward an Earth system approach [1]. When forecasts of medium-range to longer range are the focus, components of the Earth system that are typically slower than the atmosphere become more important. This is both in terms of their presence in the model and an accurate specification of their initial conditions [2]. Such components include not only the ocean and sea ice but also the land surface, waves, and aerosols, as well as their interactions with each other and the atmosphere [3]. Figure 1 represents the various components present in the system.

The land surface and waves are fully established components of the ECMWF systems. Aerosols are treated separately within the Integrated Forecasting System (IFS) as part of the Copernicus Atmosphere Monitoring Service (CAMS). The ocean has been used in the IFS for seasonal applications since 1997 [4], for monthly forecast since 2002 [5] and in the Ensemble Forecasts (ENS) from the initial time step since 2013 [6,7]. In late 2016, an interactive sea ice model was added to the ENS. Only now are these components starting to interact with the atmospheric analyses.

Figure 1. Components of the European Centre for Medium-Range Weather Forecasts' (ECMWF's) Integrated Forecasting System (IFS) Earth system. Along with the atmosphere, there are the ocean, wave, sea ice, land surface and lake models.

In order to make the most of the Earth system approach in a forecast, the components should be somehow consistent with one another. If the different components are not internally consistent, they are sometimes referred to as *unbalanced*. This lack of balance can lead to fast adjustments in the system in the initial stages of the forecast in a phenomenon known as *initialisation shock* [8]. Initialisation shock can be reduced by initialising the various components together via coupled data assimilation [9].

Much of the literature on coupled assimilation has focused on the initialisation of forecasts for seasonal to decadal timescales. For example, the Japan Agency for Marine-Earth Science and Technology (JAMSTEC) has a fully coupled four-dimensional variational data assimilation (4D-Var) system used for experimental seasonal and decadal predictions [10–12]. The National Oceanic and Atmospheric Administration Geophysical Fluid Dynamics Laboratory (NOAA/GFDL) has a coupled assimilation system based on the ensemble Kalman filter (specifically, the ensemble adjustment Kalman filter) to initialise decadal predictions [13,14]. The NOAA National Centers for Environmental Prediction (NOAA/NCEP) has a coupled assimilation system [15] for subseasonal and seasonal predictions, as well as reanalysis [16]. A prototype system built in March 2016 in the Japan Meteorological Agency Meteorological Research Institute (JMA/MRI) was designed to replace the ocean-only observation assimilation approach. The atmosphere component is updated every 6 h by 4D-Var with a TL159L100 uncoupled inner loop model, while the ocean component runs on a 10-day cycle using 3D-Var with an incremental analysis update. Experimentation with this system for coupled reanalysis and NWP is underway [17].

The U.S. Naval Research Laboratory (NRL), having a different focus from most centres, runs a coupled model, with most resources dedicated to the ocean component. Its global coupled model goes up to an ocean resolution of $1/25^{\circ}$ and is initialised by separate assimilation systems. In the near future, the organisation plans to implement the interface solver of Frolov et al. (2016) [18] to allow more coupling within the analysis.

Environment and Climate Change Canada (ECCC) has recently begun producing its global deterministic NWP forecasts using a coupled ocean–atmosphere model. This model is currently initialised separately in each of its components. The UK Met Office has a system to initialise global coupled NWP forecasts where the background for each component in a 6 hour assimilation window comes from the coupled model [19], although this is not yet operational.

Under the ERA-CLIM2 project, ECMWF has piloted techniques for coupled ocean–atmosphere data assimilation that were applied in the context of reanalysis. These are the Coupled European Reanalysis of the 20th century (CERA-20C) [20] and the CERA-SAT [21] reanalysis using the modern-day satellite observation system. The assimilation method developed for CERA involved "outer loop" coupling within the 4D-Var algorithm of the atmosphere and the ocean. This method has a high level of coupling in the analysis, which, as Figure 2 shows, can mean that the whole NWP system can be degraded by model biases in the ocean component of the coupled model.

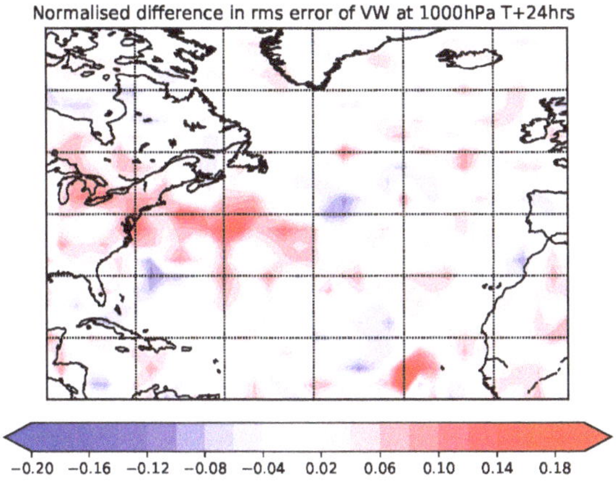

Figure 2. Impact of first implementation of outer loop coupling (quasi strongly coupled data assimilation) on high-resolution global numerical weather prediction (NWP). The presence of model bias in the western boundary currents of the ocean is evident as degradations (red) to the 24 h forecast scores of vector winds (VW) at 1000 hPa. Results were obtained from IFS cycle 45R1 at a resolution of 25 km (TCo399) with a 0.25° ocean and are based on global outer loop coupling over the period from 1 June 2017 to 2 July 2017. See Section 4 for details of the error diagnostic.

As the first steps toward coupled ocean–atmosphere data assimilation for NWP at ECMWF, we have chosen to adopt a weaker form of coupled assimilation than in the CERA system.

Following a World Meteorological Organisation (WMO) meeting on coupled assimilation, Penny et al. (2017) [17] defined *weakly* and *strongly* coupled data assimilation (and variations thereof). Their definitions were as follows:

- "Quasi Weakly Coupled DA (QWCDA): assimilation is applied independently to each of a subset of components of the coupled model. The result may be used to initialize a coupled forecast."
- "Weakly Coupled DA (WCDA): assimilation is applied to each of the components of the coupled model independently, while interaction between the components is provided by the coupled forecasting system."
- "Quasi Strongly Coupled DA (QSCDA): observations are assimilated from a subset of components of the coupled system. The observations are permitted to influence other components during the analysis phase, but the coupled system is not necessarily treated as a single integrated system at all stages of the process."

- "Strongly Coupled DA (SCDA): assimilation is applied to the full Earth system state simultaneously, treating the coupled system as one single integrated system. In most modern DA systems this would require a cross-domain error covariance matrix be defined."

Hence, QWCDA might be thought of as uncoupled assimilation to initialise a coupled model. Observations in one component never influence the analysis of the other component. In WCDA, an observation of one component is not able to directly influence the analysis of the other component in the valid assimilation window. However, as a coupled forecast is used, the observational information gets propagated to the background used for subsequent analysis cycles; hence, there is a lag by which observations can influence different components. The CERA system falls under the QSCDA category, where observations from each component can influence the analysis of the other within a single analysis window. SCDA is simply treating a coupled system as a multivariate assimilation problem, and no special terminology or mathematical analysis is necessary.

In this paper, we introduce a form of weakly coupled data assimilation which allows for the different timescales in the ocean and atmospheric analysis windows. The atmosphere and the ocean are coupled implicitly at a frequency of 24 h, determined by the frequency of the slowest component to update.

The remainder of this paper is organised as follows. In Section 2, we describe the various components of the IFS and describe both uncoupled and weakly coupled ocean–atmosphere assimilation strategies. The experimental design is described in Section 3. Section 4 shows and discusses the experimental results and gives a detailed examination of local impacts to sea ice. Finally, in Section 5, we look to future developments of the weakly coupled data assimilation system at ECMWF.

2. IFS

The ECMWF Integrated Forecasting System consists of multiple components. The main component for it all is the upper atmospheric analysis. The dynamical model which propagates the analysis from one cycle to the next contains a limited number of components: the *atmospheric model* [22,23], the *land* model [24], the *lake* model [25] and the *wave* model [26]. The atmosphere is represented on a 3D reduced Gaussian grid, and its analysis is deduced by using 4D-Var in incremental form [27]. A number of outer loops are used, and the minimisation is performed at increasingly high resolution. The number of outer loops and resolution of the inner loops are dependent on the resolution of the nonlinear model. The land data assimilation component is weakly coupled to the atmosphere. They share the same model to produce the first guess, and the state of the land surface and the state of the atmosphere are modified separately [28]. Subsequent forecasts are initialised using the latest analysis of the atmosphere and the land surface. This is archetypal *weakly coupled* assimilation, as defined previously. Currently, the land surface has various components. The *snow analysis* is performed using two-dimensional optimal interpolation (2D-OI), as is the *soil temperature* analysis. *Soil moisture* is analysed using a simplified extended Kalman filter (SEKF). Similar to the land, the *wave analysis* is weakly coupled to the atmosphere and uses 2D-OI. However, the first guess used for the wave analysis is not the same first guess that is used for the atmosphere. The first guess is the nonlinear trajectory of one of the outer loops of the atmospheric 4D-Var. Currently, the final trajectory is used. This means that, in a given cycle, observations of the atmosphere update the surface wind fields and thus will influence the wave analysis in that given cycle. The opposite is not true: wave observations will not modify the atmospheric state during that cycle. These observations will only modify the atmospheric state at the subsequent cycles due to the interactions in the forecasts that cycle the analysis.

The model that cycles the analysis does not contain a dynamical ocean model or sea ice model. For the purposes of this paper, we say it is *uncoupled*, referring to ocean–atmosphere interactions. The lower boundary of the atmosphere needs to be supplied; the sea-surface temperature (SST) field and sea ice concentration (CI) field are required.

2.1. Observations

Over 40 million observations are processed and used daily, with the vast majority of these coming from satellites. These include polar orbiting and geostationary, infrared and microwave imagers, scatterometers, altimeters, and GPS radio occultations [29]. In addition to the satellite observations, there are in situ observations used from aircraft, radiosondes and dropsondes, as well as observations from ships, buoys, land-based stations and radar-derived rainfall [30].

For the sea surface, L4 gridded products are used to give global coverage of sea-surface temperatures and sea ice concentrations. The L4 product used is the Operational Sea Surface Temperature and Sea Ice Analysis (OSTIA) [31], a 0.05° resolution dataset that is solely observation based. For its SST product, OSTIA combines satellite data from the Group for High Resolution Sea Surface Temperature (GHRSST) and in situ observations to produce a daily analysed field of foundation sea-surface temperature. Sea ice concentration fields in OSTIA are derived from the EUMETSAT Ocean Sea Ice Satellite Application Facility (OSI SAF) L3 OSI-401-b observations of sea ice concentration [32]. Lake ice concentration observations that are outside of the domain of OSI-401-b are taken from an NCEP sea ice concentration product [33].

2.2. 4D-Var, HRES and the EDA

The above observations are assimilated with the 4D-Var methodology (see, e.g., Rabier et al., 2000 [27]), which uses, amongst other details, hybrid-B and a weak constraint term. A single high-resolution (HRES) analysis and forecast are produced. The flow-dependent component of the background error covariance matrix B comes from an Ensemble of Data Assimilations (EDA) that solves similar 4D-Var problems but at a lower resolution and with stochastically perturbed observations [34]. The EDA currently runs with 25 members. The HRES analysis is performed twice daily over a 12 hour analysis window from 2100Z (0900Z) to 0900Z (2100Z). The 4D-Var is solved in incremental form with three outer loops, such that each inner loop minimisation is performed on a lower-resolution grid. From each analysis, a 10-day coupled ocean–atmosphere forecast is produced. For more details on the configuration, see Haseler (2004) [35].

2.3. OCEAN5

OCEAN5 is a reanalysis–analysis system with two streams—behind real-time (BRT) and real-time (RT) [36]. The three-dimensional ocean Nucleus for European Modelling of the Ocean (NEMO) model and the Louvain-la-Neuve 2 (LIM2) sea ice model are coupled and used as the model within OCEAN5. The OCEAN5 analysis is initialised from a behind-real-time ocean and sea ice coupled reanalysis, known as ORAS5 [37] (see purple boxes in Figure 3). The variables temperature, salinity, and horizontal currents (T, S, U, V) are analysed using the 3D-Var First Guess at Appropriate Time (FGAT) assimilation technique. The length of the assimilation window varies from 8 to 12 days and is split into two *chunks* (see blue boxes in Figure 3), the first of which is 5 days long. In parallel, a separate minimisation is performed to analyse sea ice concentration using the same 3D-Var FGAT method.

Observations that are assimilated currently are in situ profiles of temperature and salinity, and satellite-derived sea level anomaly and sea ice concentration observations. For SST, a relaxation is performed toward the OSTIA operational SST product in the OCEAN5 RT analysis.

The ocean and sea ice analysis system requires forcing fields in the form of a surface wind field, surface temperature and humidity fields, as well as surface fluxes. These come from the HRES analysis (and forecast for the final day of the OCEAN5 assimilation window). The surface fluxes consist of downward solar radiation, thermal radiation downwards, total precipitation, and snowfall. From the wave model, the ocean requires forcing fields of significant wave height, mean wave period, coefficient of drag with waves, 10 metre neutral windspeed, normalized energy flux into the ocean, normalized wave stress into the ocean, and Stokes drift. A full description is given in Zuo et al. (2018) [37].

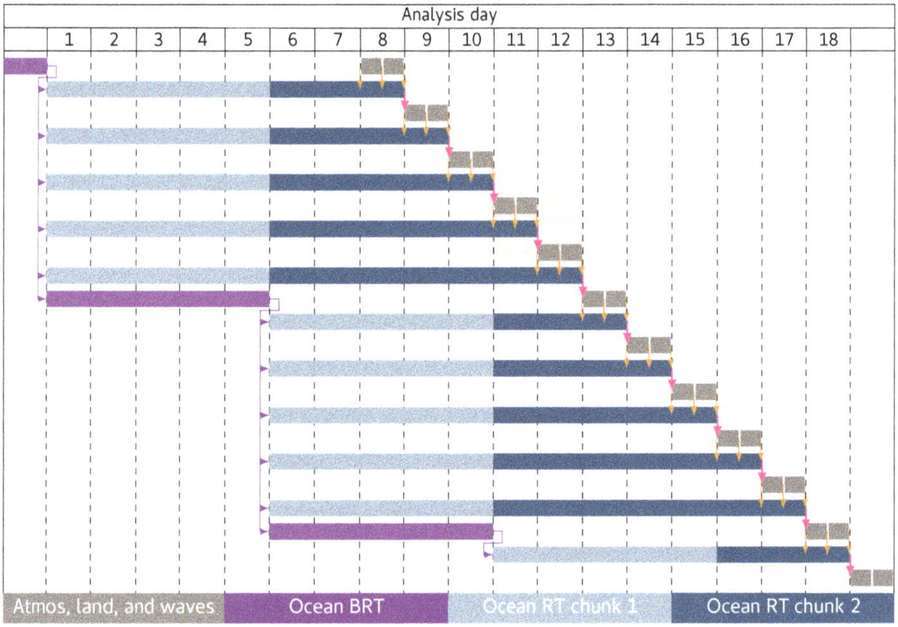

Figure 3. Weakly coupled assimilation system information flow. Horizontal bars represent the analysis window for the various different components of the system. This is a simplified plot ignoring a 3 h offset of the systems. Orange arrows show the existing transfer of forcing from the atmosphere to the ocean. Magenta arrows show the addition using the OCEAN5 fields as the lower-boundary condition for the atmospheric analysis, thus forming the weakly coupled data assimilation (WCDA) system. The highlighted region is discussed in an example in the text.

2.4. Uncoupled Approach/Workflow

From the above description, we can see that the HRES system can stand alone. It does not require any information from the OCEAN5 analysis. OCEAN5, on the other hand, requires forcing fields from an atmospheric analysis to operate.

Under this system, observations in the atmosphere will modify the atmospheric state. This change in atmospheric state will lead to a change in the forcing fields by which the ocean analysis is driven. This will lead to a change in the ocean analysis.

Observations of the ocean (unused by OSTIA, such as observations of currents) will not modify the atmospheric state, as no information from the ocean model is propagated back to the atmosphere. This system as a whole can be thought of as a *"one-way"* coupled assimilation system. The flow of information from the atmosphere to the ocean is depicted in the diagram in Figure 3 by orange arrows.

2.5. WCDA

We have seen that the atmospheric analysis requires the provision of an SST field and a sea ice field for use as its lower boundary condition. Similarly, the ocean analysis requires a set of atmospheric forcing fields to drive the ocean-only analysis.

To form a weakly coupled ocean–atmosphere data assimilation system, fields from the OCEAN5 analysis are used as the lower-boundary conditions for the atmospheric analysis over the ocean, rather than taking fields directly from the external OSTIA product. This will mean that observations of the ocean and the sea ice which previously would only influence the ocean analyses will also modify the atmospheric analysis via the lower-boundary conditions. The effect is not realized within a given assimilation cycle, but it is delayed.

Figure 3 shows the information flow between atmosphere and ocean. Consider an atmospheric observation on day 11 (within the highlighted region). This will change the analysis fields of the atmosphere on day 11, which will lead to the forcing fields that drive the ocean to change. Hence, this observation will have an effect on the latest ocean analysis valid at the start of day 12 (which, in this diagram, has an 11-day window from day 1 to day 12).

Now, instead, consider an ocean or sea ice observation on day 11. This directly changes the ocean/sea ice analysis for that day, but, as there is no feedback to the atmosphere until the end of the window, the atmospheric analysis for day 11 is unchanged. The impact of that ocean or sea ice observation is only detected by the atmosphere on day 12, when the updated sea ice analysis is seen as the lower-boundary condition for the atmosphere (magenta arrow in Figure 3).

Recall the definition of Weakly Coupled DA (WCDA): assimilation is applied to each of the components of the coupled model independently, while interaction between the components is provided by the coupled forecasting system. Clearly, the assimilation is applied independently to the atmosphere and the ocean/sea ice. Interaction between the components is not provided by a coupled *model* (i.e., a single parallel task on the supercomputer) but by the coupled *forecasting system*. That is, over a 24 hour period, the forecasting system passes forcings from the ocean to the atmosphere, and lower-boundary conditions are passed from the ocean/sea ice to the atmosphere. Hence, we categorise this as a weakly coupled data assimilation system for the ocean–atmosphere interaction.

Partial Coupling

The analysis of SST and sea ice concentration from OCEAN5 may not always be better than the OSTIA product. In particular, there are known deficiencies in the OCEAN5 analysis that can lead to degradations in forecast performance. For example, in the extratropics, the position of western boundary currents, such as the Gulf Stream, are known to be less accurate in the OCEAN5 analyses compared with OSTIA [37]. Hence, for WCDA, as with the model, flexibility has been developed so that ocean fields can be taken only over specific regions and not globally.

The sub-optimality of the ocean analyses are due to a well-known model bias in the ocean model which has been recognised already in the coupled model used to produce the 10-day forecasts [38–41]. The solution to this problem has been to use, for the model, a "partial coupling" approach, where the tendencies, rather than the absolute values, of SST are passed to the atmosphere. Partial coupling is required at latitudes (>25°) where the ocean model is unable to resolve eddies. Partial coupling can be described by the following equations.

$$SST_{\mathrm{IFS}}(t) = SST_{\mathrm{NEMO}}(t) + \alpha(t)\left(SST_{\mathrm{REF}}(0) - SST_{\mathrm{NEMO}}(0)\right) \quad (1a)$$

where $\alpha(t)$ is a function of lead time, with $\alpha(0) = 1$ decreasing to 0 by the end of the forecast. The reference field at initial time, $SST_{\mathrm{REF}}(0)$, is given by

$$SST_{\mathrm{REF}}(0) = \beta SST_{\mathrm{OSTIA}}(0) + (1-\beta) SST_{\mathrm{NEMO}}(0), \quad (1b)$$

where β is spatially varying and is depicted in Figure 4.

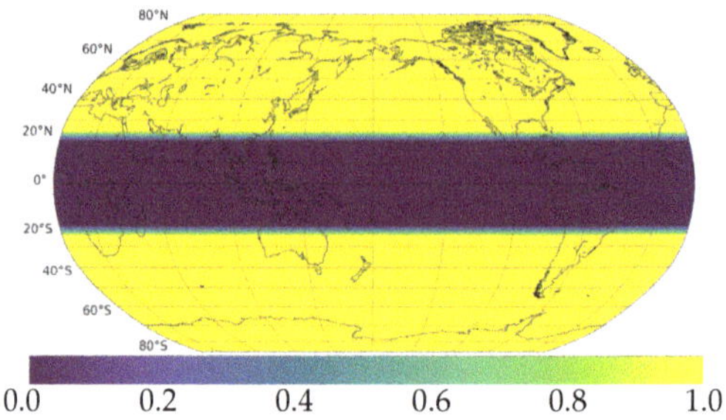

Figure 4. β coefficient from Equation (1b) indicating that initial sea-surface temperature (SST) is taken from the ocean analysis in the tropics (from 20° S to 20° N) and from Operational Sea Surface Temperature and Sea Ice Analysis (OSTIA) in the extratropics. The 5° transition regions can be seen as the bands consisting of intermediate colours.

Equation (1) evaluated at $t = 0$ gives an initial SST field of

$$SST_{\text{IFS}}(0) = \beta SST_{\text{OSTIA}}(0) + (1 - \beta) SST_{\text{NEMO}}(0) \tag{2}$$

which is the SST from NEMO in the tropics and the SST from OSTIA in the extratropics. For WCDA, we have therefore chosen to align the SST in the analysis with that used to initialise the forecast, i.e., following Equation (2).

3. Experimental Setup

A set of two experiments were conducted: the first is a control which does not use weakly coupled assimilation, and the second is an experiment with weakly coupled assimilation. Both experiments are based on IFS cycle 45R1. They were run at a 9 km global resolution (TCo1279), and both used the same uncoupled atmospheric EDA. The time period of the investigation is from 9 June 2017 to 21 May 2018. From each analysis, a *10-day coupled ocean–atmosphere forecast* is produced.

3.1. Control—Uncoupled Assimilation

The initial conditions for the ocean component were taken from an ocean-only analysis, which takes its forcing fields from the operational HRES system, i.e., the same resolution but atmosphere only and driven by OSTIA boundary conditions. The atmospheric analysis used OSTIA SST and CI fields globally as its lower-boundary conditions.

3.2. Experiment—WCDA

An ocean analysis was run alongside the atmospheric analysis. The sea ice concentration field used for the lower boundary of the atmospheric analysis comes from the ocean analysis. Similarly, the sea-surface temperature field comes from the ocean analysis, although this is restricted to the tropics only, as described in Equation (2) and Figure 4. That is, the SST comes from the ocean analysis between 20° S and 20° N, OSTIA outside of 25° N(S), and a linear interpolation of the two in the 5° band from 20° N(S) to 25° N(S).

The atmospheric analysis is otherwise identical to the control run. Similarly, except for the forcing fields coming from the atmospheric analysis rather than the operational HRES system, the ocean analysis is identical to the ocean analysis used in the control experiment. A comparison of the WCDA experiment and the uncoupled control setup is shown in Figure 5.

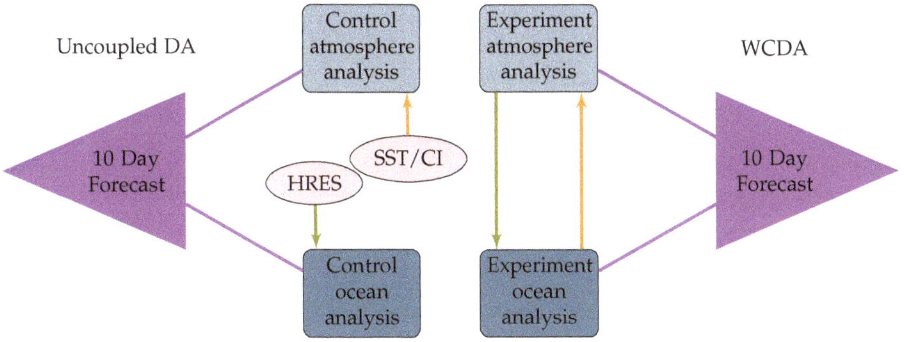

Figure 5. Schematic of experimental design. On the right is the experiment, with the weakly coupled DA passing information between atmosphere and ocean analyses. On the left is the control atmospheric analysis getting its SST and sea ice concentration (CI) from the external OSTIA product, with the forcing field for the control ocean analysis coming from the operational HRES system.

4. Results and Discussion

For brevity, we show normalised differences in root-mean-square error (RMSE) as a measure of forecast errors [42]. Normalised RMSE differences (dRMSE) for an experiment e compared with a control experiment c is defined as

$$\text{dRMSE} = \frac{||\boldsymbol{x}_f^e(T:T+t) - \boldsymbol{x}_a^e(T+t)|| - ||\boldsymbol{x}_f^c(T:T+t) - \boldsymbol{x}_a^c(T+t)||}{||\boldsymbol{x}_f^c(T:T+t) - \boldsymbol{x}_a^c(T+t)||},$$

where $\boldsymbol{x}_f(T:T+t)$ and $\boldsymbol{x}_a(T+t)$ refer to forecasts of length t and analyses valid at time $T+t$, and the norm $||\cdot||$ is the root mean square throughout the number of samples through time.

4.1. Atmospheric Performance

Figure 6 shows the impact of WCDA on atmospheric humidity and temperature. There are three distinct regions of impact—the tropics and the poles—coming from the separate influences of WCDA through tropical SST and CI, respectively. The areas of hashed shading indicate that the differences in forecast errors are statistically significant. It is clear that the impact of WCDA does not have long-range impacts on the upper troposphere or on the spatial regions where WCDA is not active.

In the tropical region of Figure 6, we can see that the impact of the tropical SST is detected from the surface up to around 850 hPa in both temperature and humidity. The hashed shading shows that these improvements, indicated by blue colours, are statistically significant. The maps in Figure 7 show clearly that the improvement from SST is restricted to the latitudinal band for which the WCDA SST is active. Within this band, there are variations in how much benefit we get from WCDA. For instance, with the temperature at 1000 hPa, the strongest positive impacts are seen in the Arabian Sea, the Eastern Atlantic and Eastern Pacific (associated with cold tongues [43]).

In the Arabian Sea, there is evidence of improvement in other variables, such as in low-level winds and significant wave heights (not shown). This indicates that the SST WCDA has improved the position of the summer monsoon, which is known to be difficult to forecast well. The regions of positive impact in the Atlantic and the equatorial Pacific are regions that tend to have high cloud cover. This persistent cloud makes observing the SST from satellites difficult, and so it may be that the use of the ocean model within the OCEAN5 analysis system is able to effectively fill the observational gap.

In the polar regions, we see significant improvements in forecast errors due to the weakly coupled assimilation. Figures 8 and 9 show that the improvements due to sea ice encompass the entire extent of the sea ice cover and are not simply confined to the ice edge. The areas of negative forecast impact in the sea ice variables (final row in Figures 8 and 9) are likely an artefact of the mask used to compute these scores. In regions where both the analysis and forecast have no sea ice concentration, the forecast errors are identically zero. Hence, in the case where the control has a smaller extent than the experiment, we can find artificially high negative scores around the ice edge, which is what is seen in Figures 8 and 9.

The small region of negative impact seen in humidity around 80° S (Figure 6a) is restricted to the areas around the Ross Ice Shelf and the Filchner-Ronne Ice Shelf (Figure 8, second row). The zonally averaged diagnostics give disproportionately large weight to areas at high latitudes. These ice shelves are mainly outside the domain of the ocean model which is currently used, and so interpolation between the ocean grid and the atmospheric grid is delicate in this area. Before operational implementation, a modified interpolation scheme was developed for this area to eliminate this negative signal.

The influence of the WCDA CI extends up to roughly 700 hPa (Figure 6). This is further vertically than the impact of SST seen in the tropics. This can be explained by considering Figure 10, which shows the usage of microwave humidity soundings in the southern hemisphere. Such soundings are rejected over sea ice, and rejecting more contaminated soundings could lead to a more consistent atmospheric state in these regions. Further area-averaged dRMSEs of forecast error are shown in Appendix A.

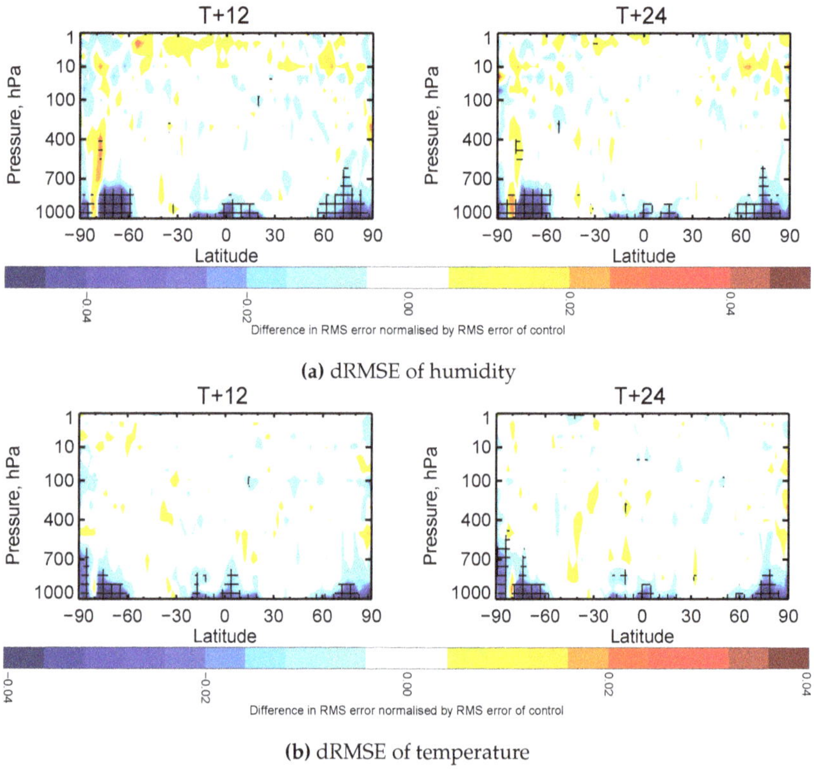

Figure 6. Latitude–pressure diagram of the zonally averaged normalised difference in root-mean-square error (RMSE) between WCDA and the control for humidity (**top**) and temperature (**bottom**) forecasts at 12 h (**left**) and 24 h (**right**) lead times, for the period from 9 June 2017 to 21 May 2018. Hashed areas indicate statistically significant differences.

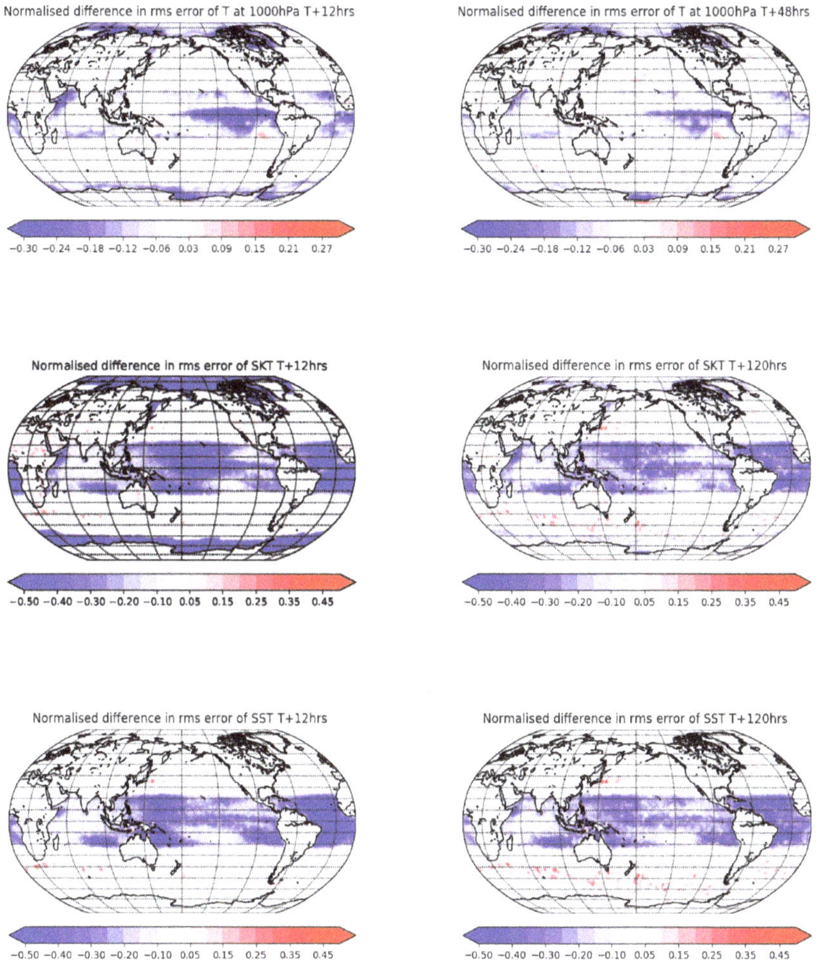

Figure 7. Spatial maps of the normalised difference in RMSE between WCDA and the control for temperature at 1000 hPa (**top**), skin temperature (**middle**) and sea-surface temperature (**bottom**) forecasts at 12 h (**left**) and 120 h (**right**) lead times (48 h lead time for temperature at 1000 hPa), for the period from 9 June 2017 to 21 May 2018.

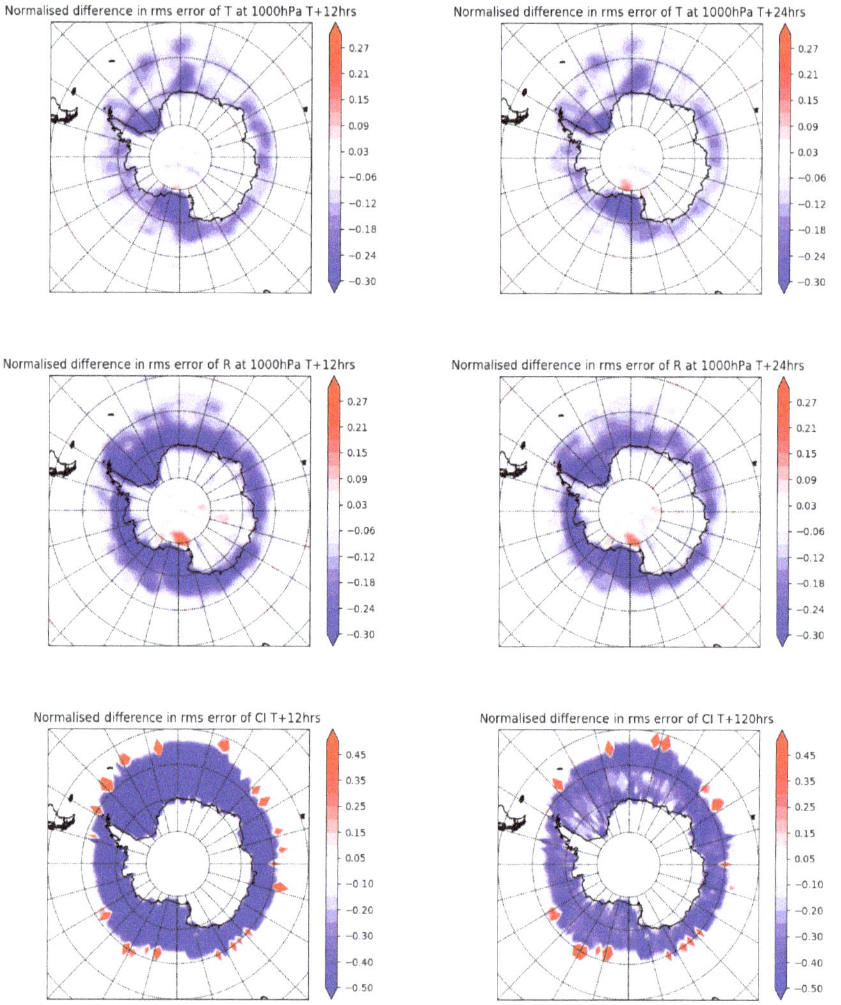

Figure 8. Spatial maps centred on the Antarctic of the normalised difference in RMSE between WCDA and the control for 1000 hPa temperature (**top**), 1000 hPa humidity (**middle**) and sea ice concentration (**bottom**) forecasts at 12 h (**left**) and 24 h (**right**) lead times (120 h lead time for sea ice concentration), for the period 9 June 2017 to 21 May 2018.

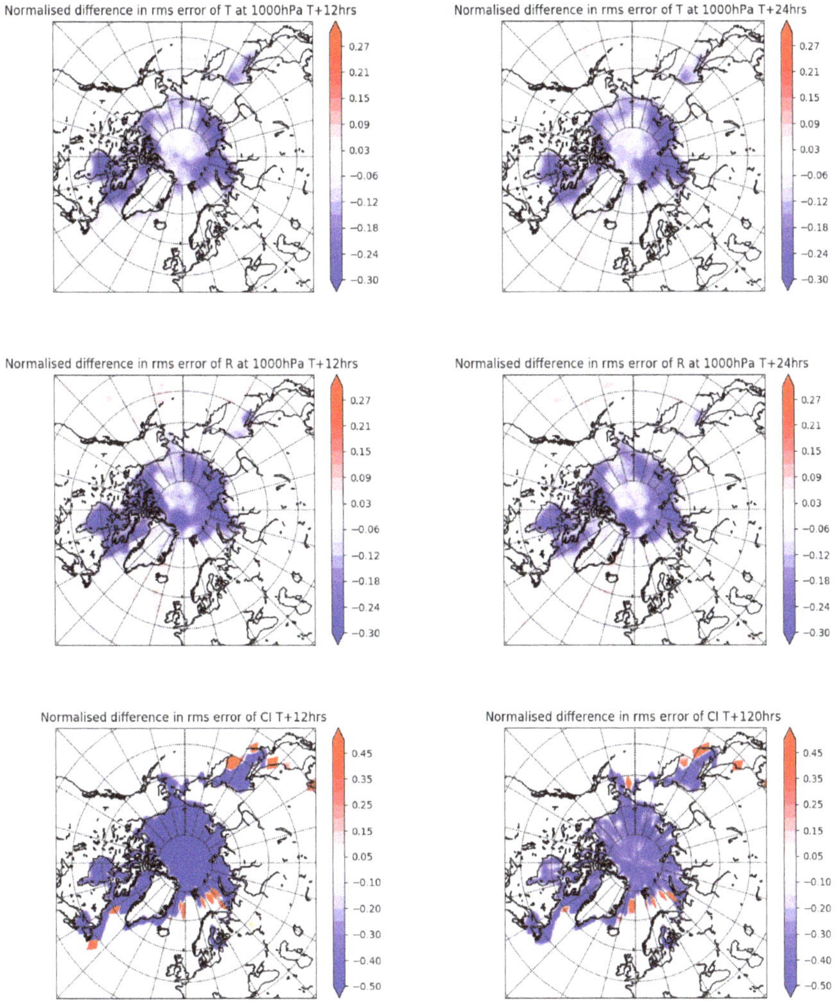

Figure 9. Similar to Figure 8 but focused on the Arctic.

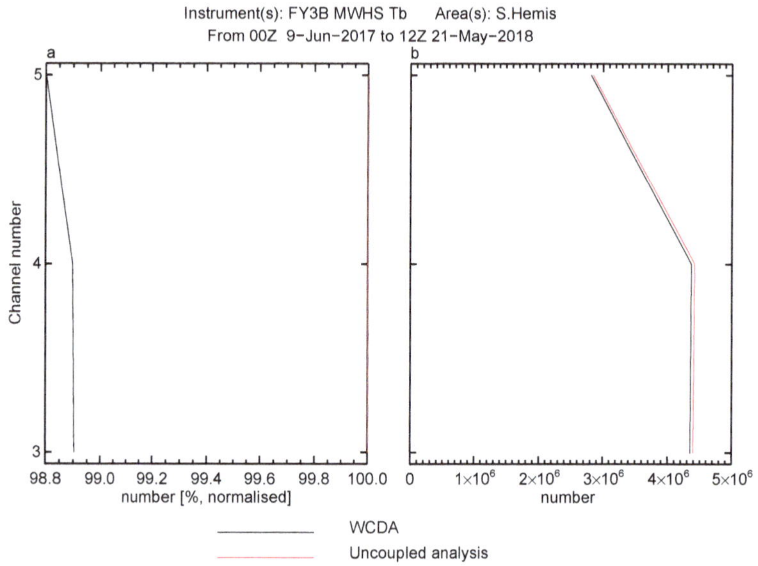

Figure 10. Observation usage of MWHS-1 in the southern hemisphere for the WCDA experiment (black) and the uncoupled control (red). Figure (**a**) shows the normalised observation usage and figure (**b**) shows the absolute numbers.

4.2. Ocean Performance

Figure 11 shows the area average of the heat flux correction, δQ, used in the ocean analyses, which is defined as

$$\delta Q = \gamma \left(SST_{\mathrm{OSTIA}} - SST_{\mathrm{NEMO}} \right),$$

where γ is a globally uniform restoration term of $-200\ \mathrm{Wm^{-2}K^{-1}}$. This heat flux correction is applied to the surface non-solar heat flux. This flux correction represents the strength of the relaxation toward the OSTIA SST product. One can see that the flux correction in the extratropics is broadly similar in both experiments, with only the northern hemisphere extratropics showing a slightly smaller flux correction in July and August in the WCDA experiment. In the tropics, the flux correction is consistently substantially smaller in WCDA than in the uncoupled. This shows that the SST coupling in the tropics is leading to an SST field that is more consistent than the uncoupled analysis, requiring less corrections. We postulate that this could be due to the improved timeliness of the the SST that the atmospheric component uses (for the uncoupled analysis, the OSTIA SST field is only available on the day after its valid time); however, this requires a more detailed investigation that is outside the scope of this paper.

Figure 12 shows the area-averaged surface temperature increments in the ocean from the uncoupled and weakly coupled assimilation experiments. Similar to Figure 11, the larger benefits from WCDA appear to be seen in the tropics with the coupling of SST, showing reduced increments compared with the uncoupled analysis. However, in the northern hemisphere extratropics, the weakly coupled assimilation increments are consistently smaller than the increments in the uncoupled system. For the southern hemisphere extratropics, the picture is mixed, and it seems like the two systems are behaving very similarly.

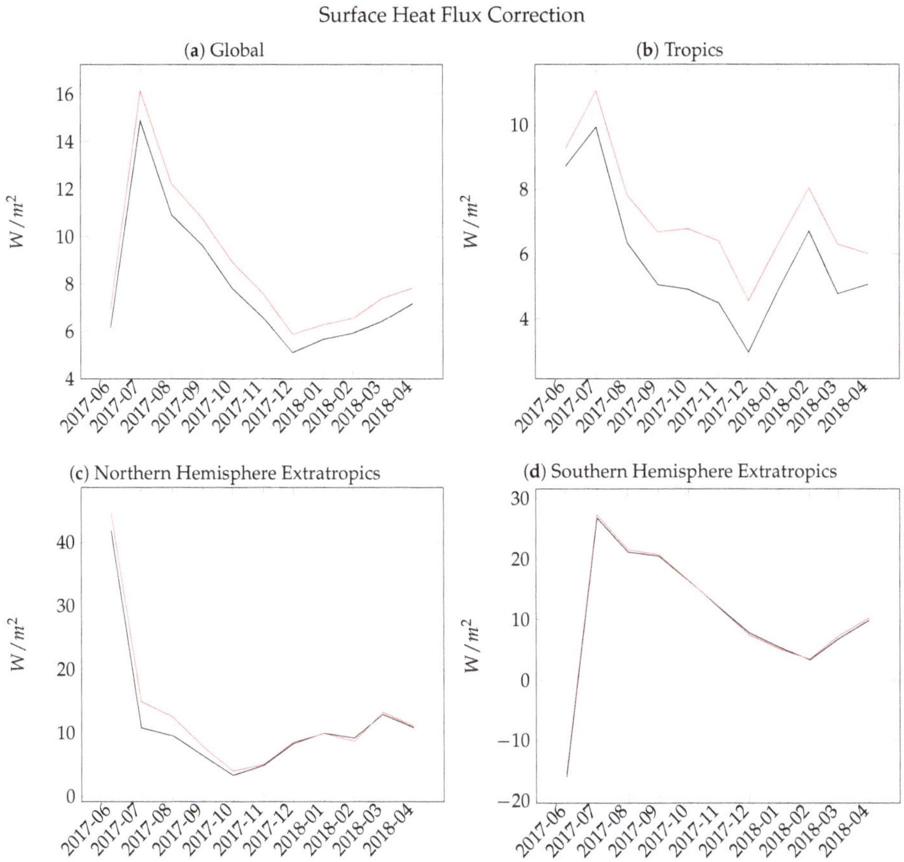

Figure 11. Monthly averaged ocean heat flux correction from the uncoupled analysis (red) and the weakly coupled analysis (black). This is split by region showing global (**top left**), the tropics (**top right**), the northern hemisphere extratropics (**bottom left**) and the southern hemisphere extratropics (**bottom right**).

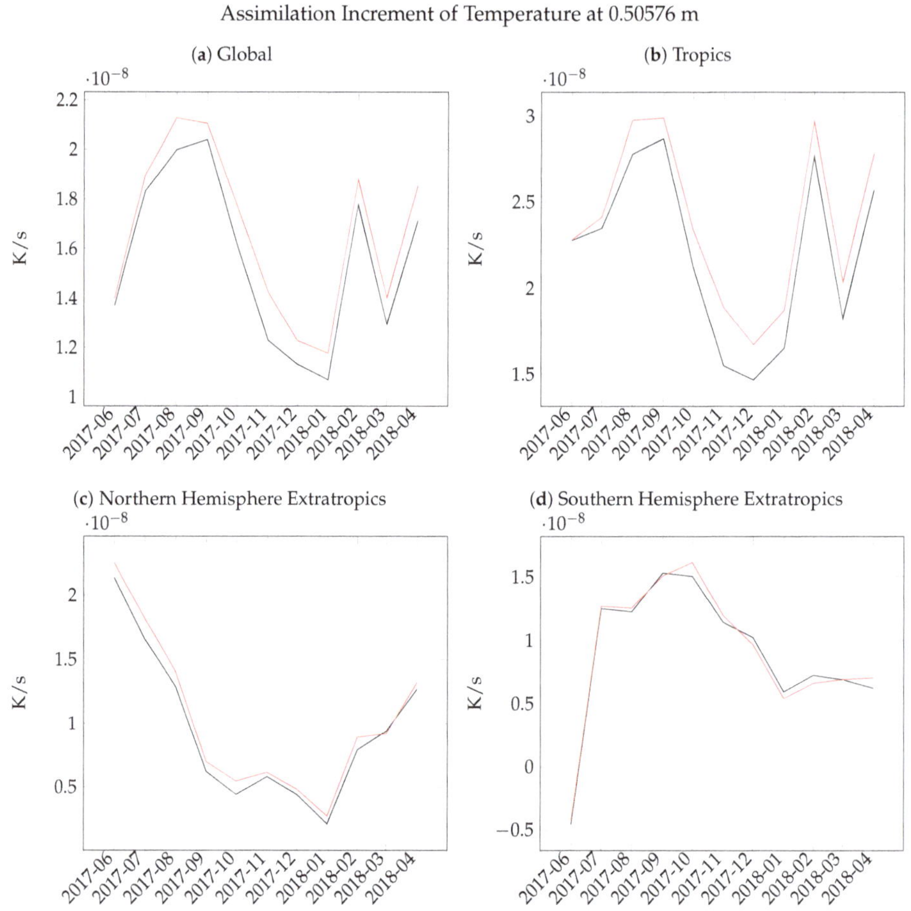

Figure 12. As in Figure 11 but for surface assimilation temperature increments.

4.3. Operational Impacts—Baltic Sea Detailed Sea Ice Investigation

Here, we take a detailed look at the spatial distribution of sea ice in the Baltic Sea, a particularly challenging area for sea ice concentration analyses. We show results for 2 separate days, as this is sufficient to highlight the behaviour of the different sea ice products in various scenarios throughout the winter season. Figure 13 shows different representations of the sea ice on 17 February 2018 from different sources. Figure 13a is a manually produced ice chart from FMI/SMHI showing sea ice concentration at the north of the Gulf of Bothnia, as well as in the eastern end of the Gulf of Finland. Figure 13b shows the available OSI SAF L3 sea ice concentration observations in the area. Note that because of the geography of the area, observations are only available in the centre of the Gulfs; coastal contamination requires that those points near the coast be masked from the product.

Figure 13c,d show the sea ice from uncoupled assimilation and WCDA experiments. One can see that the uncoupled analysis effectively smooths out the L3 observations and does not capture the high ice concentrations along the northern coastlines. WCDA, on the other hand, does a much better job at capturing the structures seen in the manual ice chart. In particular, the use of the background information coming from the dynamical model gives a much more realistic spatial distribution of the ice field. Figure 14 is similar but on 5 March 2018, the date of maximal sea ice extent in the Baltic Sea for the 2017/2018 season.

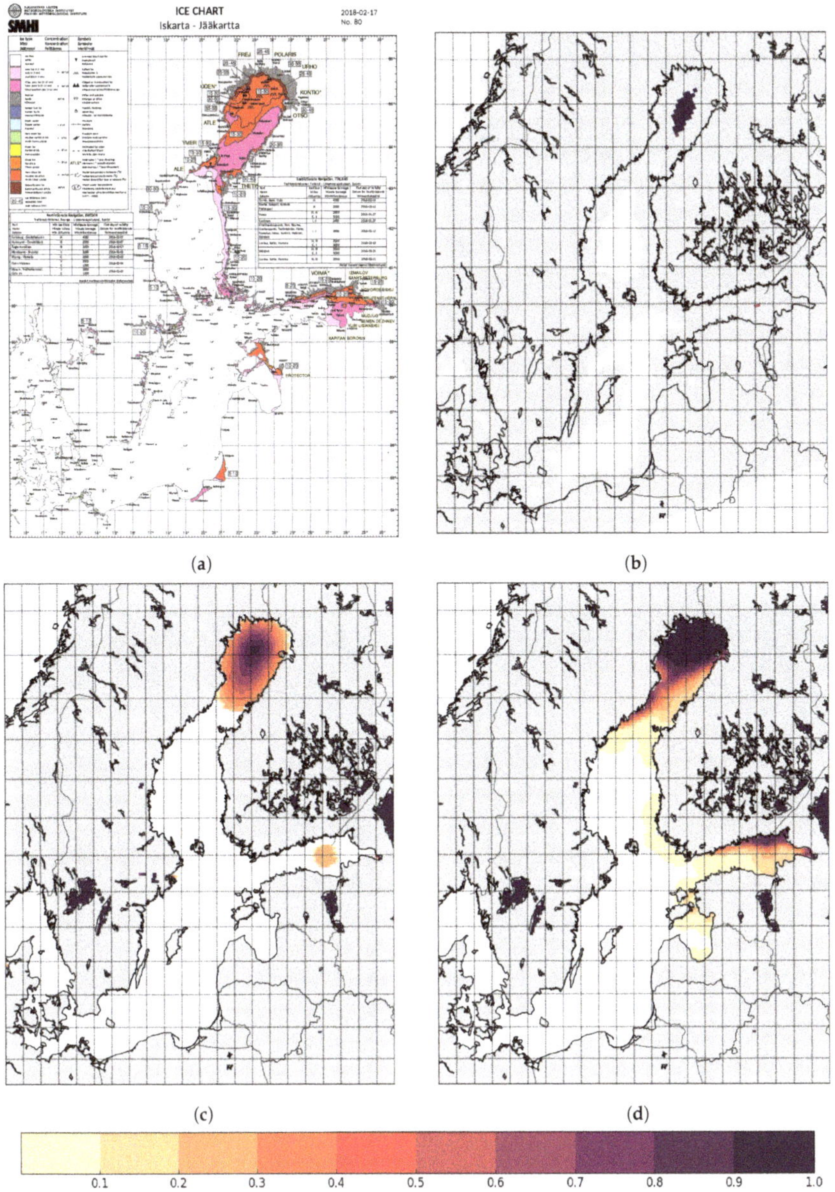

Figure 13. (**a**) Manually produced Finnish–Swedish ice chart of the Baltic Sea. © [44] Reproduced with permission. (**b**) OSI SAF 401-b product. Note missing data (grey) around coastlines due to coastal contamination in satellite retrievals of sea ice concentration. (**c**) Uncoupled analysis. Note the Gaussian nature of the ice field centred on the available OSI SAF L3 observations. (**d**) WCDA. Note the much more realistic structure and the good agreement with the manual ice chart. A manual ice chart (**a**) and sea ice concentration values in the Baltic sea on 17 February 2018 from OSI SAF L3 observations (**b**), uncoupled analysis (**c**) and WCDA (**d**).

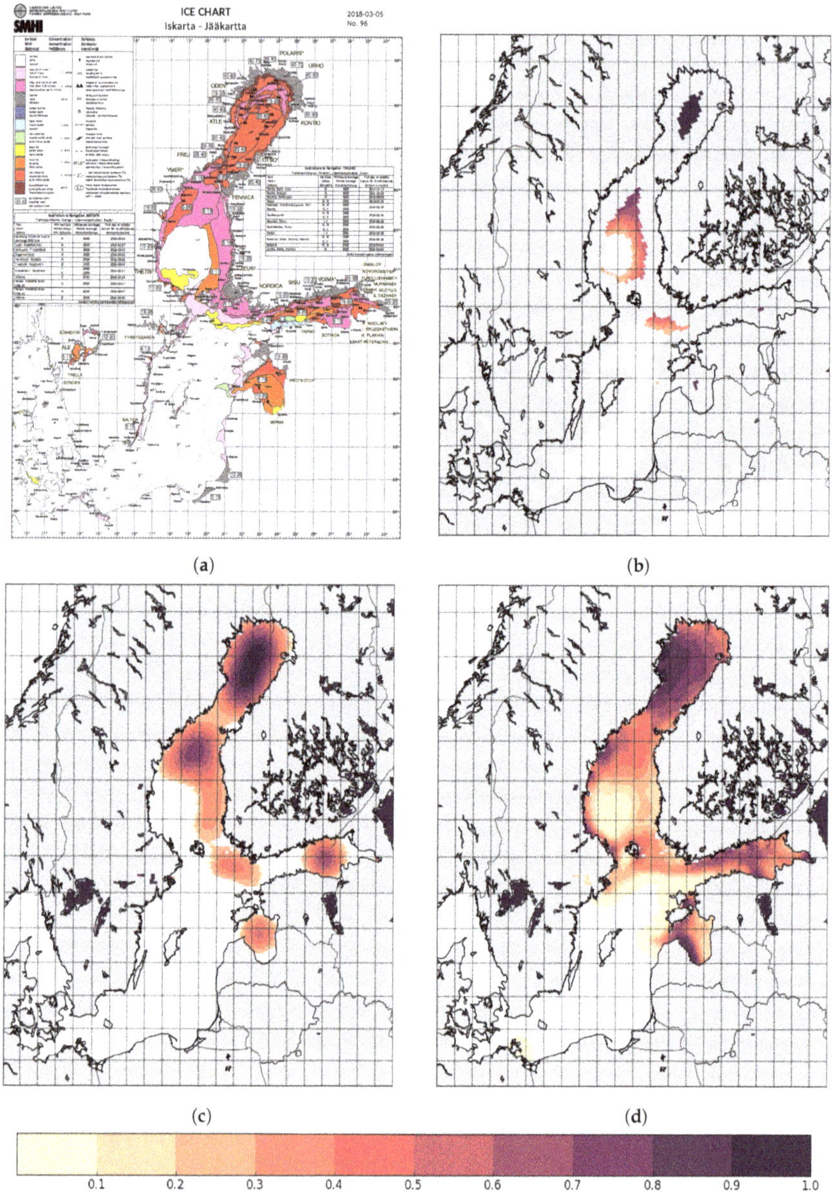

Figure 14. (**a**) Manually produced Finnish–Swedish ice chart of the Baltic Sea. © [44] Reproduced with permission. (**b**) OSI SAF 401-b product. Note missing data (grey) around coastlines due to coastal contamination in satellite retrievals of sea ice concentration. (**c**) Uncoupled analysis. Note the Gaussian nature of the ice field centred on the available OSI SAF L3 observations. (**d**) WCDA. Note the much more realistic structure and the good agreement with the manual ice chart. As in Figure 13 but on 5 March 2018, the date of maximum sea ice extent in the region for the season.

5. Conclusions and Future Plans

This study investigated the impact of weakly coupled ocean–atmosphere data assimilation on the ECMWF forecasts. The WCDA approach allows components of the Earth system with different timescales and assimilation methods to be linked together. As an alternative to using purely observation-based L4 products for the lower boundary of the atmosphere, with their associated latencies, WCDA allows dynamical models of the ocean and sea ice to fill the gaps in observations and propagate fields to the appropriate time. The results presented show that the use of WCDA improves the coupled forecasts in the regions near the interface of the variables being coupled.

In particular, it was shown that near-surface temperatures and humidities have smaller forecast errors in the WCDA system than the control experiment. Due to changes in the sea ice field, it was shown that the usage of satellite microwave sounder data within 4D-Var changes, as this data is screened based on the presence of sea ice. Statistics from the ocean analyses were compared and showed reduced analysis increments in the WCDA system, an indication of a more consistent analysis.

ECMWF's operational upgrade to cycle 45R1 in June 2018 saw the introduction of WCDA through sea ice concentration. The forthcoming upgrade to cycle 46R1 is scheduled to also couple tropical SST, as per the experiments shown in this paper.

In addition to surface temperature and ice concentration information, the ocean analysis system can provide surface current information. Surface currents can be important for the assimilation of scatterometer data. Scatterometers measure the backscattering coefficient of the ocean surface. This coefficient is a function of wind velocity relative to the ocean current. In the current usage of scatterometers at ECMWF, we assume zero ocean currents, and so a WCDA system that has knowledge of the ocean currents should be able to make better use of scatterometer data.

There is plenty of scope to improve the partial coupling approach and its geospatial structure that determines the extent of SST coupling in the WCDA system. At the moment, it is a simple function of latitude. It may be beneficial for this to be basin dependent. Given the model biases in the western boundary currents, it may be beneficial to differ in the west and east of each basin.

These are the first steps in operational coupled ocean–atmosphere assimilation at ECMWF. A progressive approach toward implementation has been adopted rather than introducing coupling in all variables in a single system upgrade. This weakly coupled data assimilation system is flexible and could be applied in areas other than NWP, such as reanalysis. Whilst the weakly coupled approach is being developed, in parallel, the outer loop coupling approach is being explored as a possible operational system which would have a more immediate impact across the various Earth system components.

Author Contributions: Conceptualization, P.A.B., P.d.R.; software, P.A.B., H.Z., A.B., A.D.; writing—original draft preparation, P.A.B.; writing—review and editing, P.A.B., P.d.R., H.Z., A.B., A.D.

Funding: This research received no external funding.

Acknowledgments: We are grateful to Patrick Eriksson of FMI for providing assistance with the FMI/SMHI sea ice charts. Our thanks go to Magdalena Balmaseda, Stephen English, Sarah Keeley, and Kristian Mogensen for their help with implementations and insightful discussion of results.

Conflicts of Interest: The authors declare no conflict of interest.

Appendix A. Area-Averaged dRMSE of Forecast Errors

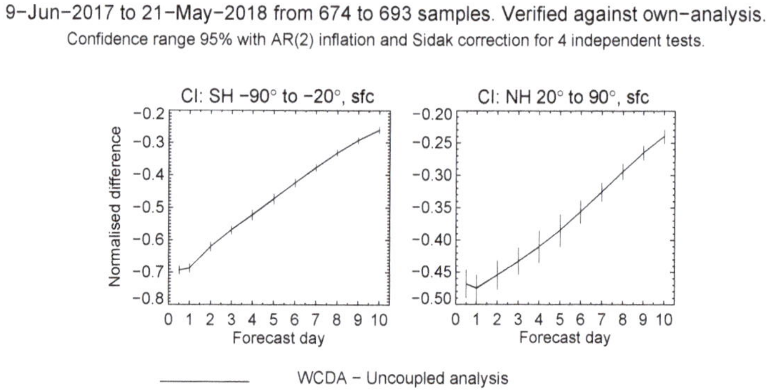

Figure A1. Area-averaged diagram of the normalised difference in RMSE between WCDA and the control for sea ice concentration in the southern hemisphere (**left**), tropics (**centre**) and northern hemisphere (**right**) for forecasts at lead times of up to 10 days.

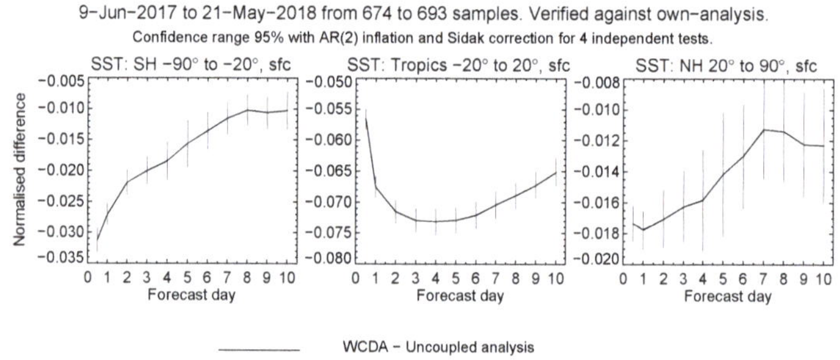

Figure A2. As in Figure A1 but for sea-surface temperature.

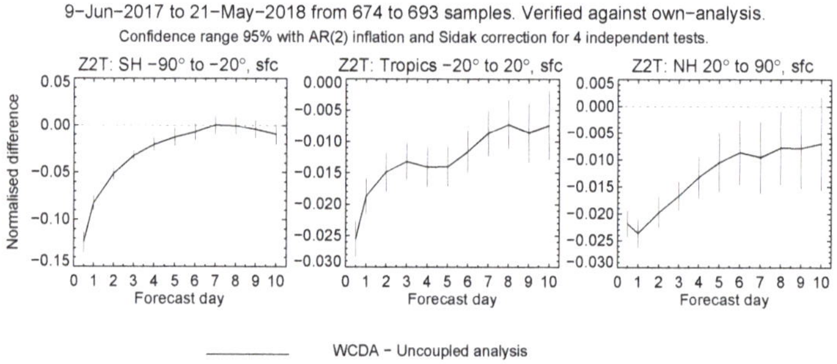

Figure A3. As in Figure A1 but for atmospheric 2-m temperature.

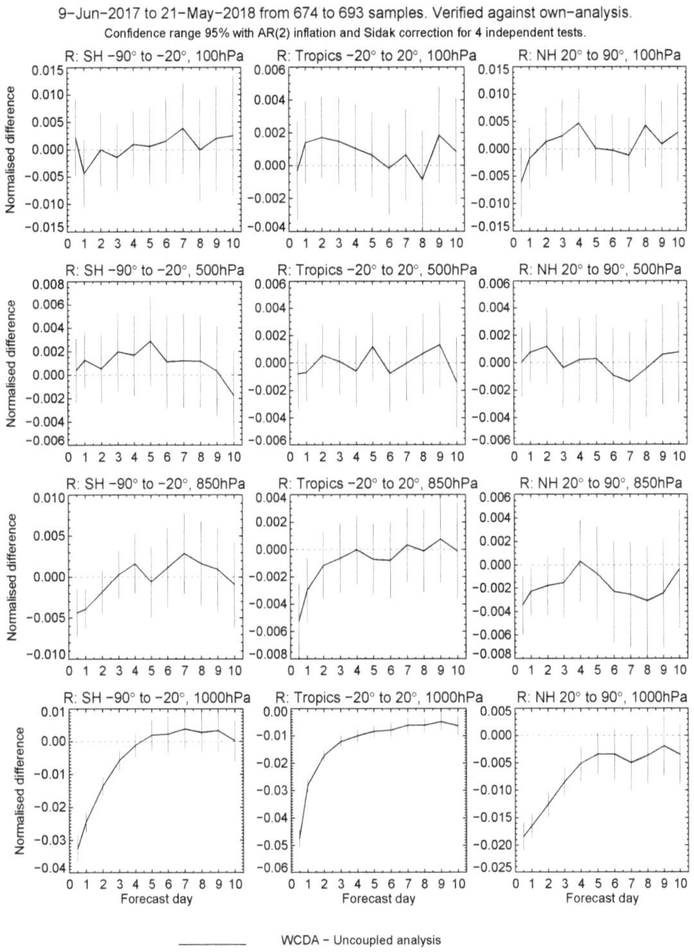

Figure A4. As in Figure A1 but for atmospheric relative humidity and stratified by level, 100 hPa (**top row**), 500 hPa (**second row**), 850 hPa (**third row**) and 1000 hPa (**bottom row**).

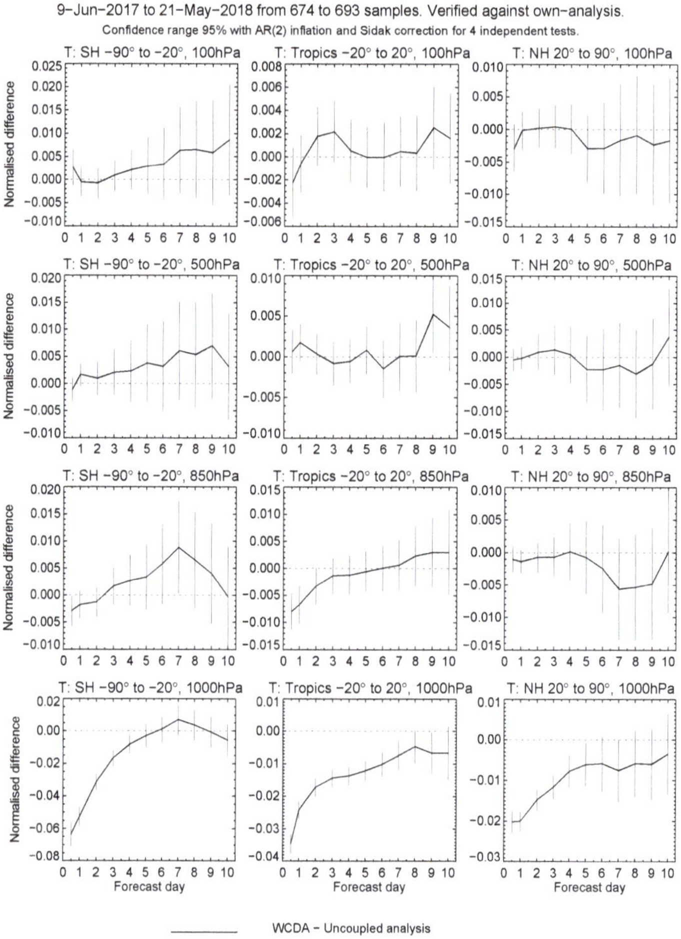

Figure A5. As in Figure A4 but for atmospheric temperature.

References

1. ECMWF. *The Strength of a Common Goal: A Roadmap to 2025*; Technical Report; ECMWF: Reading, UK, 2016.
2. Bauer, P.; Thorpe, A.; Brunet, G. The quiet revolution of numerical weather prediction. *Nature* **2015**, *525*, 47–55. [CrossRef] [PubMed]
3. Maclachlan, C.; Arribas, A.; Peterson, K.A.; Maidens, A.; Fereday, D.; Scaife, A.A.; Gordon, M.; Vellinga, M.; Williams, A.; Comer, R.E.; et al. Global Seasonal forecast system version 5 (GloSea5): A high-resolution seasonal forecast system. *Q. J. R. Meteorol. Soc.* **2015**, *141*, 1072–1084. [CrossRef]
4. Stockdale, T.N.; Anderson, D.L.; Alves, J.O.; Balmaseda, M.A. Global seasonal rainfall forecasts using a coupled ocean-atmosphere model. *Nature* **1998**, *392*, 370–373. [CrossRef]
5. Vitart, F. Monthly Forecasting at ECMWF. *Mon. Weather Rev.* **2004**, *132*, 2761–2779. [CrossRef]
6. Balmaseda, M.A.; Mogensen, K.; Weaver, A.T. Evaluation of the ECMWF ocean reanalysis system ORAS4. *Q. J. R. Meteorol. Soc.* **2013**, *139*, 1132–1161. [CrossRef]
7. Bauer, P.; Richardson, D. *New Model Cycle 40r1*; ECMWF Newsletter No. 138, Winter 2013/2014; ECMWF: Reading, UK, 2014; p. 3.
8. Rahmstorf, S. Climate drift in an ocean model coupled to a simple, perfectly matched atmosphere. *Clim. Dyn.* **1995**, *11*, 447–458. [CrossRef]

9. Mulholland, D.P.; Laloyaux, P.; Haines, K.; Alonso Balmaseda, M. Origin and Impact of Initialization Shocks in Coupled Atmosphere-Ocean Forecasts. *Mon. Weather Rev.* **2015**, *143*, 4631–4644. [CrossRef]
10. Sugiura, N.; Awaji, T.; Masuda, S.; Mochizuki, T.; Toyoda, T.; Miyama, T.; Igarashi, H.; Ishikawa, Y. Development of a four-dimensional variational coupled data assimilation system for enhanced analysis and prediction of seasonal to interannual climate variations. *J. Geophys. Res. Oceans* **2008**, *113*, 1–21. [CrossRef]
11. Masuda, S.; Philip Matthews, J.; Ishikawa, Y.; Mochizuki, T.; Tanaka, Y.; Awaji, T. A new Approach to El Niño Prediction beyond the Spring Season. *Sci. Rep.* **2015**, *5*, 1–9. [CrossRef]
12. Mochizuki, T.; Masuda, S.; Ishikawa, Y.; Awaji, T. Multiyear climate prediction with initialization based on 4D-Var data assimilation. *Geophys. Res. Lett.* **2016**, *43*, 3903–3910. [CrossRef]
13. Yang, X.; Rosati, A.; Zhang, S.; Delworth, T.L.; Gudgel, R.G.; Zhang, R.; Vecchi, G.; Anderson, W.; Chang, Y.S.; DelSole, T.; et al. A predictable AMO-like pattern in the GFDL fully coupled ensemble initialization and decadal forecasting system. *J. Clim.* **2013**, *26*, 650–661. [CrossRef]
14. Zhang, S.; Chang, Y.S.; Yang, X.; Rosati, A. Balanced and coherent climate estimation by combining data with a biased coupled model. *J. Clim.* **2014**, *27*, 1302–1314. [CrossRef]
15. Saha, S.; Moorthi, S.; Wu, X.; Wang, J.; Nadiga, S.; Tripp, P.; Behringer, D.; Hou, Y.T.; Chuang, H.Y.; Iredell, M.; et al. The NCEP climate forecast system version 2. *J. Clim.* **2014**, *27*, 2185–2208. [CrossRef]
16. Saha, S.; Moorthi, S.; Pan, H.L.; Wu, X.; Wang, J.; Nadiga, S.; Tripp, P.; Kistler, R.; Woollen, J.; Behringer, D.; et al. The NCEP climate forecast system reanalysis. *Bull. Am. Meteorol. Soc.* **2010**, *91*, 1015–1057. [CrossRef]
17. Penny, S.G.; Akella, S.; Alves, O.; Bishop, C.; Buehner, M.; Chevallier, M.; Counillon, F.; Draper, C.; Frolov, S.; Fujii, Y.; et al. *Coupled Data Assimilation for Integrated Earth System Analysis and Prediction: Goals, Challenges and Recommendations*; Technical Report; World Meteorological Organisation: Geneva, Switzerland, 2017.
18. Frolov, S.; Bishop, C.H.; Holt, T.; Cummings, J.; Kuhl, D. Facilitating Strongly Coupled Ocean-Atmosphere Data Assimilation with an Interface Solver. *Mon. Weather Rev.* **2016**, *144*, 3–20. [CrossRef]
19. Lea, D.J.; Mirouze, I.; Martin, M.J.; King, R.R.; Hines, A.; Walters, D.; Thurlow, M. Assessing a New Coupled Data Assimilation System Based on the Met Office Coupled Atmosphere–Land–Ocean–Sea Ice Model. *Mon. Weather Rev.* **2015**, *143*, 4678–4694. [CrossRef]
20. Laloyaux, P.; Balmaseda, M.; Dee, D.; Mogensen, K.; Janssen, P. A coupled data assimilation system for climate reanalysis. *Q. J. R. Meteorol. Soc.* **2016**, *142*, 65–78. [CrossRef]
21. Schepers, D.; Boisséson, E.D.; Eresmaa, R.; Lupu, C.; Rosnay, P.D. CERA-SAT: A coupled satellite-era reanalysis. *ECMWF Newslett.* **2018**, *155*, 32–37. [CrossRef]
22. ECMWF. Part III: Dynamics and numerical procedures. In *IFS Documentation CY43R3*; Number 3 in IFS Documentation; ECMWF: Reading, UK, 2017.
23. ECMWF. Part IV: Physical processes. In *IFS Documentation CY43R3*; Number 4 in IFS Documentation; ECMWF: Reading, UK, 2017.
24. Balsamo, G.; Beljaars, A.; Scipal, K.; Viterbo, P.; van den Hurk, B.; Hirschi, M.; Betts, A.K. A Revised Hydrology for the ECMWF Model: Verification from Field Site to Terrestrial Water Storage and Impact in the Integrated Forecast System. *J. Hydrometeorol.* **2009**, *10*, 623–643. [CrossRef]
25. Dutra, E.; Stepanenko, V.M.; Balsamo, G.; Viterbo, P.; Miranda, P.M.A.; Mironov, D.; Schär, C. Impact of Lakes on the ECMWF Surface Scheme. In *ECMWF Technical Memorandum*; ECMWF: Reading, UK, 2009; pp. 1–15.
26. ECMWF. Part VII: ECMWF wave model. In *IFS Documentation CY43R3*; Number 7 in IFS Documentation; ECMWF: Reading, UK, 2017.
27. Rabier, F.; Järvinen, H.; Klinker, E.; Mahfouf, J.F.; Simmons, A. The ECMWF operational implementation of four-dimensional variational assimilation. Part I: Experimental results and diagnostics wiht operational configuration. *Q. J. R. Meteorol. Soc.* **2000**, *126*, 1143–1170. [CrossRef]
28. De Rosnay, P.; Balsamo, G.; Albergel, C.; Muñoz-Sabater, J.; Isaksen, L. Initialisation of Land Surface Variables for Numerical Weather Prediction. *Surv. Geophys.* **2014**, *35*, 607–621. [CrossRef]
29. English, S.; McNally, A.; Borman, N.; Salonen, K.; Matricardi, M.; Horányi, A.; Rennie, M.; Janisková, M.; Di Michele, S.; Geer, A.; et al. Impact of Satellite Data. In *Technical Memoradum ECMWF*; ECMWF: Reading, UK, 2013; p. 46.

30. Haiden, T.; Dahoui, M.; Ingleby, B.; Rosnay, P.D.; Prates, C.; Kuscu, E.; Hewson, T.; Isaksen, L.; Richardson, D.; Zuo, H.; Jones, L. Use of In Situ Surface Observations at ECMWF. In *Technical Memoradum ECMWF*; ECMWF: Reading, UK, 2018.
31. Donlon, C.J.; Martin, M.; Stark, J.; Roberts-Jones, J.; Fiedler, E.; Wimmer, W. The Operational Sea Surface Temperature and Sea Ice Analysis (OSTIA) system. *Remote Sens. Environ.* **2012**, *116*, 140–158. [CrossRef]
32. OSI SAF. OSI-401-b. Available online: http://www.osi-saf.org/?q=content/global-sea-ice-concentration-ssmis (accessed on 14 January 2019).
33. NCEP. MMAB Sea Ice Analysis. Available online: http://polar.ncep.noaa.gov/seaice/Analyses.shtml (accessed on 14 January 2019).
34. Bonavita, M.; Isaksen, L.; Hólm, E. On the use of EDA background error variances in the ECMWF 4D-Var. *Q. J. R. Meteorol. Soc.* **2012**, *138*, 1540–1559. [CrossRef]
35. Haseler, J. *The Early-Delivery Suite*; Technical Memorandum 454; ECMWF: Reading, UK, 2004.
36. Zuo, H.; Balmaseda, M.A.; Tietsche, S.; Mogensen, K.; Mayer, M. The ECMWF operational ensemble reanalysis-analysis system for ocean and sea-ice: A description of the system and assessment. *Ocean Sci. Discuss.* **2019**, under review. [CrossRef]
37. Zuo, H.; Balmaseda, M.A.; Mogensen, K.; Tietsche, S. *OCEAN5: The ECMWF Ocean Reanalysis System and Its Real-Time Analysis Component*; Technical Report 823; ECMWF: Reading, UK, 2018.
38. ECMWF. Part V: Ensemble prediction system. In *IFS Documentation CY40R1*; Number 5 in IFS Documentation; ECMWF: Reading, UK, 2014.
39. Mogensen, K.; et al. The ECMWF Coupled Atmosphere-Wave-Ocean-Ice Model as Implemented in CY45R1: Part 1: Technical Implementation. In *ECMWF Technical Memorandum*; ECMWF: Reading, UK, 2019, in prep.
40. Mogensen, K.; et al. The ECMWF Coupled Atmosphere-Wave-Ocean-Ice Model as Implemented in CY45R1: Part 2: Ocean Model Performance. In *ECMWF Technical Memorandum*; ECMWF: Reading, UK, 2019, in prep.
41. Keeley, S.; et al. The ECMWF Coupled Atmosphere-Wave-Ocean-Ice Model as Implemented in CY45R1: Part 3: Ice model performance. In *ECMWF Technical Memorandum*; ECMWF: Reading, UK, 2019, in prep.
42. Geer, A.J. Significance of changes in medium-range forecast scores. *Tellus A* **2016**, *68*, 1–21. [CrossRef]
43. Siedler, G.; Gould, J.; Church, J. *Ocean Circulation and Climate: Observing and Modelling the Global Ocean*; International Geophysics, Elsevier Science: Amsterdam, The Netherlands, 2001.
44. FMI; SMHI. Baltic sea ice chart. Available online: http://cdn.fmi.fi/marine-observations/products/ice-charts/20180217-full-color-ice-chart.png (accessed on 10 July 2018).

remote sensing

MDPI

Article

Assimilation of GPSRO Bending Angle Profiles into the Brazilian Global Atmospheric Model

Ivette H. Banos [1,*], Luiz F. Sapucci [1], Lidia Cucurull [2], Carlos F. Bastarz [1] and Bruna B. Silveira [1]

1 Center for Weather Forecast and Climate Studies, National Institute for Space Research, Cachoeira Paulista, São Paulo 12630-000, Brazil; luiz.sapucci@inpe.br (L.F.S.); carlos.bastarz@inpe.br (C.F.B.); brunabs.silveira@gmail.com (B.B.S.)

2 NOAA Atlantic Oceanographic and Meteorological Laboratory, Miami, FL 33149, USA; lidia.cucurull@noaa.gov

* Correspondence: ivette.banos@inpe.br

Received:17 September 2018; Accepted: 27 November 2018; Published: 28 January 2019

Abstract: The Global Positioning System (GPS) Radio Occultation (RO) technique allows valuable information to be obtained about the state of the atmosphere through vertical profiles obtained at various processing levels. From the point of view of data assimilation, there is a consensus that less processed data are preferable because of their lowest addition of uncertainties in the process. In the GPSRO context, bending angle data are better to assimilate than refractivity or atmospheric profiles; however, these data have not been properly explored by data assimilation at the CPTEC (acronym in Portuguese for Center for Weather Forecast and Climate Studies). In this study, the benefits and possible deficiencies of the CPTEC modeling system for this data source are investigated. Three numerical experiments were conducted, assimilating bending angles and refractivity profiles in the Gridpoint Statistical Interpolation (GSI) system coupled with the Brazilian Global Atmospheric Model (BAM). The results highlighted the need for further studies to explore the representation of meteorological systems at the higher levels of the BAM model. Nevertheless, more benefits were achieved using bending angle data compared with the results obtained assimilating refractivity profiles. The highest gain was in the data usage exploring 73.4% of the potential of the RO technique when bending angles are assimilated. Additionally, gains of 3.5% and 2.5% were found in the root mean square error values in the zonal and meridional wind components and geopotencial height at 250 hPa, respectively.

Keywords: radio occultation data; GPSRO; bending angle; data assimilation; GSI; numerical weather prediction

1. Introduction

The development of advanced computer modeling techniques, the increase in the density of ground and satellite-based observation networks, as well as the enhancement of measuring instruments, data processing techniques and new methodologies (in particular, those based on satellites), have led to an improvement in weather and climate forecasts [1]. However, real observations and numerical models are not perfect and the atmosphere is chaotic by nature, which imposes a finite limit of predictability on the forecasts [2]. Data assimilation algorithms emerged to increase the forecast skill by reducing the uncertainties in the initial conditions for Numerical Weather Prediction (NWP) models. Initial conditions from data assimilation are generated by a statistical combination between a short-term forecast (called first guess/background) and the available observations over a time window [3]. The quantity, quality and distribution of observational data over the entire model domain, as well as a correct characterization of the errors associated with both the observations and the NWP model, are essential elements for a successful data assimilation process.

Over the last two decades, the Global Positioning System (GPS) Radio Occultation (RO) technique (hereinafter GPSRO) has provided valuable information of the thermodynamic state of the Earth's atmosphere (e.g., Kursinski et al. [4]), improving initial conditions and weather forecasts (e.g., Eyre [5], Cardinali and Healy [6]). The GPSRO technique is based on the transmission of GPS signals and their reception by a receiver on a Low Earth Orbit (LEO) satellite. GPS signals are delayed and bent along the ray path due to the refraction caused by the vertical variation of the atmospheric molecular concentration [7]. Measurements of the bending angle of the signals and the atmospheric profiles of refractivity, electrons content, temperature, pressure and water vapor pressure can be retrieved from GPSRO at various processing levels. Observations from the limb-view GPSRO technique offer complete global coverage, being independent of radiosonde calibration, with high accuracy and relatively high vertical resolution compared to nadir-view satellite radiances. Furthermore, as GPS signals go through clouds and droplets of rain without being greatly affected, the atmospheric information can be retrieved under all weather conditions [8].

Assessing the impact of GPSRO observations in the operational European Centre for Medium-Range Weather Forecasts (ECMWF) assimilation and forecast system, Cardinali and Healy [9] found that the information content of GPSRO observations was quite noticeable, being fourth in the satellite Degree of Freedom for Signal (DFS) ranking with 7%. Using sensitivity techniques from the adjoint method, the authors pointed out that GPSRO provides 10% of the 24-h forecast error reduction together with the Infrared Atmospheric Sounding Interferometer (IASI) and Atmospheric Infrared Sounder (AIRS). Cucurull and Anthes [10] confirmed the role of GPSRO data as an anchor observation, reducing the global forecast bias. A more effective use of satellite radiances was produced and a greater number of these observations passed through quality control procedures by assimilating GPSRO data. The impact of the loss of microwave observations from the National Oceanic and Atmospheric Administration (NOAA) and Aqua satellites and all RO soundings was assessed in Cucurull and Anthes [11]. A much larger negative impact on the forecasts was observed when losing RO observations, with an increase of 0.4 K in the cold bias in the upper stratosphere.

A more optimal use of GPSRO observations can be achieved when less processed data are assimilated, further exploiting the information contained in the prior data [12]. However, some centers such as the Center for Weather Forecast and Climate Studies of the Brazilian National Institute for Space Research (CPTEC/INPE) firstly assimilated temperature and humidity profiles to generate the analysis. Phase measurements and precise knowledge of the positions and velocities of the GPS and LEO satellites make up the level-0 of the standard products derived from the GPSRO technique. However, the complexity of implementing an observation operator for its assimilation in NWP models continues to be a challenge for the scientific community [13]. Therefore, most NWP centers directly assimilated refractivity or bending angle data in an operational framework. Several studies have shown an increase in the predictive skill of NWP models assimilating bending angles when compared to the assimilation of refractivity profiles. Rennie [14] pointed out that the Met Office was the first center to operationally assimilate GPSRO (refractivity profiles) data and investigate the impact of refractivity assimilation on bending angle assimilation. The results showed that, even after modifying the weight of the refractivity matrix error, the assimilation of bending angles was superior with more positive impacts. At the ECMWF, Healy and Thepaut [15] found statistically significant improvements in the Southern Hemisphere in the temperature field by assimilating GPSRO bending angle data from the CHAllenging Minisatellite Payload (CHAMP) mission when compared to radiosondes measurements. Bonavita [16] confirmed that the assimilation of GPSRO bending angle data reduced the effect of model bias in the upper troposphere and the stratosphere mostly in the Southern Hemisphere. Various studies were conducted at the Environmental Modeling Center (EMC) of the National Centers for Environmental Prediction (NCEP) to use GPSRO observations in an operational framework (e.g., Cucurull et al. [17,18]). In May 2012, the operational assimilation of refractivity ([19]) was replaced with the implementation of the NCEP's Bending Angle Method (NBAM) [20]. With the new method, the top of the profiles was extended from 30 to 50 km for the assimilation of bending angles and the quality control procedures and error characterization were tuned up to 50 km. Besides, to locate a GPSRO observation with higher precision within the model vertical grid, NBAM includes the capability of using compressibility factors in the computation of the geopotential heights of the model layers. The assimilation of bending angles with the NBAM showed improvements in the weather forecasting skill for all levels and variables.

Although it is well known that better results are obtained by assimilating GPSRO bending angle data, these observations have not been explored at the CPTEC/INPE. The main focus of the CPTEC is to constantly improve the reliability and quality of its operational weather and climate forecasts, especially over South America. This center uses the Brazilian Atmospheric Model (BAM) as the operational global circulation model and its performance for tropical precipitation forecast has already been evaluated [21]. However, the role of the BAM model as an essential part in the NWP system (i.e., a numerical model coupled to a data assimilation system) is yet to be researched. The present study aims to evaluate the impact of assimilating less processed data as GPSRO bending angle profiles on the improvement of the quality of the analysis and forecasts performed using the Gridpoint Statistical Interpolation (GSI) system coupled to the BAM model (GSI/BAM system). Possible deficiencies may mask benefits or detriments when assimilating bending angles into the data assimilation cycle and need to be investigated.

Previous studies at the CPTEC have assessed the assimilation of retrieved profiles of refractivity, temperature and humidity from GPSRO. Sapucci et al. [22] studied the impact of assimilating GPSRO refractivity profiles on the improvement of the CPTEC's Atmospheric Global Circulation Model (AGCM/CPTEC) performance. The refractivity profiles were from the Constellation Observing System for Meteorology, Ionosphere, and Climate (COSMIC) satellites within an experimental version of the Local Ensemble Transform Kalman Filter (LETKF) system coupled to the AGCM/CPTEC. Results indicated a positive impact of the geopotential height at 500 hPa over the Southern Hemisphere and South America, and a significant positive impact was also found over the Tropical region. Azevedo et al. [23] identified among observation systems, such as radiosondes, satellite radiances, and GPSRO refractivity profiles, which had the greatest impact on the CPTEC's analysis and forecasts. Several numerical experiments were performed employing the three-dimensional variational method (3DVar) based on the GSI coupled with the AGCM/CPTEC model (G3DVar). A reduction of the root mean square error (RMSE) was found over the Southern Hemisphere for the geopotential height at 500 hPa when the refractivity profiles were added, very close to results obtained with the assimilation of satellite radiances. It was shown that the assimilation of refractivity profiles allowed more radiance observations to be assimilated, which confirmed the anchoring role of GPSRO observation.

One of the deficiencies found in the previous version of the CPTEC data assimilation system was the lack of a proper observation operator for the assimilation of bending angle profiles. This was a motivation that led us to consider the potential benefits of these data for the skill of the CPTEC's global analysis and forecasts. We addressed this issue in this study through a coupling between the version 3.3 of the GSI system (in its global 3DVar application) and the BAM model, which was called the GSI/BAM system.

In Section 2 the materials and methodology are presented, including a description of the database assimilated, the numerical experiments setup, as well as the main characteristics of the GSI and the BAM model. The analysis and discussion of the obtained results are provided in Section 3. Additional comments and conclusions of this study are presented in Section 4.

2. Materials and Methods

In this work, we are using an updated version of the GSI system from the Developmental Testbed Center (DTC distribution, version 3.3), and the Brazilian atmospheric global model from the CPTEC (BAM) [21]. In this section, we give a brief review of the system components detailing its most relevant aspects.

2.1. Brazilian Global Atmospheric Model (BAM)

The BAM model is the global atmospheric circulation model developed in Brazil. It is an upgraded version of the AGCM/CPTEC, although it remains as a hydrostatic spectral model in which a shallow atmosphere is considered. A detailed description of the AGCM/CPTEC is provided in Cavalcanti et al. [24]. In the BAM model [21], the primitive equations are written using a pure sigma coordinate in the vertical and spherical coordinates in the horizontal domain. Spurious gravity waves are controlled through explicit diffusion mechanisms. The physical space is discretized in an Arakawa-A grid and a semi-implicit Eulerian method is used for the temporal integration with an Asselin filter. BAM includes a Eulerian dynamic core, which was used in this study. The model resolution used was TQ299L64 representing a spectral triangular truncation in the 299 zonal wavenumber (approximately 40 km around the equator line), with 64 levels in the vertical domain. The model top in this resolution is located at 0.3253 hPa (approximately 57 km). For the purpose of numeric stability, the integration time step was 200 s. In the beginning of the integration model, climatological values are used for the surface variables (such as soil moisture, snow depth, surface albedo and land surface temperature), which are adjusted during the integration. Observations of sea surface temperature and snow cover with a spatial resolution of $1^\circ \times 1^\circ$ are introduced in each integration. In addition, an initialization is performed using diabatic normal modes. A quadratic and not reduced grid (900×450 horizontal grid points) is employed in the post-processing of forecasts. Table 1 outlines the physical parameterizations used in this study.

Table 1. Parameterizations of physical processes in the BAM model used in this study.

Long-wave radiation	Harshvardhan et al. [25]
Short-wave radiation	CliRAD (Chou and Suarez [26])
Deep convection	Grell and Dévényi [27]
Shallow convection	Tiedtke [28]
Surface scheme	SSIB (Xue et al. [29])
Boundary layer top	Holtslag and Boville [30]
Boundary layer bottom	Mellor and Yamada [31]

2.2. Gridpoint Statistical Interpolation (GSI) Setup

Since the CPTEC began operational data assimilation activities, the GSI has been the system used to generate the analyses. In this study, we used GSI version 3.3 (v3.3) for implementation in the GSI/BAM, as it includes the most recent improvements for the assimilation of GPSRO data in GSI [32]. It has been verified that within GSI v3.3, the NBAM code presented in Cucurull et al. [20] is implemented for the assimilation of bending angles and, the observation operator described in Cucurull [19] is used for the assimilation of refractivity profiles. The results provided in Cucurull et al. [20] were taken as reference since they show the performance of the NBAM in the GSI.

The 3DVar algorithm in GSI v3.3 was used for the generation of the analyses. It contains information of the control variables at each grid point and vertical model level and it is used as the initial condition for the model integration. In this method, the analysis is conducted by minimizing a cost function (J), which represents the weighted distance between the analysis and the background and the weighted distance between the analysis and the observations. The minimization of J is solved by iterative numerical algorithms [3]. The minimization was performed in one outer loop with 100 inner loops using the conjugate gradient algorithm. This number of inner and outer loop iterations was considered enough to reach the convergence condition. The humidity constraints were activated inside the minimization process of J to control negative and supersaturated moisture values. This procedure is important to penalize the solutions where not-physical humidity values are generated by the numerical and statistical process. Relative pseudo-humidity was the variable chosen to control humidity. With this variable, it is ensured that the relative humidity control variable can only change through changes in specific humidity [32]. The choice of this variable modifies the impact of the data, especially for humidity fields [33]. Since NBAM is implemented in GSI v3.3, the system offers the capability to use compressibility factors to calculate the geopotential heights of the model layers in both observation operators. In addition, the refractivity factors provided by Bevis et al. [34] and Rüeger [35] are included (see Table 2). Cucurull et al. [20] showed that the use of Rüeger coefficients and compressibility factors together did not lead to modification in the results. Therefore, in order to make more suitable the comparison of the results obtained in this study with those reported by Cucurull et al. [20], the Rüeger coefficients and compressibility factors were used in our experiments.

Table 2. Values of the refractivity coefficients available in GSI v3.3.

	Rüeger [35]	Bevis et al. [34]	Unit
k_1	77.6890	77.60	K mb^{-1}
k_2	3.75463×10^5	3.739×10^5	K^2 mb^{-1}
k_3	71.2952	70.4	K mb^{-1}

2.3. Experimental Design

Three numerical experiments were performed for August 2014, considering a spin-up period from 17 until 31 July 2014. The first experiment conducted included the assimilation of all conventional and unconventional data available for this period at the CPTEC. Conventional data assimilated included observations of zonal and meridional winds; temperature; specific humidity; and surface pressure from radiosondes, dropsondes, continental and maritime surface stations, aircraft sensors, balloons and profilers. Atmospheric motion vectors obtained by satellite images were included. Satellite observations included radiances from the sensors: Microwave Humidity Sounder (MHS) on board the NOAA-18 and 19 satellites and from the Meteorological Operational (MetOp) A and B satellites; High Resolution Infrared Radiation Sounder (HIRS/4) on board the MetOp-A satellite; AIRS on board the Aqua satellite; Advanced Microwave Sounding Unit (AMSU-A) on board the NOAA-15, 18 and 19 satellites, the MetOp-A and B satellites and the Aqua satellite; and the IASI on board the MetOp-A and B satellites. In this experiment, any GPSRO data were excluded and was taken as the control run (CNT). CNT is used as a reference to compare the results by assimilating refractivity profiles or bending angles separately. A second experiment was performed adding GPSRO refractivity profiles to the dataset in the CNT experiment. This is called the experiment with refractivity profiles (REF). The last experiment (BND) was set up to include GPSRO bending angle data and the entire dataset of the CNT experiment. GPSRO refractivity and bending angle profiles assimilated were from the missions COSMIC, TerraSAR-X, MetOp-A and B. GPSRO data from COSMIC satellites are processed at COSMIC Data Analysis and Archive Center (CDAAC), from MetOp-A and B satellites are delivered by the Radio Occultation Meteorology Satellite Application Facility (ROM SAF), and the German Research Centre for Geosciences (*GeoForschungsZentrum*; GFZ) provides GPSRO observations from the TerraSAR-X satellite. All data used are distributed in near-real-time by the Global Telecommunication System (GTS) and are received at the CPTEC via File Transfer Protocol (FTP). Table 3 shows a summary of the experiments executed.

Table 3. Configuration of the experiments conducted in this study.

CNT	REF	BND
Conventional data Unconventional data (Radiances) (No GPSRO data)	Conventional data Unconventional data (Radiances) (GPSRO refractivity profiles)	Conventional data Unconventional data (Radiances) (GPSRO bending angles)

According to the literature, refractivity profiles above 30 km are heavily weighted with climatological data during the retrieval process. Thus, in the NBAM, all refractivity data above this height are excluded before quality control procedures. As the bending angle does not suffer from this issue and the model top is at approximately 57 km, the bending angle assimilation has been extended beyond 30 km. However, the upper limit was cut off at 50 km since it is not recommended to assimilate bending angles close to 60 km due to the possible influence of ionospheric noise [36]. Other quality control measures included procedures recommended by each satellite processing center (see details in [32]). Regarding the assigned observation errors, the globally constant GPSRO observation matrices from the NCEP were used, which are distributed into the GSI system. The latitudinal and vertical variations of these matrices were as in Cucurull et al. [20] which in turn followed Desroziers et al. [37].

In all the experiments, cycling analysis and forecasts were performed. The first set of forecasts was obtained by running the BAM model using an analysis from the NCEP. Next, the BAM's forecast was used as background to calculate the next analysis using GSI v3.3. Available data in a time window of ± 3 h were assimilated around the synoptic times (i.e., 00, 06, 12 and 18 UTC). The resulting analysis was used as initial condition for the BAM model in the next step. The First Guess at Appropriate Time (FGAT) [38] approach was used. The forecasts were then generated for a 9 h interval where the forecasts for 3, 6, and 9 h were used as background to assimilate the new dataset of observations and calculate the analysis. Finally, forecasts for a 120-h interval were performed by BAM integration using each analysis obtained as an initial condition.

3. Results and Discussion

The forecast variables used in the analysis of results included integrated content of precipitated water (AGPL); profiles of geopotential height (ZGEO); zonal and meridional wind components (UVEL and VVEL, respectively); specific humidity (UMES); as well as temperature (TEMP) and virtual temperature (VTMP). Variables such as ZGEO, UVEL, VVEL and TEMP were evaluated at levels 250, 500 and 850 hPa while UMES and VTMP were evaluated at levels 500, 850 and 925 hPa. Different regions were considered in the evaluations: the global region from 80°N to 80°S; extratropical Southern Hemisphere (SH), between 80°S and 20°S; extratropical Northern Hemisphere (NH), between 80°N and 20°N and Tropical region (EQ), between 20°S and 20°N. As South America (SA) is the area of major interest for the CPTEC, the results were also focused on this region between 50°S and 10°N and 80°W and 30°W.

3.1. Assimilated GPSRO Data

Table 4 summarizes the number of available and assimilated observations in the REF and BND experiments for August 2014. The total GPSRO observations available corresponds to observations in the BND experiment because bending angles are a prior product. Please note that 2.1% of bending angle observations seems to not be converted in refractivity during the retrieval process. It could be related to the reference point, being the impact parameter in the bending angles and the geometric height in the refractivity profiles, which implies that some observations of bending angles at elevated heights were not correctly retrieved in refractivity. A greater amount of GPSRO observations were assimilated in BND when compared with REF, which means that much more data were able to pass through quality controls in BND. When assimilating bending angles, 73.4% of the data is used, while in REF 42.1% is used. In the range between 0 and 30 km, the assimilated data in BND exceed 3.4% of the total of assimilated observations in REF and the total number of non-assimilated observations in each experiment indicates that 55.8% is not used in REF.

Table 4. Number of available observations and those assimilated in each experiment.

	BND	%	REF	%
Number of available observations	21,385.736	100	20,935.518	97.9
Assimilated observations	15,688.745	73.4	9002.718	42.1
0–30 km	9736.151	45.5	9002.718	42.1
30–50 km	5952.594	27.8	-	-
Not Assimilated observations	5696.991	26.6	11,932.800	55.8

3.2. Observation-Minus-First Guess (OmF)

Figure 1 shows the heights that received a larger contribution from GPSRO observations, as in Cucurull et al. [20] but for the global domain. Statistics of the fractional differences between the observed and modeled refractivity are presented in Figure 1a, and between bending angle observations and those simulated from the forecast model in Figure 1b, respectively. The light gray solid curve represents the mean values and the standard deviation is provided by the light gray dashed curve. The count of assimilated observations is shown through the dark gray dotted curve. Simultaneously, Figure 1 presents the results of normalizing the mean (aven, black solid curve) and standard deviation (stddevn, black dashed curve) of the fractional difference values by the mean observation error. Fractional differences were normalized by the mean refractivity error in REF (Figure 1a) and by the mean bending angle error (Figure 1b) in BND. Fractional differences were quantified by layers of 0.5 km in both experiments. The number of assimilated GPSRO observations in each layer is greater in BND than in the REF experiment, mostly between 8 and 15 km where, on average, the high troposphere and low stratosphere are located, respectively. At around 10 km, the number of assimilated GPSRO data in BND increased to approximately 1000 observations more than in REF. The assimilation of less processed data allows more data to be accepted during quality control procedures. Between 30 and 50 km, the extension of the vertical model domain adjusted by the bending angle observations is observed. The curve of the mean remains very close to zero in both experiments indicating that the analyses were highly influenced by the observations. However, a positive increase in the mean and standard deviation (light gray curves) is observed between 45 and 50 km in BND. Few observational systems are able to sample the atmosphere at those heights, thus, the BAM model as probably many others, may have a poor representation of atmospheric systems at this level. In the lower troposphere, at around 2.5 km, an increase in the standard deviation is also observed although with negative mean values. This increase may be related to the high content and horizontal gradients of water vapor, which render the retrieval of GPSRO observations difficult. The vertical distribution of the water vapor can cause simultaneous multiple paths between the transmitter and the receiver, with GPS signals arriving in different ways and data retrieval becoming more difficult [36]. Super-refraction conditions can also lead to a greater bend and delay of the GPS signals; they are sometimes never received at LEO satellites. Refractivity observations suffer more from these situations during data retrieval [39], but bending angles can be affected by super-refraction when calculating the modeled bending angles in the data assimilation process. On the other hand, the bending angle operator in the NBAM assumes spherical symmetry, neglecting horizontal gradients [20], which could limit the results at 2.5 km and below. After normalizing, it is observed that the mean values are closer to zero in BND than in REF. The normalized standard deviation no longer shows an exponential behavior in BND with values of approximately 1.5% over the entire vertical domain, similar to the results in REF. The influence of horizontal gradients remains, to a certain extent, in the troposphere represented by a small increase in the deviation at around 5 km in both experiments. Despite the stratosphere not being directly involved in the development of daily weather systems, stratospheric conditions impose limitations or restrictions on weather and climate variability. An adequate representation of the stratosphere in the forecasting model can increase its predictability, as also achieved when using the sea surface temperature or sea ice cover data [40]. The assimilation of bending angles data is shown to be suitable to improve the predictability of the high-level systems.

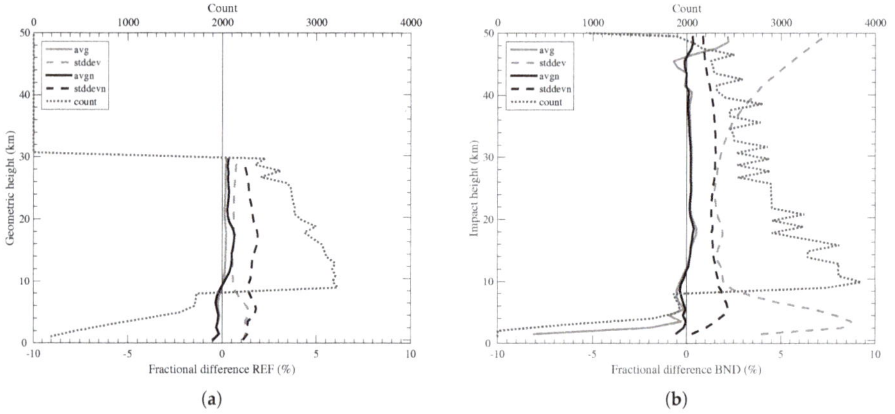

Figure 1. Statistics of the fractional differences ((O-B)/B) (avg and stddev) and normalized by the error observation (avgn and stddevn), between (**a**) the observations and modeled refractivity profiles as a function of the geometric height and (**b**) the observations of bending angles and those simulated from the model as a function of the impact height. Count refers to the total refractivity (**a**) and bending angle (**b**) observations assimilated in each experiment.

Figure 2 shows frequency histograms of the incremental differences in REF and BND stratified in the vertical domain, for 10 August 2014 at 1200 UTC. This result follows what was proposed in Cucurull et al. [20], but for different ranges of height and showing the normal distribution fit curve. Refractivity profiles were analyzed for heights between 0 and 15 km and 15 and 30 km, including the second value of each interval. As the assimilation of bending angles is extended up to 50 km, results between 30 and 50 km were also analyzed. Both types of observations appear positively deviated from the modeled observations, although refractivity profiles show a less deviated probability density function. The bars in BND are heavier than in REF for all the analyzed heights, indicating a larger amount of differences for each class in the former. Between 0 and 15 km, almost 600 observations are concentrated per bin in BND, whereas 450 are concentrated in REF. At heights between 15 and 30 km, this amount decreases to 330 observations per bin concentrated close to zero in BND and a reduction is also observed in REF to around 220 observations per bin. Between 30 and 50 km, the number of bending angle observations is higher than that in the layers below, up to more than 400 observations per bin around zero. However, at this height, a great number of differences are located in the tail of the histogram, agreeing with the highest values of standard deviation observed in Figure 1b. The standard deviation in the lower atmosphere is also observed in the first heights range in BND through a wider Gaussian curve. The behavior in each interval shows a Gaussian shape with mean difference values located closest to zero, which in turn means that the first-guess is closer to the observations. These results are slightly different from Cucurull et al. [20], where the observations in the last interval showed a clearly non-Gaussian shape curve.

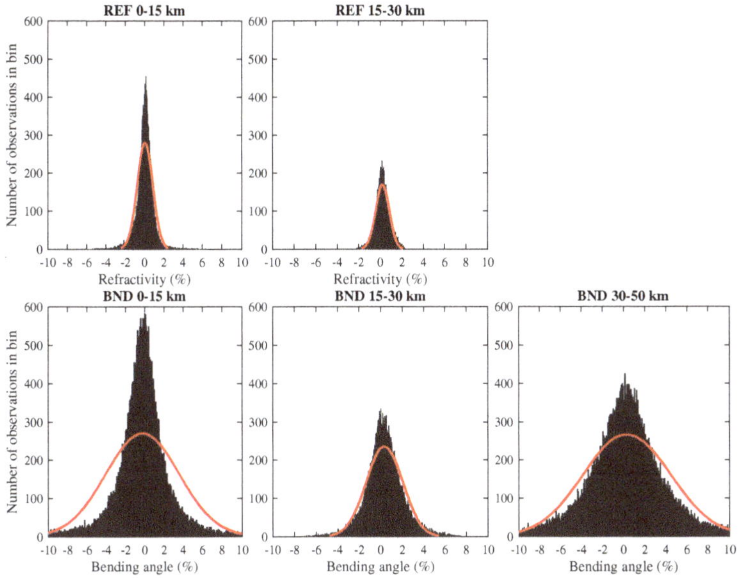

Figure 2. Frequency histograms of the differences between the observations and background by height intervals, in the REF (**upper** panels) and BND (**lower** panels) experiments, referring to the analysis generated on 10 August 2014 at 1200 UTC. Red curves represent the normal distribution fit curve in each height range.

3.3. Cost Function Minimization

Figure 3 presents the mean and standard deviation of the cost functions and gradient norms in each experiment (CNT, REF e BND). The results of all the analyses are shown, generated at 1200 UTC since a greater number of data were available; therefore, a higher efficiency of the process was required. An increase in the values of J is observed from CNT to REF and BND which is directly related to the total of assimilated observations in each case. The highest values of J are reached in BND, with the standard deviation slightly higher than those obtained in the REF and CNT experiments. However, the standard deviation of the gradient norm indicates that similar values are obtained in all the experiments performed. It is noticeable that even with a significant increase in the number of observations in BND (almost twice that in REF), the system converged to the minimum value in the first 50 iterations as in CNT and REF. The obtained results are consistent with our objective since the statistics for all the experiments are quite similar. As the focus in this study is to assess the impact of assimilating less processed GPSRO data in our NWP system, it was configured in a simple form to obtain explicit results, as we can observe in Figure 3. A further study to optimize the minimization process may be conducted in which the execution of more inner and outer loops is assessed.

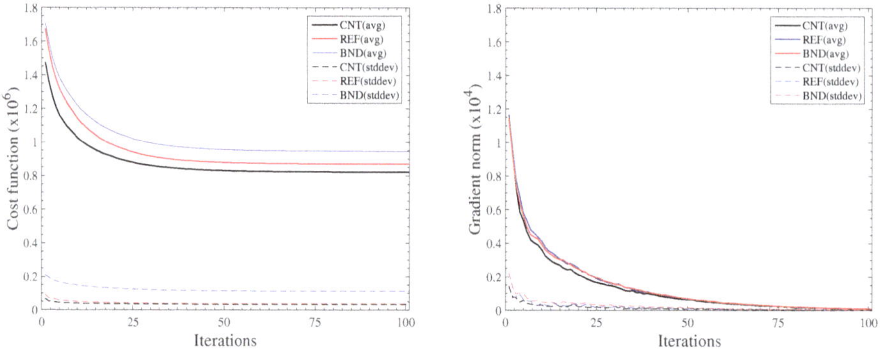

Figure 3. Statistics of the cost functions and gradient norms for the CNT, REF, BND experiments at 1200 UTC.

The fraction of reduction of the initial cost function provided by a group of type **x** observations during the minimization process was computed as in Cucurull et al. [20]. This analysis is very useful to determine which GPSRO observation type is more costly in the minimization of J. The calculation was carried out using the following formula:

$$Reduction_x = \frac{J_{0_x} - J_{f_x}}{J_0 - J_f} \tag{1}$$

where J_0 and J_f are the initial (from the first guess) and final (from the analysis x_a) total observation cost functions, respectively, and J_{0_x} and J_{f_x} are the cost function components of the group of type **x** observations, which are as follows: surface pressure (P_s); temperature (T); wind (W); humidity; GPS (refractivity or bending angle data); and Radiance.

Figure 4 shows the results, in percentage form, for 10 August 2014 at 1200 UTC (Figure 4a) and normalized by the total of assimilated observations of each group (Figure 4b). The largest contribution reducing J_0 is accomplished by radiance observations with 64% in BND and 63% in REF. Refractivity profiles indicated 17% reduction and bending angles indicated 17.5% reduction, whereas wind observations contributed to 16% of reduction in BND and 16.5% in REF. Surface pressure, temperature and humidity observations contributed up to 7%. Globally, the high percentage value is found in the radiances because the higher number of assimilated data is from this observation system. However, the greatest contribution is given by the GPS when normalizing by the total of each observation, as shown in Figure 4b where the values are relative. A contribution of 2.5×10^{-6} was obtained by each refractivity data and 1.4×10^{-6} by each bending angle observation, while each radiance observation contributes 0.25×10^{-6}. For temperature, humidity and radiance observations, the greatest reduction was found when assimilating bending angles. However, a somewhat larger contribution remains when assimilating refractivity profiles for wind and pressure surface observations, as well as for GPS themselves where the highest reduction value was achieved in REF. These results are due to the smaller amount of assimilated refractivity data compared to the amount of bending angles being assimilated in BND. Observation operators to assimilate less processed data are more complex than for direct observations. For bending angles, the operator requires the projection of the modeled refractivity into bending angle values and its location [20].

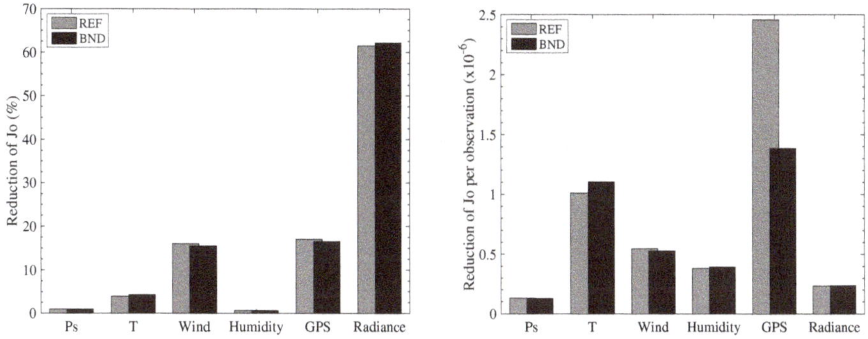

(**a**) (**b**)

Figure 4. Contribution of (**a**) each type of observation (in %) and (**b**) each type of observation normalized by the number of observations used in each case, in the reduction of the total cost function in the analysis generated for 10 August 2014 at 1200 UTC.

3.4. Analyses Differences

Figure 5 shows the global mean analysis temperature difference between BND and REF at 10 hPa and 850 hPa, respectively. It is observed that at 850 hPa, temperature analyses are slightly cooler in BND than in REF over the tropical latitudes and warmer in the Antarctic region. However, the mean differences are very small varying from −1 to 1 K. At 10 hPa, the differences increase to values between −3 and 3 K, with positive differences over the central Pacific Ocean and the west of the Atlantic Ocean including the tropical region of SA. High latitudes in the SH also show a temperature analysis slightly warmer in BND than in REF. At both levels, it is observed that over the southern high latitudes, on average, temperature analysis is warmer when assimilating bending angle observations. It could indicate that the model is probably routinely cooling the South Pole region and warming tropical latitudes. Figueroa et al. [21] (Figure 1) shows the surface latent heat fluxes averaged for December, January and February in the BAM and Era-Interim reanalysis, where BAM reproduced cooler surface latent heat fluxes in the southern high latitudes with slightly high values over tropical regions for that period. Results in this study suggest that, although the BAM model still needs to be improved, by assimilating bending angles into the NWP system of the CPTEC, it is possible to generate a more realistic initial condition.

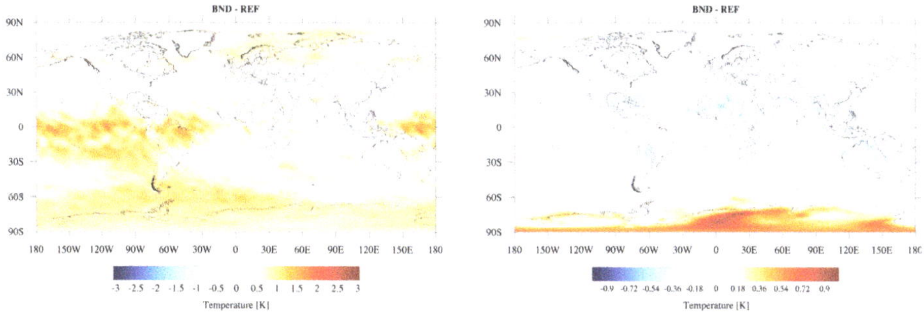

(a) (**b**)

Figure 5. Mean difference between BND and REF temperature analysis at (**a**) 10 hPa and (**b**) 850 hPa, for August 2014.

3.5. Balance in the First Guess

The initial condition after the data assimilation process should be balanced so as not to degrade weather forecasts. According to Lynch and Huang [41], the unbalance generated from the assimilation process can be measured by computing the mean absolute surface pressure tendency. Wang et al. [42] also used the mean absolute tendency calculation as a measure of high-frequency noise in the forecasts generated using different types of initialization. In this study, we use the mean absolute surface pressure tendency to analyze the surface pressure forecasts that are used as the first guess in the assimilation cycle. Because a FGAT approach was used, the first 9-h forecasts were analyzed. Following Lynch and Huang [41], the mean absolute tendency (N) was calculated as:

$$N = \left(\frac{1}{MN}\right) \sum_{m=1}^{M} \sum_{n=1}^{N} \left|\frac{\partial P_s}{\partial t}\right| \tag{2}$$

where M and N are the points of the entire global domain, P_s is the surface pressure and t is the forecast time. As an indication of balance, the stable oscillation of tendency values around a determined value was considered. Figure 6 shows the mean absolute tendencies in each experiment. Please note that N in BND, REF, and CNT has an initial value of 0.42, 0.39 and 0.3 hPa/h, respectively. Afterwards, the values fall in the second hour of forecasts in all the experiments and then oscillate around approximately 0.25 hPa, which indicates that the 3-, 6- and 9-h forecasts, used as the first guess, are balanced. Although BND shows the larger initial value, which is due to the largest amount of assimilated data in this experiment, the tendency values also fluctuate between 0.25 and 0.27 hPa/h after two forecast hours. The results suggest that the model is not creating or losing mass during the forecast step in the analysis cycle, which contributes to the balanced analysis being obtained.

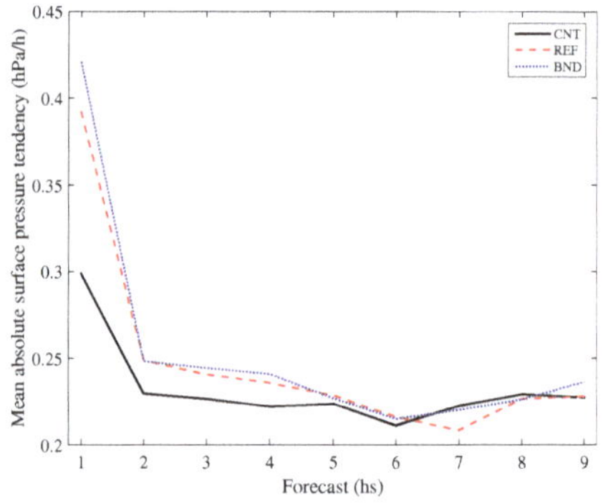

Figure 6. Mean absolute surface pressure tendency in each experiment over the global region for the period under study.

3.6. Forecasts Skill

The anomaly correlation coefficient (ACC) and RMSE were calculated to evaluate the 120-h forecast. For a better interpretation of the RMSE and ACC results, a Gain Coefficient (GC) was calculated following Sapucci et al. [22]. The GC is very useful to show how important the results were in the forecasts when adding bending angle or refractivity data to the CNT experiment, respectively. This gain is relative to the experiment taken as a control and was calculated using the formulas below:

$$\text{GAIN}_{v_t}^{\text{RMSE}} = \frac{\text{RMSE}_{v_t}^{E_i} - \text{RMSE}_{v_t}^{C}}{\text{RMSE}_{perfect} - \text{RMSE}_{v_t}^{C}} \times 100\% \tag{3}$$

$$\text{GAIN}_{v_t}^{\text{ACC}} = \frac{\text{ACC}_{v_t}^{E_i} - \text{ACC}_{v_t}^{C}}{\text{ACC}_{perfect} - \text{ACC}_{v_t}^{C}} \times 100\% \tag{4}$$

where v corresponds to any of the variables evaluated at each integration time t, while E indicates the results of the i experiments in which some GPSRO data set was incremented (REF and BND) and C represents the results of the CNT experiment. $\text{RMSE}_{perfect}$ represents the RMSE value in the case of predictions of a perfect theoretical model, which would imply that it is equal to 0. For the ACC gain, the perfect value corresponds to 1, representing a 100% of correlation between the anomalies of the predicted fields and the analysis in each experiment with respect to the climatology. For both measures, a positive gain value indicates that the experiment adding some GPSRO-type observations benefited the weather forecasts, that is, RMSE values were reduced and ACC values increased. Otherwise, negative values indicate that the addition of this data degrades the forecasts. To highlight where the gains were concentrated, the difference in the RMSE and ACC gain values between BND and REF experiments was calculated for each forecast time, variable and analyzed level.

Figure 7 presents results for the mean difference in RMSE gain values for the 12-h temperature forecast at 250, 500, and 850 hPa. Please note that when assimilating bending angles, a noticeable gain in the RMSE values is concentrated in the tropical latitudes between 20°S and 20°N in the vertical atmospheric column. A reduction in the RMSE is noticeable at 250 hPa where gain values up to 60% are observed. Although, at this level, high values of losses are also reached, throughout the middle and lower atmosphere, the losses are highly reduced showing small values and minor areas at 850 hPa. A gain in the RMSE in BND is also observed over the North and South Poles at 500 and 850 hPa, indicating the positive influence of the bending angle in the temperature analysis. The results from temperature forecasts are representative of the other evaluated variables, which indicate that, probably, the vertical model resolution used (64 vertical sigma levels) may not yet be adequate to assimilate this type of observation that has a high vertical resolution. Furthermore, to fully explore these data, the model should correctly represent atmospheric systems at high levels; however, since there are not many other types of meteorological observations that reliably provide measurements at high atmospheric heights (above 30 km), the model may not appropriately characterize the errors at those heights of the domain. An additional study should be carried out to adjust the background error covariance matrix for the BAM model.

The results of the RMSE and ACC differences are shown in Figure 8. The highest impact in the RMSE gain values is observed when assimilating bending angles during the 5-day forecast for most variables. The RMSE is decreased by 3.5% in the zonal and meridional wind components at 250 hPa, and by 2.5% in the geopotential height also at 250 hPa. These results corroborate the role of GPSRO data assimilation, indirectly impacting mass field forecasts [14]. Virtual and absolute temperature forecasts at 500 and 850 hPa show improvements of about 1 to 2% until the 72-h forecast. For 96- and 120-h forecasts, the enhancements still remain with gains of 0.5 and 1.5%, respectively. The specific humidity forecasts also present gains in the RMSE values when bending angles are assimilated. Gains of 2.5% are observed at 500 and 925 hPa and 1% at 850 hPa for the 24-h forecast. In this variable, a decrease in the RMSE persists for the 120-h lead time in the three evaluated levels with values of 0.5 to 1.8%. The improvements for the specific humidity are remarkable since the BAM has some deficiencies in the precipitation forecast over some regions such as the Amazon and La Plata [21]. The results from the absolute temperature forecast show a degradation in the RMSE gain values at 250 hPa, reaching 2% in the 24-h forecast and diminishing as the forecast time advances. Otherwise, the ACC differences results also show noticeable gain values in the zonal and meridional wind components, geopotential height, and specific humidity by assimilating bending angles. Gain values reach 5.5% and 7% in the wind components at 250 hPa, and specific humidity and geopotential height at 250 hPa, respectively. On average, the forecasts of virtual and absolute temperature show a degradation of 0.5 to 1.5% in BND. Degradation of 2.5% is noted in the absolute temperature at 500 hPa for the 24-h forecast, which becomes neutral in the 48-, 72-, and 96-h forecasts and is again degraded in the 120-h forecast with a loss of 0.5%. Although losses are also found when assimilating bending angles, the results are quite noticeable and suggest the operational use of these data.

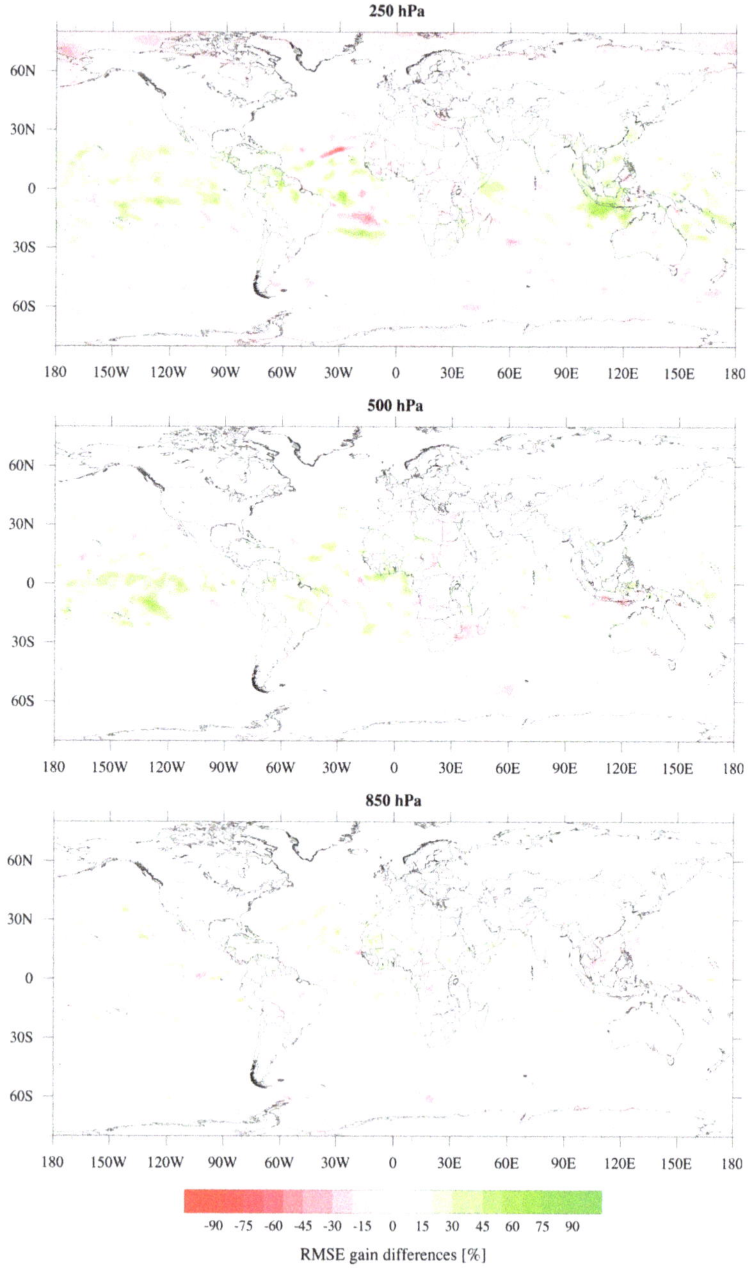

Figure 7. Mean difference in the RMSE gain between BND and REF for the 12-h temperature forecast at 250, 500 and 850 hPa (from **top** to **bottom**), for August 2014. Green values indicate a gain in the RMSE values, while red values indicate a loss.

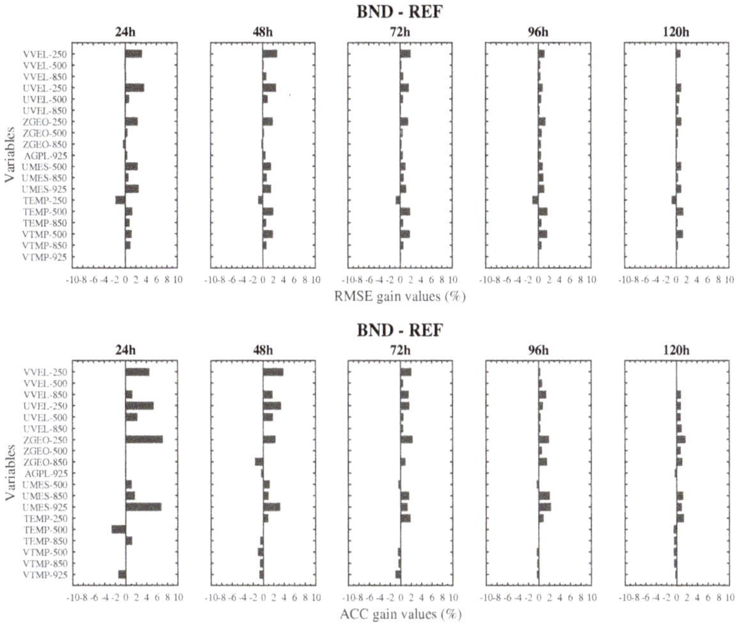

Figure 8. Differences in RMSE and ACC gain values for all variables (represented on the ordinate axis) for: 24-, 48-, 72-, 96- and 120-h forecasts (from **left** to **right**, respectively) over the global region.

Considering the mission of the CPTEC to improve forecasts for the SA region, the Fractional Change (FC) was calculated to highlight the gains in the RMSE over this region. Both the GC and the FC measures are positively oriented, with positive values indicating improvements or positive impacts in the forecasts made. The FC was computed by Equation (5) as recommended in Anthes et al. [43]. The nomenclature used has the same meaning as for the GC formulation.

$$FC = 1 - \frac{RMSE_{E_i}}{RMSE_C} \tag{5}$$

Figure 9 shows the FC values calculated for REF and BND experiments in the SA region, where the zero line is highlighted. The results of FC indicate that in BND the values are more positive than when assimilating the refractivity profiles in most variables. The results obtained for the global domain are also observed over SA, where at 500, 850, and 925 hPa most of the variables show improvements when assimilating bending angles (more FC results reaching the zero value). Temperature forecasts for the 120 h of integration present a low performance at 250 hPa, although small improvements are observed until the 96-h forecast. As mentioned before, the results obtained in this work suggest the need for a study in which the skill of the BAM to represent the atmospheric systems at high levels is investigated and documented; the background error covariance matrix of the BAM should also be appropriately adjusted. Moreover, an increase in the vertical resolution of the model would allow us to further explore the data usage and the precision of the modeled observations from the background.

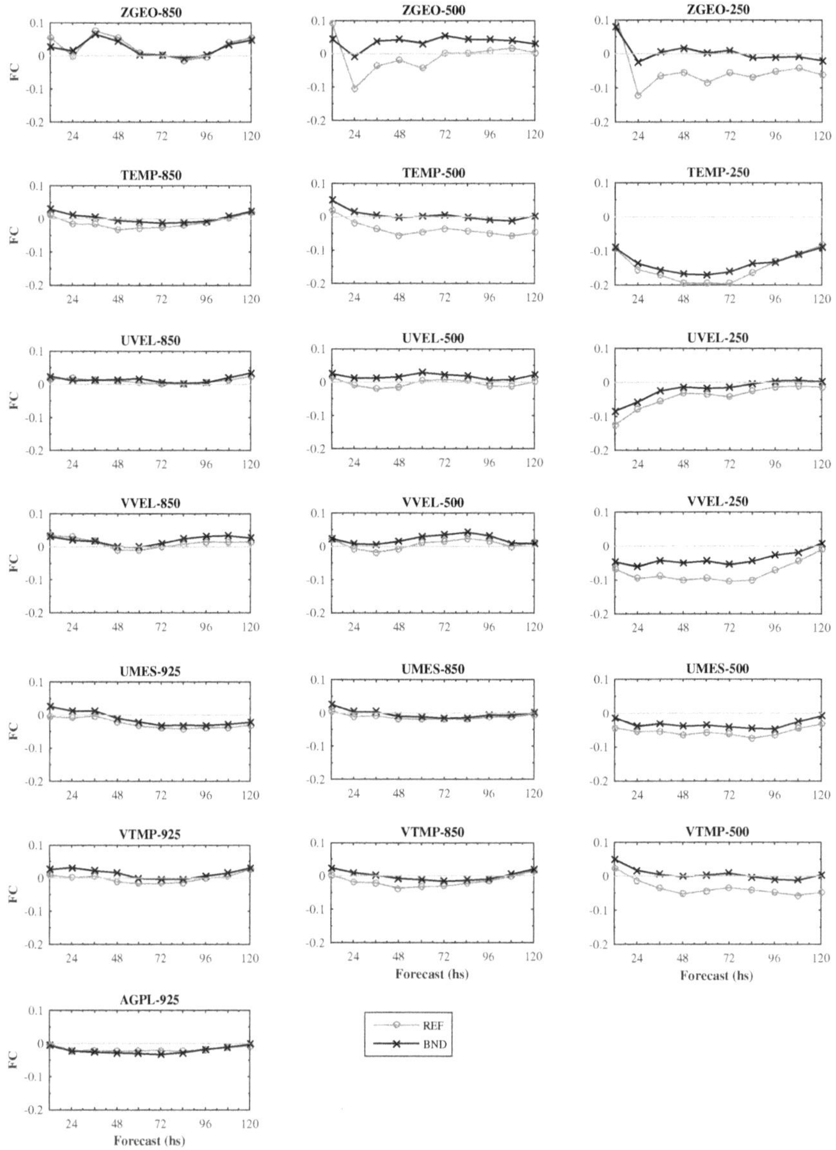

Figure 9. Fractional change of the RMSE values for all variables evaluated and the 120 h of model integration over the South America region.

4. Conclusions

This study presents the first results of the impact of assimilating less processed GPSRO data into the NWP system at the CPTEC, specifically assimilating bending angles over observations of refractivity profiles. Three numerical experiments were conducted—BND, REF and CTN—for August 2014, in which the GPSRO observation used was modified, that is, bending angles compared with refractivity profiles and not using any GPSRO data, respectively. The GSI system was used, coupled to the BAM model. With the assimilation of bending angle data, a greater extension of the vertical domain of the model was impacted, exploring 73.4% of the potential of the RO technique, while with the assimilation of refractivity profiles, less than half of this potential was explored. Despite the exponential behavior of the standard deviation of the fractional differences in BND, it was observed that standard deviation values are similar for the two observation types when normalizing by the mean of the error value in each layer. Although the number of observations of GPSRO is almost doubled using bending angles, the analyses do not require extra computational cost. A further study should be developed to optimize the minimization process in the data assimilation system at the CPTEC.

Regarding the impact on the forecasts, a high impact was clearly observed with the assimilation of bending angles, throughout the model integration time evaluated. The greatest impact of assimilating those data over the refractivity profiles was the RMSE global reduction in almost all the evaluated variables throughout the 5-day forecast. RMSE gain values achieved 3.5% and 2.5% in the zonal and meridional wind components and geopotencial height at 250 hPa, respectively.

The results of the fractional change over South America indicate that the most impacted atmospheric levels were 500 and 850 hPa; however, in BND, at 250 hPa, more positive results were obtained than in REF. The slightly lower performance at the 250 hPa level may be related to the need for a higher vertical resolution of the model and a better representation of the atmosphere at high levels, which should be explored in future research. On the other hand, since there are not many other types of observation systems that reliably provide measurements at higher heights, it is necessary to carry out studies that explore the representation of meteorological systems at those heights and adjust the background error covariance matrix currently in use.

In summary, this study established that the highest gain—after assimilating bending angles into the global data assimilation system of the CPTEC—was the significant increase of the assimilated data throughout the vertical atmospheric column, and especially at the higher levels of the atmosphere. The results reported here are meaningful for the modeling activities at the CPTEC and give clear indications about the improvements in the results due to the best use of GPSRO data, which can guide daily operational assimilation.

Author Contributions: L.F.S. and L.C. had an important role in the design of the study, results discussion and review of the manuscript. C.F.B. and B.B.S. provided scientific and technical support for the coupling of the GSI/BAM system, interpretation of the results and writing of the manuscript. I.H.B. performed the experiments, analyzed the data, generated the figures and wrote the document.

Funding: The first author acknowledges the financial support of the CNPq (acronym in Portuguese for National Council for Scientific and Technological Development) and CAPES (acronym in Portuguese for Coordination for the Improvement of Higher Education Personnel) during her Master and ongoing PhD studies. The APC was funded by the Graduate Program in Meteorology of INPE.

Acknowledgments: The authors thank the CPTEC for providing access to the computational resources used for the development of this research. The Group on Data Assimilation Development and the BAM model developers are also acknowledged for its technical support. We also thank the three anonymous reviewers and the Academic Editor whose comments and suggestions contributed significantly to the improvement of this article.

Conflicts of Interest: The authors declare no conflict of interest.

References

1. Bauer, P.; Thorpe, A.; Brunet, G. The quiet revolution of numerical weather prediction. *Nature* **2015**, *525*, 47–55. [CrossRef] [PubMed]
2. Palmer, T.; Hagedorn, R. *Predictability of Weather and Climate*; Cambridge University Press: Cambridge, UK, 2006; p. 702.
3. Kalnay, E. *Atmospheric Modeling, Data Assimilation, and Predictability*; Cambridge University Press: Cambridge, UK, 2003; Volume 54, p. 341.
4. Kursinski, E.R.; Hajj, G.A.; Bertiger, W.I.; Leroy, S.S.; Meehan, T.K.; Romans, L.J.; Schofield, J.T.; McCleese, D.J.; Melbourne, W.G.; Thornton, C.L.; et al. Initial results of radio occultation observations of Earth's atmosphere using the Global Positioning System. *Science* **1996**, *271*, 1107–1110. [CrossRef]
5. Eyre, J. Assimilation of Radio Occultation measurements into a numerical weather prediction system. In *ECMWF Technical Memoranda*; ECMWF: Reading, UK, 1994; p. 22.

6. Cardinali, C.; Healy, S. Impact of GPS radio occultation measurements in the ECMWF system using adjoint-based diagnostics. *Q. J. R. Meteorol. Soc.* **2014**, *140*, 2315–2320. [CrossRef]
7. Fussen, D.; Tétard, C.; Dekemper, E.; Pieroux, D.; Mateshvili, N.; Vanhellemont, F.; Franssens, G.; Demoulin, P. Retrieval of vertical profiles of atmospheric refraction angles by inversion of optical dilution measurements. *Atmos. Meas. Tech.* **2015**, *8*, 3135–3145. [CrossRef]
8. Kursinski, E.; Hajj, G.; Schofield, J.T.; Linfield, R.P.; Hardy, K.R. Observing Earth's atmosphere with radio occultation measurements using the Global Positioning System. *J. Geophys. Res.* **1997**, *102*, 23429–23465. [CrossRef]
9. Cardinali, C.; Healy, S. GPS-RO at ECMWF. In Proceedings of the ECMWF Seminar on Data Assimilation for Atmosphere and Ocean, Reading, UK, 6–9 September 2011; pp. 323–336.
10. Cucurull, L.; Anthes, R.A. Impact of Infrared , Microwave , and Radio Occultation Satellite Observations on Operational Numerical Weather Prediction. *Mon. Weather Rev.* **2014**, *142*, 4164–4186. [CrossRef]
11. Cucurull, L.; Anthes, R.A. Impact of Loss of U.S. Microwave and Radio Occultation Observations in Operational Numerical Weather Prediction in Support of the U.S. Data Gap Mitigation Activities. *Weather Forecast.* **2015**, *30*, 255–269. [CrossRef]
12. Cucurull, L.; Derber, J.C. Operational Implementation of COSMIC Observations into NCEP's Global Data Assimilation System. *Weather Forecast.* **2008**, *23*, 702–711. [CrossRef]
13. Cucurull, L.; Derber, J.C.; Treadon, R.; Purser, R.J. Assimilation of Global Positioning System Radio Occultation Observations into NCEP's Global Data Assimilation System. *Mon. Weather Rev.* **2007**, *135*, 3174–3193. [CrossRef]
14. Rennie, M.P. The impact of GPS radio occultation assimilation at the Met Office. *Q. J. R. Meteorol. Soc.* **2010**, *136*, 116–131. [CrossRef]
15. Healy, S.B.; Thepaut, J.N. Assimilation experiments with CHAMP GPS radio occultation measurements. *Q. J. R. Meteorol. Soc.* **2006**, *132*, 605–623. [CrossRef]
16. Bonavita, M. On some aspects of the impact of GPSRO observations in global numerical weather prediction. *Q. J. R. Meteorol. Soc.* **2014**, *140*, 2546–2562. [CrossRef]
17. Cucurull, L.; Kuo, Y.H.; Barker, D.; Rizvi, S.R.H. Assessing the Impact of Simulated COSMIC GPS Radio Occultation Data on Weather Analysis over the Antarctic: A Case Study. *Mon. Weather Rev.* **2006**, *134*, 3283–3296. [CrossRef]
18. Cucurull, L.; Derber, J.C.; Treadon, R.; Purser, R.J. Preliminary Impact Studies Using Global Positioning System Radio Occultation Profiles at NCEP. *Mon. Weather Rev.* **2008**, *136*, 1865–1877. [CrossRef]
19. Cucurull, L. Improvement in the Use of an Operational Constellation of GPS Radio Occultation Receivers in Weather Forecasting. *Weather Forecast.* **2010**, *25*, 749–767. [CrossRef]
20. Cucurull, L.; Derber, J.C.; Purser, R.J. A bending angle forward operator for global positioning system radio occultation measurements. *J. Geophys. Res. Atmos.* **2013**, *118*, 14–28. [CrossRef]
21. Figueroa, S.N.; Bonatti, J.P.; Kubota, P.Y.; Grell, G.A.; Morrison, H.; Barros, S.R.M.; Fernandez, J.P.R.; Ramirez, E.; Siqueira, L.; Luzia, G.; et al. The Brazilian Global Atmospheric Model (BAM): Performance for Tropical Rainfall Forecasting and Sensitivity to Convective Scheme and Horizontal Resolution. *Weather Forecast.* **2016**, *31*, 1547–1572. [CrossRef]
22. Sapucci, L.F.; Diniz, F.L.R.; Bastarz, C.F.; Avanço, L.A. Inclusion of Global Navigation Satellite System radio occultation data into Center for Weather Forecast and Climate Studies Local Ensemble Transform Kalman Filter (LETKF) using the Radio Occultation Processing Package as an observation operator. *Meteorol. Appl.* **2016**, *23*, 328–338. [CrossRef]
23. Azevedo, H.B.D.; Gonçalves, L.G.G.D.; Bastarz, C.F.; Silveira, B.B. Observing System Experiments in a 3DVAR Data Assimilation System at CPTEC/INPE. *Weather Forecast.* **2017**, *32*, 873–880. [CrossRef]
24. Cavalcanti, I.F.; Marengo, J.A.; Satyamurty, P.; Nobre, C.A.; Trosnikov, I.; Bonatti, J.P.; Manzi, A.O.; Tarasova, T.; Pezzi, L.P.; D'Almeida, C.; et al. Global climatological features in a simulation using the CPTEC-COLA AGCM. *J. Clim.* **2002**, *15*, 2965–2988. [CrossRef]
25. Davies, R.; Randall, D.A.; Corsetti, T.G. A fast radiation parameterization for atmospheric circulation models. *J. Geophys. Res.* **1987**, *92*, 1009–1016.
26. Chou, M.D.; Suarez, M.J. A Solar Radiation Parameterization Atmospheric Studies. In *Technical Report Series on Global Modeling and Data Assimilation*; NASA: Washington, DC, USA, 1999.

27. Grell, G.A.; Dévényi, D. A generalized approach to parameterizing convection combining ensemble and data assimilation techniques. *Geophys. Res. Lett.* **2002**, *29*, 38-1–38-4. [CrossRef]
28. Tiedtke, M. The sensitivity of the time-mean large-scale flow to cumulus convection in the ECMWF model. In Proceedings of the ECMWF Workshop on Convection in Large-Scale Models, Shinfield Park, Reading, 28 November–1 December 1983; European Centre for Medium-Range Weather Forecasts: Reading, UK, 1983; pp. 297–316.
29. Xue, Y.; Sellers, P.; Kinter, J.; Shukla, J. A Simplified Biosphere Model for Global Climate Studies. *J. Clim.* **1991**, *4*, 345–364. [CrossRef]
30. Holtslag, A.A.M.; Boville, B.A. Local versus nonlocal boundary-layer diffusion in a global climate model. *J. Clim.* **1993**, *6*, 1825–1842. [CrossRef]
31. Mellor, G.; Yamada, T. Development of a turbulence closure for geophysical fluid problems. *Rev. Geophys. Space Phys.* **1982**, *20*, 851–875. [CrossRef]
32. Developmental Testbed Center. Gridpoint Statistical Interpolation (GSI) User's Guide for Version 3.3. In *Community Gridpoint Statistical Interpolation System*; Developmental Testbed Center: Boulder, CO, USA, 2014; p. 108.
33. Campos, T.B.; Sapucci, L.F.; Lima, W.; Ferreira, D.S. Sensitivity of Numerical Weather Prediction to the Choice of Variable for Atmospheric Moisture Analysis into the Brazilian Global Model Data Assimilation System. *Atmosphere* **2018**, *9*, 123. [CrossRef]
34. Bevis, M.; Businger, S.; Chiswell, S.; Herring, T.A.; Anthes, R.A.; Rocken, C.; Ware, R.H. GPS Meteorology: Mapping Zenith Wet Delays onto Precipitable Water. *J. Appl. Meteorol.* **1994**, *33*, 379–386. [CrossRef]
35. Rüeger, J.M. Refractive Index Formulae for Radio Waves. In Proceedings of the JS28 Integration of Techniques and Corrections to Achieve Accurate Engineering, FIG Proceedings XXII International Congress, Washington, DC, USA, 19–26 April 2002; pp. 19–23.
36. Jin, S.; Cardellach, E.; Xie, F. *GNSS Remote Sensing*; Springer: Berlin, Germany, 2014; Volume 19, pp. 215–239.
37. Desroziers, G.; Berre, L.; Chapnik, B.; Poli, P. Diagnosis of observation, background and analysis-error statistics in observation space. *Q. J. R. Meteorol. Soc.* **2005**, *131*, 3385–3396. [CrossRef]
38. Lee, M.s.; Barker, D.; Huang, W.; Kuo, Y.H. First Guess at Appropriate Time (FGAT) with WRF 3DVAR. *J. Korean Meteorol. Soc.* **2005**, *41*, 495–505.
39. Syndergaard, S. On the ionosphere calibration in GPS radio occultation measurements. *Radio Sci.* **2000**, *35*, 865–883. [CrossRef]
40. Karpechko, A.; Tummon, F.; Secretariat WMO. Climate Predictability in the Stratosphere. *Bull. World Meteorol. Organ.* **2016**, *65, 54–57*
41. Lynch, P.; Huang, X.Y. Initialization of the HIRLAM Model Using a Digital Filter. *Mon. Weather Rev.* **1992**, *120*, 1019–1034. [CrossRef]
42. Wang, H.G.; Wu, Z.S.; Kang, S.F.; Zhao, Z.W. Monitoring the marine atmospheric refractivity profiles by ground-based GPS occultation. *IEEE Geosci. Remote Sens. Lett.* **2013**, *10*, 962–965. [CrossRef]
43. Anthes, R.A.; Ector, D.; Hunt, D.C.; Kuo, Y.H.; Rocken, C.; Schreiner, W.S.; Sokolovskiy, S.V.; Syndergaard, S.; Wee, T.K.; Zeng, Z.; et al. The COSMIC/FORMOSAT-3 Mission: Early Results. *Bull. Am. Meteorol. Soc.* **2008**, *89*, 313–333. [CrossRef]

remote sensing

MDPI

Article

Data-Driven Interpolation of Sea Level Anomalies Using Analog Data Assimilation

Redouane Lguensat [1,*], Phi Huynh Viet [2], Miao Sun [3], Ge Chen [4], Tian Fenglin [4], Bertrand Chapron [5] and Ronan Fablet [2]

1 IGE, Université Grenoble Alpes, CNRS, IRD, Grenoble INP, 38000 Grenoble, France
2 IMT Atlantique, Lab-STICC UMR CNRS 6285, UBL, 29200 Brest, France; Vietphi3892@gmail.com (P.H.V.); ronan.fablet@imt-atlantique.fr (R.F.)
3 Key Laboratory of Digital Ocean, National Marine Data and Information Service, Tianjin 300171, China; miaomiao_1987qq@126.com
4 Department of Marine Information Technology, Ocean University of China, Qingdao 266100, China; gechen@ouc.edu.cn (G.C.); tianfenglin@ouc.edu.cn (T.F.)
5 Laboratoire d'Océanographie Physique et Spatiale, IFREMER, 29200 Brest, France; bertrand.chapron@ifremer.fr
* Correspondence: redouane.lguensat@univ-grenoble-alpes.fr

Received: 21 January 2019; Accepted: 4 April 2019; Published: 9 April 2019

Abstract: From the recent developments of data-driven methods as a means to better exploit large-scale observation, simulation and reanalysis datasets for solving inverse problems, this study addresses the improvement of the reconstruction of higher-resolution Sea Level Anomaly (SLA) fields using analog strategies. This reconstruction is stated as an analog data assimilation issue, where the analog models rely on patch-based and Empirical Orthogonal Functions (EOF)-based representations to circumvent the curse of dimensionality. We implement an Observation System Simulation Experiment (OSSE) in the South China Sea. The reported results show the relevance of the proposed framework with a significant gain in terms of Root Mean Square Error (RMSE) for scales below 100 km. We further discuss the usefulness of the proposed analog model as a means to exploit high-resolution model simulations for the processing and analysis of current and future satellite-derived altimetric data with regard to conventional interpolation schemes, especially optimal interpolation.

Keywords: analog data assimilation; sea level anomaly; sea surface height; interpolation; data-driven methods

1. Introduction

Over the last two decades, ocean remote sensing data has benefited from numerous remote earth observation missions. These satellites measured and transmitted data about several ocean properties, such as sea surface height, sea surface temperature, ocean color, ocean current, sea ice, etc. This has helped building big databases of valuable information and represents a major opportunity for the interplay of ideas between the ocean remote sensing community and the data science community. Exploring machine learning methods in general and non-parametric methods in particular is now feasible and is increasingly drawing the attention of many researchers [1,2].

More specifically, analog forecasting [3] which is among the earliest statistical methods explored in geoscience benefits from recent advances in data science. In short, analog forecasting is based on the assumption that the future state of a system can be predicted throughout the successors of past (or simulated) similar situations (called analogs). The amount of currently available remote sensing and simulation data offers analog methods a great opportunity to catch up their early promises. Several recent works involving applications of analog forecasting methods in geoscience fields contribute in the revival of these methods, recent applications comprise the prediction of soil moisture anomalies [4], the prediction of sea-ice anomalies [5], rainfall nowcasting [6], numerical weather prediction [7–9], etc. One may also cite methodological developments such as dynamically-adapted kernels [10] and novel parameter estimation schemes [11]. Importantly, analog strategies have recently been extended to address data assimilation issues within the so-called analog data assimilation (AnDA) [12,13], where the dynamical model is stated as an analog forecasting model and combined to state-of-the-art stochastic assimilation procedures such as Ensemble Kalman filters.

Producing time-continuous and gridded maps of Sea Surface Height (SSH) is a major challenge in ocean remote sensing with important consequences on several scientific fields from weather and climate forecasting to operational needs for fisheries management and marine operations (e.g., [14]). The reference gridded SSH product commonly used in the literature is distributed by Copernicus Marine Environment Monitoring Service (CMEMS) (formerly distributed by AVISO+). This product relies on the interpolation of irregularly-spaced along-track data using an Optimal Interpolation (OI) method [15,16]. While OI is relevant for the retrieval of horizontal scales of SSH fields up to $\approx$100 km, the prescribed covariance priors lead to smoothing out finer-scales. Typically, horizontal scales from a few tens of kilometers to $\approx$100 km may be poorly resolved by OI-derived SSH fields, while they may be partially revealed by along-track altimetric data. This has led to a variety of research studies to improve the reconstruction of the altimetric fields. One may cite both methodological alternatives to OI, for instance locally-adapted convolutional models [17] and variational assimilation schemes using model-driven dynamical priors [18], as well as studies exploring the synergy between different sea surface tracers, especially the synergy between SSH and SST (Sea Surface Temperature) fields and Surface Quasi-Geostrophic dynamics [17,19–23].

In this work, we build upon our recent advances in analog data assimilation and its application to high-dimensional fields. While the works in [12,13] presented the AnDA framework by combining the analog forecasting method and stochastic filtering, these works have only shown applications to geophysical toy models. It was not until the work in [24] that the AnDA methodology was applied to realistic high dimensional fields, namely, Sea Surface Temperature (SST). Dealing with the curse of dimensionality is a critical challenge, in [24] we have shown that the use of patch-based representations (a patch is a term used by the image processing community to refer to smaller image parts of a given global image [25]) combined with EOF-based representations (EOF stands for Empirical Orthogonal Function, a classic dimensionality reduction technique also known as Principal Component Analysis (PCA)) leads to a computationally-efficient interpolation of missing data in SST maps outperforming classical OI-based interpolation schemes. Another development in AnDA applied to high dimensional fields was the introduction of conditional and physically-derived operators [26], where the analog forecasting operators account for the theoretical studies relating to synergies between ocean variables (e.g., SSH and SST) and those highlighting the importance of inter-scale dependencies. In this paper, we make use of these previously developed methodologies and tools and apply the AnDA to Sea Level Anomaly fields. The contribution of this work is two-fold: (i) Confronting AnDA to the reconstruction of an ocean tracer with scarce observations compared to SST (due to the nature of the available altimeters); (ii) designing an Observation System Simulation Experiment (OSSE) based on numerical simulation data to build the archived datasets used for the analog search; (iii) Reconstructing Sea Level Anomaly (SLA) by using SST or large scale SLA as auxiliary variables embedded in the analog regression techniques as shown in Section 4.

Using OFES (Ocean General Circulation Model (OGCM) for the Earth Simulation) numerical simulations [27,28], we design an Observation System Simulation Experiment (OSSE) for a case-study in the South China Sea using real along-track sampling patterns of spaceborne altimeters. Several particular mesoscale variation patterns characterizing this region were studied in the literature, we refer the reader to [29] and references therein. We also note that our method is not region specific and can be applied to any region of interest. Using the resulting groundtruthed dataset, we perform a qualitative and quantitative evaluation of the proposed scheme, including comparisons to state-of-the-art schemes.

The remainder of the paper is organized as follows: Section 2 presents the different datasets used in this paper to design an OSSE, Section 3 gives insights on the classical methods used for mapping SLA from along track data, Section 4 introduces the proposed analog data assimilation model. Experimental settings are detailed in section 5 and results for the considered OSSE are shown in Section 6. Section 7 further discusses the key aspects of this work.

2. Data: OFES (OGCM for the Earth Simulator)

An Observation System Simulation Experiment (OSSE) based on numerical simulations is considered to assess the relevance of the proposed analog assimilation framework. Our OSSE uses these numerical simulations as a groundtruthed dataset from which simulated along-track data are produced. We describe further the data preparation setup in the following sections.

2.1. Model Simulation Data

The Ocean General Circulation Model (OGCM) for the Earth Simulator (OFES) is considered in this study as the true state of the ocean. The simulation data is described in [27,28]. The coverage of the model is 75°S–75°N with a horizontal resolution of $1/10^{\circ}$. 34 years (1979–2012) of 3-daily simulation of SSH maps are considered, we proceed to a subtraction of a temporal mean to obtain SLA fields. In this study, our region of interest is located in the South China Sea (105°E to 117°E, 5°N to 25°N). This dataset is split into a training dataset corresponding to the first 33 years (4017 SLA maps) and a test dataset corresponding to the last year of the time series (122 SLA maps).

2.2. Along Track Data

We consider a realistic situation with a high rate of along track data. More precisely we use along-track data positions registered in 2014 where four satellites (Jason2, Cryosat2, Saral/AltiKa, HY-2A) were operating. Data is distributed by Copernicus Marine and Environment Monitoring Service (CMEMS).

From the reference 3-daily SLA dataset and real along-track data positions, we generate simulated along-track data from the sampling of a reference SLA field: More precisely, for a given along-track point, we sample the closest position of the $1/10^{\circ}$ regular model grid at the closest time step of the 3-daily model time series. As we consider a 3-daily assimilation time step (see Section 2.1 for details), we create a 3-daily pseudo-observation field, to be fed directly to the assimilation model. For a given time t, we combine all along-track positions for times $t-1$, t and $t+1$ to create an along-track pseudo-observation field at time t. We denote by $s3dAT$ the simulated 3-daily time series of along-track pseudo-observation fields. An example of these fields is given in Figure 1.

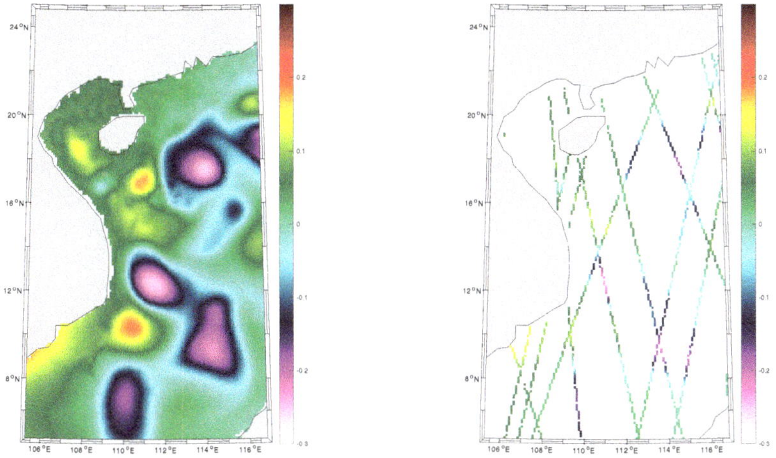

Figure 1. An example of a ground-truth Sea Level Anomaly (SLA) field (meters) in the considered region and its associated simulated pseudo-along track.

3. Problem Statement and Related Work

3.1. Data Assimilation and Optimal Interpolation

Data assimilation consists in estimating the true state of a physical variable $\mathbf{x}(t)$ at a specific time t, by combining (i) equations governing the dynamics of the variable, (ii) available observations $\mathbf{y}(1,...,T)$ measuring the variable and (iii) a background or first guess on its initial state $\mathbf{x}^b$. The estimated state is generally called the analyzed state and noted by $\mathbf{x}^a$. Data assimilation is a typical example of inverse problems, and similar formulations are known to the statistical signal processing community through optimal control and estimation theory [30]. We adopt here the unified notation of [31] and formulate the problem as a stochastic system in the following:

$$\begin{cases} \mathbf{x}(t) = \mathcal{M}(\mathbf{x}(t-1)) + \eta(t), & (1) \\ \mathbf{y}(t) = \mathcal{H}(\mathbf{x}(t)) + \epsilon(t). & (2) \end{cases}$$

Equation (1) represents the dynamical model governing the evolution of state $\mathbf{x}$ through time, while η is a Gaussian centered noise of covariance $\mathbf{Q}$ that models the process error. Equation (2) explains the relationship between the observation $\mathbf{y}(t)$ and the state to be estimated $\mathbf{x}(t)$ through the operator $\mathcal{H}$. The uncertainty of the observation model is represented by the ϵ error, considered here to be Gaussian centered and of covariance $\mathbf{R}$. We assume that ϵ and η are independent and that $\mathbf{Q}$ and $\mathbf{R}$ are known. Two main approaches are generally considered for the mathematical resolution of the system (1) and (2), namely, variational data assimilation and stochastic data assimilation. They differ in the way they infer the analyzed state $\mathbf{x}^a$, the first is based on the minimization of a certain cost function while the latter aims to obtain an optimal a posteriori estimate. We encourage the reader to consider the book of [32] for detailed insights on the various aspects and methods of data assimilation.

A popular data assimilation algorithm that is largely used in the literature to grid sea level anomalies from along-track data is called Optimal Interpolation (OI) (e.g., [15,33]), this algorithm is also the method adopted in CMEMS altimetry product. Optimal Interpolation (OI) aims at finding the Best Linear Unbiased Estimator (BLUE) of a field $\mathbf{x}$ given irregularly sampled observations $\mathbf{y}$ in space and time and a background prior $\mathbf{x}^b$. The multivariate OI equations were derived in [34] for meteorology and numerous applications in oceanography have been reported since the early work of [16]. Supposing that the background state $\mathbf{x}^b$ has covariance $\mathbf{B}$, and the observation operator is linear $\mathcal{H} = \mathbf{H}$, the analyzed state $\mathbf{x}^a$ and the analyzed error covariance $\mathbf{P}^a$ can be calculated using the following OI set of equations:

$$\mathbf{K} = \mathbf{BH}(\mathbf{R} + \mathbf{HBH}^T)^{-1} \quad \text{called the Kalman gain,} \tag{3}$$

$$\mathbf{x}^a = \mathbf{x}^b + \mathbf{K}(\mathbf{y} - \mathbf{Hx}^b), \tag{4}$$

$$\mathbf{P}^a = (\mathbf{I} - \mathbf{KH})\mathbf{B}. \tag{5}$$

It is worth mentioning that [35] showed that OI is closely related to the 3D-Var variational data assimilation algorithm which obtains $\mathbf{x}^a$ by minimizing the following cost function:

$$J(\mathbf{x}) = (\mathbf{x} - \mathbf{x}^b)^T\mathbf{B}^{-1}(\mathbf{x} - \mathbf{x}^b) + (\mathbf{y} - \mathbf{Hx})^T\mathbf{R}^{-1}(\mathbf{y} - \mathbf{Hx}). \tag{6}$$

An important limitation of OI is that the Gaussian-like covariance priors lead to smoothing out the small-scale information (e.g., mesoscale eddies), more specifically, this is a limitation due to the use of a static climatological **B** matrix. For satellite-derived altimetry fields, this usually results in over-smoothing altimetry fields for structures below ≈100 km [18]. This limitation may be even more critical in the context of future high-resolution altimetry missions, which supports the development of new OI-based methods (e.g., multi-scale OI schemes as in [36]) or alternatives as addressed by our work.

3.2. *Analog Data Assimilation*

Endorsed by the recent development in data-driven methods and data storage capacities, the Analog Data Assimilation (AnDA) was introduced as an alternative to classical model-driven data assimilation under one or more of the following situations [13]:

- The model is inconsistent with observations.
- The cost of the model integration is high computationally.
- (mandatory) The availability of a represenative (large) dataset of the dynamics of the state variables to be estimated. These datasets are hereinafter called catalogs and denoted by $\mathcal{C}$. The catalog is organized in a two-column dictionary where each state of the system is associated with its successor in time, forming a set of couples $(\mathcal{A}_i, \mathcal{S}_i)$ where $\mathcal{A}_i$ is called the analog and $\mathcal{S}_i$ its successor.

Given the considerations above, AnDA resorts to evaluating filtering, respectively smoothing, posterior likelihood, i.e., the distribution of the state to be estimated $\mathbf{x}(t)$ at time t, given past and current observations $\mathbf{y}(1, ..., t)$, respectively given all the available observation $\mathbf{y}(1, ..., T)$. This evaluation relies on the following state-space model:

$$\begin{cases} \mathbf{x}(t) = \mathcal{F}(\mathbf{x}(t-1)) + \eta(t), & (7) \\ \mathbf{y}(t) = \mathcal{H}(\mathbf{x}(t)) + \epsilon(t). & (8) \end{cases}$$

The difference between AnDA and classical data assimilation resides in the transition model Equation (7). The counterpart of a model-driven operator $\mathcal{M}$ of Equation (1) is here the operator $\mathcal{F}$ which refers to the considered data-driven operator, so called, the analog forecasting operator. This operator makes use of the available catalog $\mathcal{C}$ and assumes that the state forecast can be inferred from similar situations in the past.

Provided the definitions of the analogs and successors given above, the derivation of this operator resorts to characterizing the transition distribution i.e., $p(\mathbf{x}(t)|\mathbf{x}(t-1))$. Following [13], a Gaussian conditional distribution is adopted:

$$p(\mathbf{x}(t)|\mathbf{x}(t-1)) = \mathcal{N}(\mu(\mathbf{x}(t-1)), \Sigma(\mathbf{x}(t-1))), \tag{9}$$

where $\mathcal{N}$ is a Gaussian distribution of mean $\mu(\mathbf{x}(t-1))$ and covariance $\Sigma(\mathbf{x}(t-1))$. These parameters of the Gaussian distribution are calculated using the result of a K nearest neighbors search. The K nearest neighbors (analogs) $\mathcal{A}_{k\in(1,...,K)}$ of state $\mathbf{x}(t-1)$ and their successors $\mathcal{S}_{k\in(1,...,K)}$, along with a weight associated to each pair $(\mathcal{A}_k, \mathcal{S}_k)$ are used to calculate $\mu(\mathbf{x}(t-1))$ and $\Sigma(\mathbf{x}(t-1))$ as we will show in the next paragraph, the forecast state $\mathbf{x}(t)$ is then sampled from $\mathcal{N}(\mu(\mathbf{x}(t-1)), \Sigma(\mathbf{x}(t-1)))$. The weights are defined using a Gaussian kernel $\mathcal{K}_G$.

$$\mathcal{K}_G(u, v) = \exp\left(-\frac{\|u-v\|^2}{\sigma}\right). \tag{10}$$

Scale parameter σ is locally-adapted to the median value of the K distances $\|x(t-1) - \mathcal{A}_k\|^2$ to the K analogs. Other types of kernels might be considered (e.g., [4,10]), investigating kernel choice is out of the scope of this paper.

The mean and the covariance of the transition distribution might be calculated following several strategies. We consider in this work the three analog forecasting operators introduced in AnDA [13], more details can be found in Appendix A:

- **Locally-constant operator:** Mean $\mu(\mathbf{x}(t-1))$ and covariance $\Sigma(\mathbf{x}(t-1)))$ are given by the weighted mean and covariance of the K successors $\mathcal{S}_{k\in(1,...,K)}$.
- **Locally-incremental operator:** Here, the increments between the K analogs and their corresponding successors are calculated $\mathcal{S}_{k\in(1,...,K)} - \mathcal{A}_{k\in(1,...,K)}$. The weighted mean of the K increments is then added to the $\mathbf{x}(t-1)$ to obtain $\mu(\mathbf{x}(t-1))$. While $\Sigma(\mathbf{x}(t-1)))$ results in the weighted covariance of these differences.
- **Locally-linear operator:** A weighted least-square estimation of the linear regression of the state at time t given the state at time $t-1$ is performed based on the K pairs $(\mathcal{A}_k, \mathcal{S}_k)$. The parameters of the linear regression are then applied to state $\mathbf{x}(t-1)$ to obtain $\mu(\mathbf{x}(t-1))$. Covariance $\Sigma(\mathbf{x}(t-1)))$ is represented by the covariance of the residuals of the fitted weighted linear regression.

We may state clearly the key difference between the AnDA and reduced-rank Kalman filters and the OI method. It lies in the fact that the AnDA introduces a dynamical operator and not a prescribed space-time covariance model, and this dynamical operator is state-dependent and globally non-linear. The proposed analog forecasting operator can be seen as a state-dependent linear Gaussian operator, meaning that it is locally Gaussian and linear at each time step with a parameterization that depends on the current state, such that globally the dynamical operator is non-linear and non-Gaussian. A special case where AnDA is equivalent to OI is when all the elements of the catalog are considered as neighbors of any state vector. This case comes to assume that the dynamical operator is linear and state-independent. It is obviously of low interest due to the computational burden resulting from using all the catalog.

The application of the AnDA framework faces the curse of dimensionality i.e., the search of analogs is highly affected by the dimensionality of the problem and can fail at finding good analogs for state vector dimensions above 20 [13]. As proposed in [24], the extension of AnDA models to high-dimensional fields may then rely on turning the global assimilation issue into a series of lower-dimensional ones. We consider here an approach similar to [24] using a patch-based and EOF-based representation of the two-dimensional (2D) fields, i.e., the 2D fields are decomposed into a set of overlapping patches, each patch being projected onto an EOF space. Analog strategies then apply to patch-level time series in the EOF space.

Overall, as detailed in the following section, the proposed analog data assimilation model for SLA fields relies on three key components: A patch-based representation of the SLA fields, the selection of a kernel to retrieve analogs and the specification of a patch-level analog forecasting operator.

4. Analog Reconstruction for Altimeter-Derived SLA

4.1. Patch-Based State-Space Formulation

As stated above, OI may be considered as an efficient model-based method to recover large-scale structures of SLA fields. Following [24], this suggests considering the following two-scale additive decomposition:

$$X = \bar{X} + dX + \xi, \tag{11}$$

where $\bar{X}$ is the large-scale component of the SLA field, typically issued from an optimal interpolation, dX the fine-scale component of the SLA field we aim to reconstruct and ξ is the remaining unresolved scales.

The reconstruction of field dX involves a patch-based and EOF-based representation. It consists in regarding field dX as a set of $P \times P$ overlapping patches (e.g., $2^\circ \times 2^\circ$), an example of patch locations is shown in Figure 2. This set of patches is referred to as $\mathcal{P}$, and we denote by $\mathcal{P}_s$ the patch centered on position s. After building a catalog $\mathcal{C}_\mathcal{P}$ of patches from the available dataset of residual fields $X - \bar{X}$ (see Section 3.2), we proceed to an EOF decomposition of each patch in the catalog. The reconstruction of field $dX(\mathcal{P}_s, t)$ at time t is then stated as the analog assimilation of the coefficients of the EOF decomposition in the EOF space given an observation series in the patch space. Formally, $dX(\mathcal{P}_s, t)$ decomposes as a linear combination of a number N_E of EOF basis functions:

$$dX(\mathcal{P}_s, t) = \sum_{k=1}^{N_E} \alpha_k(s,t) EOF_k, \tag{12}$$

with EOF_k the kth EOF basis and $\alpha_k(s,t)$ the corresponding coefficient for patch $\mathcal{P}_s$ at time t. Let us denote by $\Phi(\mathcal{P}_s, t)$ the vector of the N_E coefficients $\alpha_k(s,t)$: $\Phi(\mathcal{P}_s, t) = \{\alpha_1(s,t), ..., \alpha_{N_E}(s,t)\}$. This vector represents the projection of $dX(\mathcal{P}_s, t)$ in the lower-dimensional EOF space.

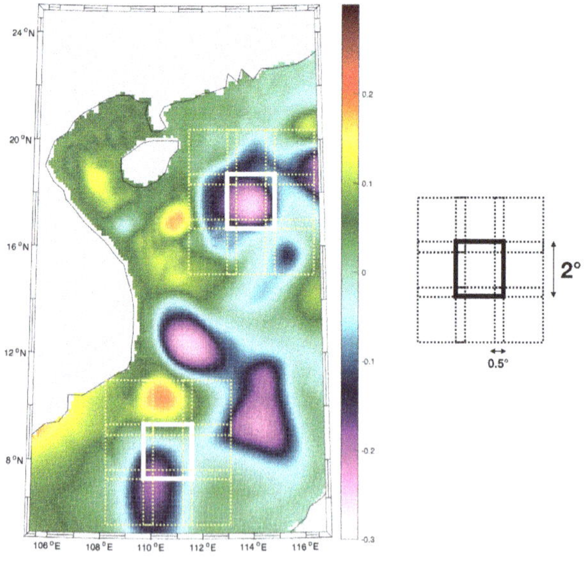

Figure 2. Two examples of patch locations and their overlapping patch neighbours.

4.2. Patch-Based Analog Dynamical Models

Given the considered patch-based representation of field dX, the proposed patch-based analog assimilation scheme involves a dynamical model stated in the EOF space. Formally, Equation (9) leads to the following Gaussian conditional distribution in the EOF space:

$$p(\Phi(\mathcal{P}_s, t)|\Phi(\mathcal{P}_s, t-1)) = \mathcal{N}(\mu(\Phi(\mathcal{P}_s, t-1)), \Sigma(\Phi(\mathcal{P}_s, t-1))), \tag{13}$$

We consider the three analog forecasting operators presented in Section 3.2, namely, the locally-constant, the locally-incremental and the locally-linear. The calculation of the weights associated to each analog-successor pair relies on a Gaussian kernel $\mathcal{K}_G$ (Equation (10)). The search for analogs in the N_E-dimensional patch space (in practice, N_E ranges from 5 to 20) ensures a better accuracy in the retrieval of relevant analogs compared to a direct search in the high-dimensional space of state dX. It also reduces the computational complexity of the proposed scheme.

Another important extension of the current study is the possibility of exploiting auxiliary variables with the state vector Φ in the analog forecasting models. Such variables may be considered in the search for analogs as well as regression variables in a locally-linear analog setting. Regarding the targeted application to the reconstruction of SSH fields and the proposed two-scale decomposition (Equation (11)), two types of auxiliary variables seem to be of interest: The low-resolution component $\bar{X}$ to take into account inter-scale relationship [17], and Sea Surface Temperature (SST) with respect to the widely acknowledged SST-SSH synergies [17,19,21]. We also apply patch-level EOF-based decompositions to include both types of variables in the considered analog forecasting models (Equation (13)).

4.3. Numerical Resolution

Given the proposed analog assimilation model, the proposed scheme first relies on the creation of patch-level catalogs from the training dataset. This step requires the computation of a training dataset of fine scale data $dX_{training}$, this is done by subtracting a large-scale component $\bar{X}_{training}$ from the original training dataset. Here, we consider the large-scale component of training data to be the result of a global (By global, we mean here an EOF decomposition over the entire case study region, by contrast to the patch-level decomposition considered in the analog assimilation setting.) EOF-based reconstruction using a number of EOF components that retains 95% of the dataset variance, which accounts for horizontal scales up to $\approx$100 km. This global EOF-based decomposition provides a computationally-efficient means for defining large-scale component $\bar{X}$. This EOF-based decomposition step is followed by the extraction of overlapping patches for all variables of interest, namely $\bar{X}_{training}$, $dX_{training}$ and potential auxiliary variables, and the identification of the EOF basis functions from the resulting raw patch datasets. This leads to the creation of a patch-level catalog $\mathcal{C}_{\mathcal{P}}$ from the EOF-based representations of each patch.

Given the patch-level catalog, the algorithm applied for the mapping SLA fields from along-track data, referred to PB-AnDA (for Patch-Based AnDA), is stated in Algorithm 1 and a sketch of the method is shown in Figure 3.

Algorithm 1 Patch-Based AnDA

1: Compute the large-scale component $\bar{X}$, here, we consider the result of optimal interpolation (OI) projected onto the global EOF basis functions.

2: Split the case study region into overlapping $P \times P$ patches, here, 20×20 patches

3: For each patch position s, use the Analog Ensemble Kalman Smoother (AnEnKS) [13], for patch $\mathcal{P}_s$ of field dX. As stated in (13), the assimilation is performed in the EOF space, i.e., for EOF decomposition $\Phi(\mathcal{P}_s, t)$, using the operator derived from EOF-based reconstruction (12) and decomposition (11) as observation model $\mathcal{H}$ in (8) and the patch-level training catalog described in the previous section. The assimilation is sequential and is performed each 3-days.

4: Reconstruct fields dX from the set of assimilated patches $\{dX(\mathcal{P}_s, \cdot)\}_s$. This relies on a spatial averaging over overlapping patches (here, a 5-pixel overlapping in both directions).

5: the reconstruction of fields X as $\bar{X} + dX$.

We may point out two important aspects in the implementation of the proposed patch-level AnDA setting:

- (step 3) In the analog forecasting setting, the search for analogs is restricted to patch exemplars in the catalog within a local spatial neighborhood (typically a patch-level 8-neighborhood), except for patches along the seashore for which the search for analogs is restricted to patch exemplars at the same location.
- (step 4) In practice, we do not apply the patch-level assimilation to all grid positions. Consequently, the spatial averaging may result in blocky artifacts. We then apply a patchwise EOF-based decomposition-reconstruction with a smaller patch-size (here, 17×17 patches) to remove these blocky artifacts.

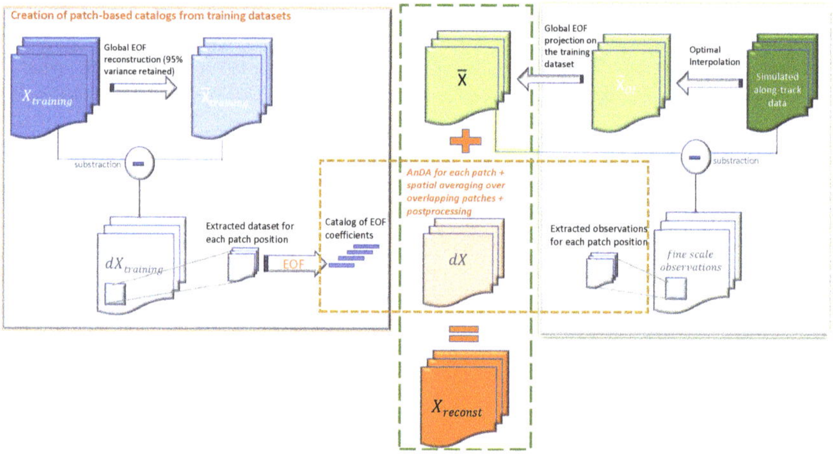

Figure 3. Sketch of the proposed patch-based Analog Data Assimilation (PB-AnDA). The left block details the construction of the patch-based catalogs from the training dataset. The right block illustrates the process of obtaining the large-scale component of the SLA reconstructed field. The orange dashed rectangle represents the application of the AnDA using the catalog and the fine-scale observations. Finally, the green dashed rectangle shows the final addition operation that yields the reconstructed SLA field.

5. Experimental Setting

We detail below the parameter setting of the models evaluated in the reported experiments, including the proposed PB-AnDA scheme:

- *PB-AnDA*: We consider 20×20 patches with 15-dimensional EOF decompositions ($N_E = 15$), which typically accounts for 99% of the data variance for the considered dataset. The postprocessing step exploits 17×17 patches and a 15-dimensional EOF decomposition. Regarding the parametrization of the AnEnKS procedure, we experimentally cross-validated the number of nearest neighbors K to 50, the number of ensemble members $n_{ensemble}$ to 100 and the observation covariance error in Equation (8) is considered to be diagonal $\mathbf{R} = \kappa^2\mathbf{I}$ and $\kappa = 0.001$m.
- *Optimal Interpolation*: We apply an Optimal Interpolation to the processed along-track data. It provides the low-resolution component for the proposed PB-AnDA model and a model-driven reference for evaluation purposes. The background field is a null field. We use a Gaussian covariance model with a spatial correlation length of 100 km and a temporal correlation length of 15 days (± 5 timesteps since our data is 3-daily). These choices result from a cross-validation experiment.
- *VE-DINEOF*: We apply a second state-of-the-art interpolation scheme using a data-driven strategy solely based on EOF decompositions, namely VE-DINEOF [37]. Using an iterative reconstruction scheme, VE-DINEOF starts by filling the missing data with a first guess, here along-track pseudo-observation field for along-track data positions and $\bar{X}$ for missing data positions. For each iteration, the resulting field is projected on the most significant EOF components calculated from the clean catalog data, then missing data positions are updated using pixels from the reconstructed new field. We run this iterative process until convergence. To make this algorithm comparable to the proposed PB-AnDA setting, we perform the reconstruction for each patch position then regroup the results as in PB-AnDA.
- *G-AnDA*: With a view to evaluating the relevance of the patch-based decomposition, we also apply AnDA at the region scale, referred to as G-AnDA. It relies on an EOF-based decomposition of the detail field dX. We use 150 EOF components, which accounts for more than 99% of the total variance of the SSH dataset. From cross-validation experiments, the associated AnEnKS procedure relies on a locally-linear analog forecasting model with $K = 500$ analogs, $n_{ensemble} = 100$ ensemble members and a diagonal observation covariance similar to as in PB-AnDA.

The patch-based experiments were run on Teralab infrastructure using a multi-core virtual machine (30 CPUs, 64G of RAM). We used the Python toolbox for patch-based analog data assimilation [24] (available at github.com/rfablet/PB_ANDA). Optimal Interpolation was implemented on Matlab using [36]. Throughout the experiments, two metrics are used to assess the performance of the considered interpolation methods: (i) Daily and mean Root Mean Square Error (RMSE) series between the reconstructed SLA fields X and the groundtruthed ones; (ii) daily and mean correlation coefficient between the fine-scale component dX of the reconstructed SLA fields and of the groundtruthed ones. These two metrics allow a good evaluation on image reconstruction capabilities and are widely used in missing data interpolation literature [38,39].

6. Results

We evaluate the proposed PB-AnDA approach using the OSSE presented in Section 2. We perform a qualitative and quantitative comparison to state-of-the-art approaches. We first describe the parameter setting used for the PB-AnDA as well as benchmarked models, namely OI, an EOF-based approach [37] and a direct application of AnDA at the region level. We then report numerical experiments for noise-free and noisy observation data as well the relevance of auxiliary variables in the proposed PB-AnDA scheme.

6.1. SLA Reconstruction from Noise-Free Along-Track Data

We first performed an idealized noise-free experiment, where the along-track observations were not contaminated with noise. The interpolation performances for this experiment are illustrated in Table 1. Our PB-AnDA algorithm significantly outperforms OI. More specifically, the locally-linear PB-AnDA results in the best reconstruction among the competing methods. We suggest that this improvement comes from the reconstruction of fine-scale features learned from the archived model simulation data. Figure 4a reports interpolated SSH fields and their gradient fields which further confirm our intuition. PB-AnDA interpolation shows an enhancement of the gradients and comes out with some fine-scale eddies that were smoothed out in OI and VE-DINEOF. This is also confirmed by the Fourier power spectrum of the interpolated SLA fields in Figure 4b.

Table 1. SLA Interpolation performance for a noise-free experiment: Root Mean Square Error (RMSE) (meters), correlation statistics for Optimal Interpolation (OI), VE-DINEOF, G-AnDA and PB-AnDA with regard to the groundtruthed SLA fields. The relative gain with regard to OI is also shown in percentage. See Section 5 for the corresponding parameter settings.

Criterion		RMSE	Correlation	$\frac{RMSE_{OI}-RMSE}{RMSE_{OI}}$
OI		0.026 ± 0.007	0.81 ± 0.08	-
VE-DINEOF		0.023 ± 0.007	0.85 ± 0.07	11.53%
G-AnDA		0.020 ± 0.006	0.89 ± 0.04	23.07%
PB-AnDA	Locally-constant	0.014 ± 0.005	0.95 ± 0.03	46.15%
	Locally-Increment	0.014 ± 0.005	0.95 ± 0.03	46.15%
	Locally-Linear	$\mathbf{0.013 \pm 0.005}$	$\mathbf{0.96 \pm 0.02}$	**50.00%**

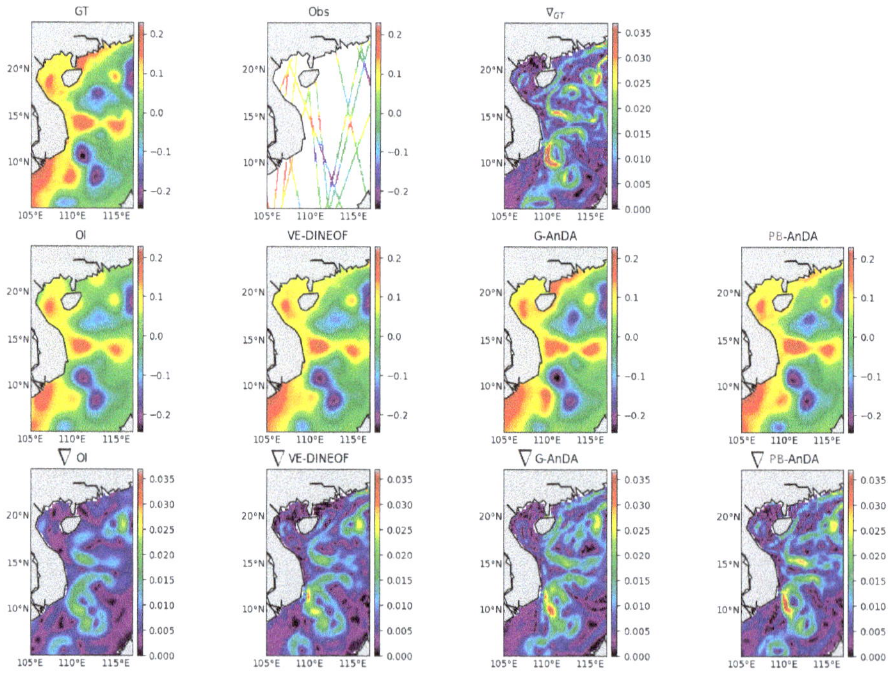

(a)

Figure 4. *Cont.*

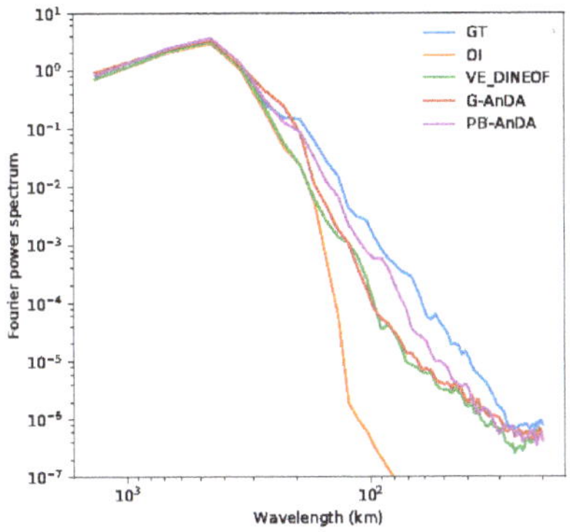

(b)

Figure 4. Reconstructed SLA fields (meters) using noise-free along-track observation using OI, VE-DINEOF, G-AnDA, PB-AnDA on the 54th day (24 February 2012): From left to right, the first row shows the ground truth field, the simulated available along-tracks for that day, the ground truth gradient field. The second and third rows show each of the reconstruction and their corresponding gradient fields, from left to right, OI, VE-DINEOF, G-ANDA and PB-AnDA. The Fourier power spectrum of the competing methods is also included.

6.2. SLA Reconstruction from Noisy Along-Track Data

We also evaluated the proposed approach for noisy along-track data. Here, we ran two experiments with an additive zero-mean Gaussian noise applied to the simulated along-track data. We considered a diagonal noise covariance of $\gamma^2\mathbf{I}$ where $\gamma = 0.01$ m (Experiment A) and $\gamma = 0.03$ m (Experiment B) which was more close to the instrumental error of conventional altimeters. Given the resulting noisy along-track dataset, we applied the same methods as for the noise-free case study.

We ran PB-AnDA using different values for κ. For Experiment A, Table 2 shows that the minimum is reached using the true value of the error $\kappa = 0.01$ m. While for Experiment B, Table 3 shows that the minimum is counter-intuitively reached again using value of the error $\kappa = 0.01$ m with a negligible margin compared to the true value.

Table 2. Impact of standard variation of observation error κ in AnDA interpolation performance using noisy along-track data (γ = 0.01 m): RMSE (meters) of AnDA interpolation for different values of parameter κ. For the same dataset, OI RMSE is 0.039.

κ	0.1	0.05	0.03	**0.01**	0.005	0.001	0.0001
$rmse_{PB-AnDA}$	0.035	0.030	0.028	**0.025**	0.025	0.029	0.044

Table 3. Impact of standard variation of observation error κ in AnDA interpolation performance using noisy along-track data (γ = 0.03 m): RMSE (meters) of AnDA interpolation for different values of parameter κ. For the same dataset, OI RMSE is 0.066.

κ	0.1	0.05	0.03	**0.01**	0.005	0.001	0.0001
$rmse_{PB-AnDA}$	0.038	0.036	0.035	**0.0349**	0.037	0.046	0.076

Our algorithm is then compared with the results of the application of the competing algorithms considered in this work. Results are shown in Table 4. PB-AnDA still outperforms OI in terms of RMSE and correlation statistics in both experiments. The locally-linear version of PB-AnDA depicts the best reconstruction performance. We report an example of the reconstruction in Figure 5. Similarly to the noise-free case study, PB-AnDA better recovers finer-scale structures in Figure 5a compared with OI,

VE-DINEOF and G-AnDA. In Figure 5b, PB-AnDA also better reconstructs a larger-scale North-East structure, poorly sampled by along-track data and hence poorly interpolated by OI.

Table 4. SLA Interpolation performance for noisy along-track data: RMSE (meters) and correlation statistics for OI, VE-DINEOF, G-AnDA and PB-AnDA w.r.t. the groundtruthed SLA fields. The relative gain with regard to OI is also shown in percentage. See Section 5 for the corresponding parameter settings.

	Criterion		RMSE	Correlation	$\frac{RMSE_{OI}-RMSE}{RMSE_{OI}}$
$\gamma = 0.01$ m	OI		0.039 ± 0.005	0.64 ± 0.09	-
	VE-DINEOF		0.035 ± 0.005	0.68 ± 0.09	10.25%
	G-AnDA		0.030 ± 0.005	0.78 ± 0.06	23.07%
	PB-AnDA	Locally constant	0.026 ± 0.005	0.82 ± 0.05	33.33%
		Increment	0.028 ± 0.005	0.81 ± 0.05	28.20%
		Local Linear	$\mathbf{0.0245 \pm 0.005}$	$\mathbf{0.83 \pm 0.05}$	**37.17%**
$\gamma = 0.03$ m	OI		0.066 ± 0.006	0.41 ± 0.09	-
	VE-DINEOF		0.060 ± 0.006	0.45 ± 0.09	9.09%
	G-AnDA		0.039 ± 0.006	0.67 ± 0.09	40.90%
	PB-AnDA	Locally constant	0.035 ± 0.006	0.688 ± 0.064	46.96%
		Increment	0.036 ± 0.006	0.656 ± 0.07	45.45%
		Local Linear	$\mathbf{0.032 \pm 0.006}$	$\mathbf{0.708 \pm 0.063}$	**51.51%**

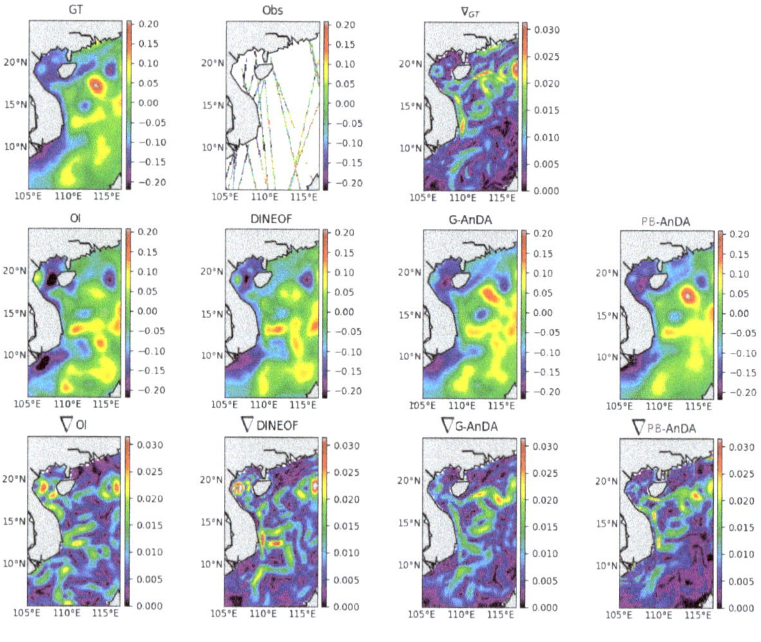

(a)

Figure 5. *Cont.*

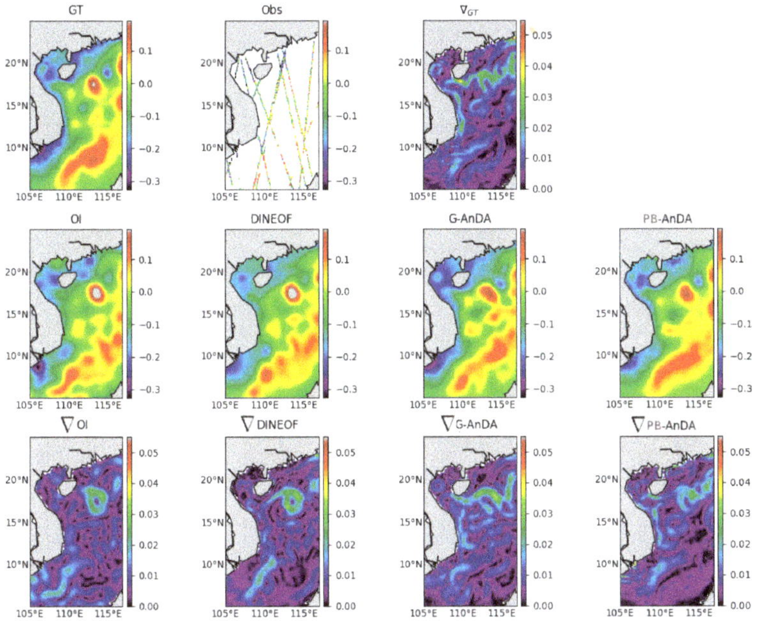

(b)

Figure 5. Reconstruction of SLA fields (meters) from noisy along-track data using OI, VE-DINEOF, G-AnDA & PB-AnDA on day 225th (**a**) & 228th (**b**).

6.3. PB-AnDA Models with Auxiliary Variables

We further explore the flexibility of the analog setting to the use of additional geophysical variable information as explained in Section 4.2. Intuitively, we expect SLA fields to involve inter-scale dependencies as well as synergies with other tracers [19,40]. The use of auxiliary variables provide the means for evaluating such dependencies and their potential impact on reconstruction performance. We consider two auxiliary variables that are used in the locally-linear analog forecasting model (7): (i) To account for the relationship between the large-scale and fine-scale component, we may consider variable $\bar{X}$; (ii) considering potential SST-SSH synergies, we consider SST fields. Overall, we consider four parameterization of the regression variables used in PB-AnDA: The sole use of dX (PB-AnDA-dX), the joint use of dX and SST fields (PB-AnDA-dX+SST), the joint use of dX and $\bar{X}$ (PB-AnDA-dX+$\bar{X}$), the joint use of dX and the groudntruthed version of $\bar{X}$ denoted by $\bar{X}^{GT}$, (PB-AnDA-dX+$\bar{X}^{GT}$). The later provides a lower-bound for the reconstruction performance, assuming the low-resolution component is perfectly estimated.

We report mean RMSE (meters) and correlation statistics for these four PB-AnDA parameterizations in Table 5 for the noisy case-study. Considering PB-AnDA-dX as reference, these results show a very slight improvement when complementing dX with SST information. Though limited, we report a greater improvement when adding the low-resolution component $\bar{X}$. Interestingly, a significantly greater improvement is obtained when adding the true low-resolution information. The mean results are in accordance with [17], which reported that large-scale SLA information was more informative than SST to improve the reconstruction of the SLA at finer scales. Though mean statistics over one year leads to rather limited improvement, daily RMSE time series (Figure 6) reveal that for some periods, for instance between day 130 and 150, relative improvements in terms of RMSE may reach 10% with the additional information brought by the large-scale component. In this respect, it may noted that PB-AnDA-dX+$\bar{X}$ always perform better than PB-AnDA-dX. An example of the reconstruction in reported in Figure 7.

Table 5. PB-AnDA reconstruction performance using noisy along-track data for different choices of the regression variables in the locally-linear analog forecasting model: PB-AnDA-dX using solely dX, PB-AnDA-dX+ SST (Sea Surface Temperature) using both dX and SST, PB-AnDA-$dX + \bar{X}$ using both dX and $\bar{X}$, and PB-AnDA-$dX + \bar{X}^{GT}$ using dX and the true large-scale component $\bar{X}^{GT}$. The table shows the RMSE (meters) and correlation statistics.

	PB-AnDA Model	RMSE	Correlation
γ = 0.01 m	PB-AnDA-dX	0.025 ± 0.005	0.83 ± 0.05
	PB-AnDA-dX + SST	0.024 ± 0.005	0.83 ± 0.05
	PB-AnDA-$dX + \bar{X}$	0.023 ± 0.005	0.84 ± 0.05
	PB-AnDA-$dX + \bar{X}^{GT}$	0.021 ± 0.004	0.87 ± 0.04
γ = 0.03 m	PB-AnDA-dX	0.032 ± 0.006	0.708 ± 0.06
	PB-AnDA-dX + SST	0.031 ± 0.006	0.710 ± 0.06
	PB-AnDA-$dX + \bar{X}$	0.029 ± 0.006	0.717 ± 0.06
	PB-AnDA-$dX + \bar{X}^{GT}$	0.026 ± 0.005	0.730 ± 0.05

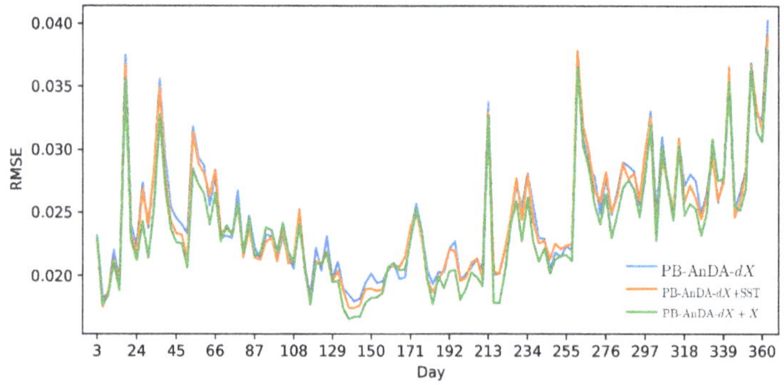

Figure 6. Daily RMSE (meters) time series of PB-AnDA SLA reconstructions using noisy along-track data for different choices of the regression variables in the locally-linear analog forecasting model: PB-AnDA-dX (light blue), PB-AnDA-dX+ SST (orange) and PB-AnDA-$dX + \bar{X}$ (green).

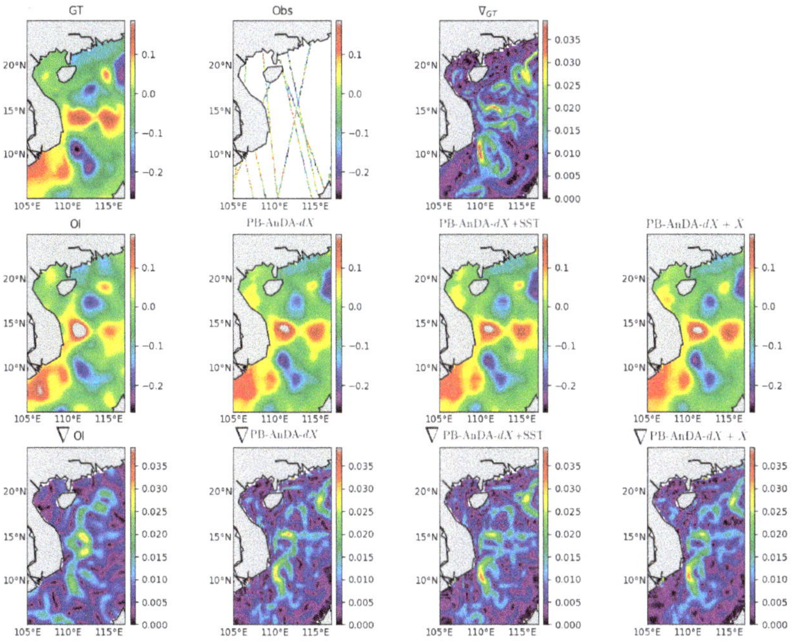

(a)

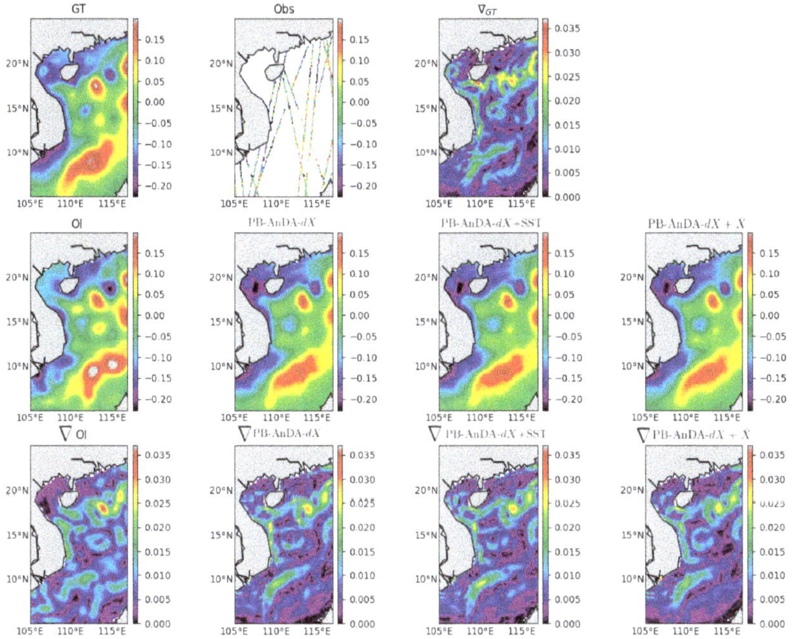

(b)

Figure 7. (Noisy observation) Reconstruction of SLA fields (meters) using PB-AnDA with different multivariate regression models on day 57th (**a**) & 237th (**b**).

7. Discussion

Analog data assimilation can be regarded as a means to fuse ocean models and satellite-derived data. We regard this study as a proof-of-concept, which opens research avenues as well as new directions for operational oceanography. Our results advocate for complementary experiments at the global scale or in different ocean regions for a variety of dynamical situations with a view to further evaluating the relevance of the proposed analog assimilation framework. Such experiments should evaluate the sensitivity of the assimilation with respect to the size of the catalog. The scaling up to the global ocean also suggests investigating computationally-efficient implementation of the analog data assimilation. In this respect, the proposed patch-based framework intrinsically ensures high parallelization performance. From a methodological point of view, a relative weakness of the analog forecasting models (9) may be their low physical interpretation compared with physically-derived priors [18]. The combination of such physically-derived parameterizations to data-driven strategies appear to be a promising research direction. While we considered an OSSE to evaluate the proposed scheme, future work will investigate applications to real satellite-derived datasets, including the use of independent observation data such as surface drifters' track data to further assess the performance of the proposed algorithm.

The analog method is at the heart of this work as it is appealing by its implementation ease and its intuitive strategy, but it must not be seen as the only data-driven method adapted to the framework we presented. As long as a data-driven forecasting operator can be derived, other data-driven methods can be investigated [41,42]. One promising path is the use of neural networks, as they sparked off a series of breakthroughs in other fields [42–46]. While neural network based approaches can lead to better performances due to their superior regressing capabilities, two clear advantages of adopting analog methods are the fact that they do not need a time consuming training phase and that analog methods are easy to understand and interpret compared to black-box approaches such as neural networks.

The Analog Data Assimilation method as used in this work relies on several hyperparameters, assumptions and design choices. These considerations are discussed in the following:

- In this work, we used all the available data for the creation of the catalog, we expect that a rich dataset is important in increasing the likelihood of finding good analogs, yet a more thorough study is needed to assess the impact of reducing the temporal resolution of the dataset versus reducing the amount of the total available data. To compensate the computational impact of using a large dataset for the search for analogs, we used the FLANN (Fast Library for Approximate Nearest Neighbors) library as in [26]. This method makes use of tree-indexing techniques and is suitable for this kind of high dimensional applications [47].
- While we used a conditional Gaussian distribution in Equation (9), another alternative is the use of a conditional multinomial distribution. This resorts to sampling one of the analogs' successors as the forecast. Adopting this alternative would mean that we rely strongly on the archived catalog, so that forecasts of each ensemble member are actual elements of the catalog. This reduces the ability of the model to generate a rich variability of forecast scenes as done by the conditional Gaussian distribution.
- While we used a Gaussian kernel in Equation (10), other alternatives include cone kernels [10] which are more adapted to finding analogs in time series. The performance of our algorithm was slightly improved at the expense of more time execution and an additional hyperparameter to tune empirically, and we decided to prioritize simpler and efficient kernels. Regarding the scale parameter, the median was chosen due to its robustness to outliers.
- Although the AnEnKS was used in this work, a possible alternative is the use of an analog based Particle Smoother which drops Gaussian assumptions, however techniques based on particle filters need more ensemble members than their Ensemble Kalman Filter counterparts, thus causing a considerable increase in computational demands which was impractical for our application.

- We encourage the reader to refer to the discussion section in [24] for more insights about the rationale behind the use of patch-based representations and EOF-based dimensionality reduction.

Through the experiments conducted in this work, it was shown that the best performance was always reached using the locally-linear analog operator, which is in line with our previous findings [13,24]. An explanation for the superiority of this approach is that it better approximates locally the true dynamical model [48].

Beyond along-track altimeter data as considered in this study, future missions such as SWOT (NASA/CNES) promise an unprecedented coverage around the globe. More specifically, the large swath is expected to provide a large number of data, urging for the inspection of the potential improvements that this new mission will bring compared to classical along-track data. In the context of analog data assimilation, the interest of SWOT data may be two-fold. First, regarding observation model (8), SWOT mission will both significantly increase the number of available observation data and enable the definition of more complex observation models exploiting for instance velocity-based or vorticity-based criterion. Second, SWOT data might also be used to build representative patch-level catalogs of exemplars. Future work should investigate these two directions using simulated SWOT test-beds [49].

8. Materials and Methods

The experimental results presented in this work were obtained using the Python PB-AnDA toolbox made available by the authors at https://github.com/rfablet/PB_ANDA.

9. Conclusions

This work sheds light on the opportunities that data science methods are offering to improve altimetry in the era of big data. Assuming the availability of high-resolution numerical simulations, we show that Analog Data Assimilation (AnDA) can outperform the Optimal Interpolation method and retrieve smoothed out structures resulting from the sole use of OI both with idealized noise-free and more realistic noisy observations for the considered case study. Importantly, the reported experiments point out the relevance for combining OI for larger scales (above 100 km) whereas the proposed patch-based analog setting successfully applies to the finer-scale range below 100 km. This is in agreement with the recent application of the analog data assimilation to the reconstruction of cloud-free SST fields [24]. We also demonstrate that AnDA can embed complementary variables in a simple manner through the regression variables used in the locally-linear analog forecasting operator. In agreement with our recent analysis [17], we demonstrate that the additional use of large-scale SLA information may further improve the reconstruction performance for fine-scale structures.

We may state here the limitations of the present work and possible research avenues for the future. The experiments presented in this work were conducted on a numerical simulation derived dataset. A major future work direction would be then to apply the same procedure on real satellite-derived SLA contaminated with more complex noise models, then investigate the contribution of the use of numerical simulation datasets as catalogs. As combining multi-source datsets can also be challenging when using auxiliary variables relationships with SLA, an interesting experiment for example would be constructing a catalog with real SST observations combined with numerical simulation SLA datasets.

More efforts should be directed to assess the quality of the catalogs (spatio-temporal resolution, total years of measurements to consider, occurence of rare events, etc.). Besides, building a good catalog can represent an opportunity for the use of neural networks based methods, and confronting these powerful regressors to our method is a promising future step. We also note that PB-AnDA can be a relevant candidate for the interpolation of other geophysical variables (e.g., Sea Surface Salinity, Chlorophyll concentrations, etc.) under the condition that they verify the set of assumptions made in this work.

Author Contributions: R.L., R.F. and G.C. conceived and designed the experiments; R.L. and M.S. created the datasets; R.L. and P.H.V. performed the experiments; R.L. and M.S. analyzed the data; T.F., B.C. contributed materials/analysis tools; R.L. and R.F. wrote the paper.

Funding: Redouane Lguensat is funded through a CNES (French Space Agency) postdoctoral grant. This work is also supported by ANR (Agence Nationale de la Recherche, grant ANR-13-MONU-0014), Labex Cominlabs (grant SEACS) and TeraLab (grant TIAMSEA).

Conflicts of Interest: The authors declare no conflict of interest. The funders had no role in the design of the study; in the collection, analyses, or interpretation of data; in the writing of the manuscript, or in the decision to publish the results.

Abbreviations

The following abbreviations are used in this manuscript:

AVISO+	Archivage, Validation et Interprétation des données des Satellites Océanographiques
CMEMS	Copernicus Marine Environment Monitoring Service
OSSE	Observation System Simulation Experiment
OGCM	Ocean General Circulation Model
OFES	OGCM for the Earth Simulation

Appendix A. Analog Forecasting Operators

In this appendix, we present the calculations needed for the three analog forecasting operators used in this work. An illustration is also given in Figure A1. Following [13], we give for each operator, the equations for μ and Σ given $\mathcal{A}$ the analogs of $x(t-1)$, their successors $\mathcal{S}$ and the corresponding weights $\mathcal{K}_G$:

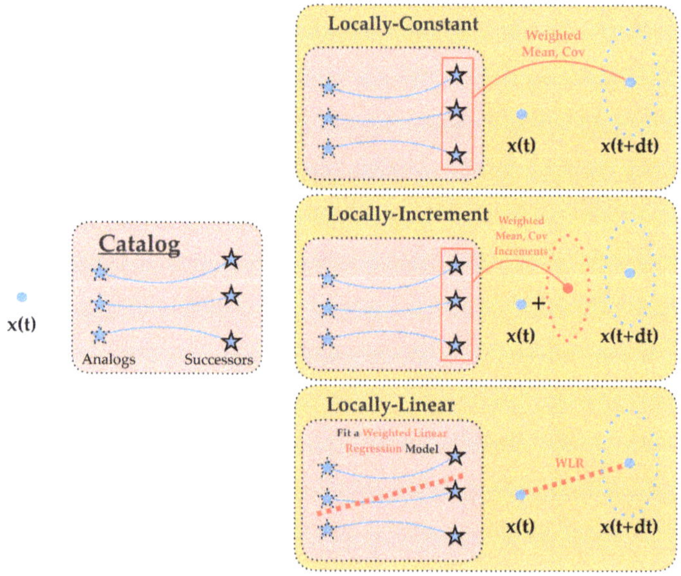

Figure A1. Illustration of the three analog forecasting operator.

Locally-Constant Operator

$\mu = \sum_{k=1}^{K} \mathcal{K}_G(x(t-1), \mathcal{A}_k)\mathcal{S}_k(x(t-1))$
$\Sigma = cov_{\mathcal{K}_G}(\mathcal{S}_k(x(t-1)))_{k\in 1,K})$ where $cov_{\mathcal{K}_G}$ is a weighted covariance.

Locally-Incremental Operator

$$\tau_k(x(t-1)) = \mathcal{S}_k(x(t-1)) - \mathcal{A}_k(x(t-1))$$
$$\mu = x(t-1) + \sum_{k=1}^{K} \mathcal{K}_G(x(t-1), \mathcal{A}_k)\tau_k(x(t-1))$$
$$\Sigma = cov_{\mathcal{K}_G}(x(t-1) + \tau_k(x(t-1))_{k \in 1,K})$$

Locally-Linear Operator

Fitting a weighted least square between the K analogs and their successors we obtain slope $\alpha(x(t-1))$ and intercept $\beta(x(t-1))$ parameters, and residuals $\xi_k(x(t-1))$ that lead us to μ and Σ:

$$\xi_k(x(t-1)) = \mathcal{S}_k(x(t-1)) - (\alpha(x(t-1))\mathcal{A}_k(x(t-1) + \beta(x(t-1)))$$
$$\mu = \alpha(x(t-1)).x(t-1) + \beta(x(t-1))$$
$$\Sigma = cov((\xi_k(x(t-1)))_{k \in 1,K})$$

References

1. Zhang, L.; Zhang, L.; Du, B. Deep learning for remote sensing data: A technical tutorial on the state of the art. *IEEE Geosci. Remote Sens. Mag.* **2016**, *4*, 22–40. [CrossRef]
2. Lary, D.J.; Alavi, A.H.; Gandomi, A.H.; Walker, A.L. Machine learning in geosciences and remote sensing. *Geosci. Front.* **2016**, *7*, 3–10. [CrossRef]
3. Lorenz, E.N. Atmospheric predictability as revealed by naturally occurring analogues. *J. Atmos. Sci.* **1969**, *26*, 636–646. [CrossRef]
4. McDermott, P.L.; Wikle, C.K. A model-based approach for analog spatio-temporal dynamic forecasting. *Environmetrics* **2015**. [CrossRef]
5. Comeau, D.; Giannakis, D.; Zhao, Z.; Majda, A.J. Predicting regional and pan-Arctic sea ice anomalies with kernel analog forecasting. *arXiv* **2017**, arXiv:1705.05228.
6. Atencia, A.; Zawadzki, I. A Comparison of Two Techniques for Generating Nowcasting Ensembles. Part II: Analogs Selection and Comparison of Techniques. *Mon. Weather Rev.* **2015**, *143*, 2890–2908. [CrossRef]
7. Delle Monache, L.; Eckel, F.A.; Rife, D.L.; Nagarajan, B.; Searight, K. Probabilistic weather prediction with an analog ensemble. *Mon. Weather Rev.* **2013**, *141*, 3498–3516. [CrossRef]
8. Yiou, P. AnaWEGE: A weather generator based on analogues of atmospheric circulation. *Geosci. Model Dev.* **2014**, *7*, 531–543. [CrossRef]
9. Hamill, T.M.; Whitaker, J.S. Probabilistic quantitative precipitation forecasts based on reforecast analogs: Theory and application. *Mon. Weather Rev.* **2006**, *134*, 3209–3229. [CrossRef]
10. Zhao, Z.; Giannakis, D. Analog Forecasting with Dynamics-Adapted Kernels. *arXiv* **2014**. arXiv: 1412.3831.
11. Horton, P.; Jaboyedoff, M.; Obled, C. Global Optimization of an Analog Method by Means of Genetic Algorithms. *Mon. Weather Rev.* **2017**, *145*, 1275–1294. [CrossRef]
12. Tandeo, P.; Ailliot, P.; Chapron, B.; Lguensat, R.; Fablet, R. The analog data assimilation: Application to 20 years of altimetric data. In Proceedings of the CI 2015: 5th International Workshop on Climate Informatics, Boulder, CO, USA, 24–25 September 2015; pp. 1–2.
13. Lguensat, R.; Tandeo, P.; Ailliot, P.; Pulido, M.; Fablet, R. The Analog Data Assimilation. *Mon. Weather Rev.* **2017**, *145*, 4093–4107. [CrossRef]
14. Hardman-Mountford, N.; Richardson, A.; Boyer, D.; Kreiner, A.; Boyer, H. Relating sardine recruitment in the Northern Benguela to satellite-derived sea surface height using a neural network pattern recognition approach. In *Progress in Oceanography: ENVIFISH: Investigating Environmental Causes of Pelagic Fisheries Variability in the SE Atlantic*; Elsevier Ltd.: Amsterdam, The Netherlands, 2003; Volume 59, pp. 241–255.
15. Le Traon, P.; Nadal, F.; Ducet, N. An improved mapping method of multisatellite altimeter data. *J. Atmos. Ocean. Technol.* **1998**, *15*, 522–534. [CrossRef]
16. Bretherton, F.P.; Davis, R.E.; Fandry, C. A technique for objective analysis and design of oceanographic experiments applied to MODE-73. In *Deep Sea Research and Oceanographic Abstracts*; Elsevier: Amsterdam, The Netherlands, 1976; Volume 23, pp. 559–582.
17. Fablet, R.; Verron, J.; Mourre, B.; Chapron, B.; Pascual, A. Improving mesoscale altimetric data from a multi-tracer convolutional processing of standard satellite-derived products. *IEEE Trans. Geosci. Remote Sens.* **2018**, *56*, 2518–2525. [CrossRef]

18. Ubelmann, C.; Klein, P.; Fu, L.L. Dynamic Interpolation of Sea Surface Height and Potential Applications for Future High-Resolution Altimetry Mapping. *J. Atmos. Ocean. Technol.* **2014**, *32*, 177–184. [CrossRef]
19. Klein, P.; Isern-Fontanet, J.; Lapeyre, G.; Roullet, G.; Danioux, E.; Chapron, B.; Le Gentil, S.; Sasaki, H. Diagnosis of vertical velocities in the upper ocean from high resolution sea surface height. *Geophys. Res. Lett.* **2009**, *36*, L12603. [CrossRef]
20. Isern-Fontanet, J.; Chapron, B.; Lapeyre, G.; Klein, P. Potential use of microwave sea surface temperatures for the estimation of ocean currents. *Geophys. Res. Lett.* **2006**, *33*, L24608. [CrossRef]
21. Isern-Fontanet, J.; Shinde, M.; Andersson, C. On the Transfer Function between Surface Fields and the Geostrophic Stream Function in the Mediterranean Sea. *J. Phys. Oceanogr.* **2014**, *44*, 1406–1423. [CrossRef]
22. Turiel, A.; Sole, J.; Nieves, V.; Ballabrera-Poy, J.; Garcia-Ladona, E. Tracking oceanic currents by singularity analysis of Microwave Sea Surface Temperature images. *Remote Sens. Environ.* **2009**, in press. [CrossRef]
23. Turiel, A.; Nieves, V.; Garcia-Ladona, E.; Font, J.; Rio, M.H.; Larnicol, G. The multifractal structure of satellite sea surface temperature maps can be used to obtain global maps of streamlines. *Ocean Sci.* **2009**, *5*, 447–460. [CrossRef]
24. Fablet, R.; Viet, P.H.; Lguensat, R. Data-driven Models for the Spatio-Temporal Interpolation of satellite-derived SST Fields. *IEEE Trans. Comput. Imaging* **2017**. [CrossRef]
25. Buades, A.; Coll, B.; Morel, J.M. A non-local algorithm for image denoising. In Proceedings of the IEEE Conference on Computer Vision and Pattern Recognition, CVPR'05, San Diego, CA, USA, 20–25 June 2005; Volume 2, pp. 60–65. [CrossRef]
26. Fablet, R.; Huynh Viet, P.; Lguensat, R.; Horrein, P.H.; Chapron, B. Spatio-Temporal Interpolation of Cloudy SST Fields Using Conditional Analog Data Assimilation. *Remote Sens.* **2018**, *10*, 310. [CrossRef]
27. Masumoto, Y.; Sasaki, H.; Kagimoto, T.; Komori, N.; Ishida, A.; Sasai, Y.; Miyama, T.; Motoi, T.; Mitsudera, H.; Takahashi, K.; et al. A fifty-year eddy-resolving simulation of the world ocean: Preliminary outcomes of OFES (OGCM for the Earth Simulator). *J. Earth Simul.* **2004**, *1*, 35–56.
28. Sasaki, H.; Nonaka, M.; Masumoto, Y.; Sasai, Y.; Uehara, H.; Sakuma, H. An eddy-resolving hindcast simulation of the quasi-global ocean from 1950 to 2003 on the Earth Simulator. In *High Resolution Numerical Modelling of the Atmosphere and Ocean*; Springer: Berlin/Heidelberg, Germany, 2008.
29. Shaw, P.T.; Chao, S.Y.; Fu, L.L. Sea surface height variations in the South China Sea from satellite altimetry. *Oceanol. Acta* **1999**, *22*, 1–17. [CrossRef]
30. Bocquet, M.; Pires, C.A.; Wu, L. Beyond Gaussian statistical modeling in geophysical data assimilation. *Mon. Weather Rev.* **2010**, *138*, 2997–3023. [CrossRef]
31. Ide, K.; Courtier, P.; Ghil, M.; Lorenc, A. Unified notation for data assimilation: operational, sequential and variational. *Practice* **1997**, *75*, 181–189.
32. Asch, M.; Bocquet, M.; Nodet, M. *Data Assimilation: Methods, Algorithms, and Applications*; Fundamentals of Algorithms; SIAM: Philadelphia, PA, USA, 2016.
33. De Mey, P.; Robinson, A.R. Assimilation of altimeter eddy fields in a limited-area quasi-geostrophic model. *J. Phys. Oceanogr.* **1987**, *17*, 2280–2293. [CrossRef]
34. Gandin, L. Objective analysis of meteorological fields. By L. S. Gandin. Translated from the Russian. Jerusalem (Israel Program for Scientific Translations), 1965. Pp. vi, 242: 53 Figures; 28 Tables. £4 1s. 0d. *Q. J. R. Meteorol. Soc.* **1966**, *92*, 447. [CrossRef]
35. Lorenc, A.C. Analysis methods for numerical weather prediction. *Q. J. R. Meteorol. Soc.* **1986**, *112*, 1177–1194. [CrossRef]
36. Escudier, R.; Bouffard, J.; Pascual, A.; Poulain, P.M.; Pujol, M.I. Improvement of coastal and mesoscale observation from space: Application to the northwestern Mediterranean Sea. *Geophys. Res. Lett.* **2013**, *40*, 2148–2153. [CrossRef]
37. Ping, B.; Su, F.; Meng, Y. An Improved DINEOF Algorithm for Filling Missing Values in Spatio-Temporal Sea Surface Temperature Data. *PLoS ONE* **2016**, *11*, e0155928. [CrossRef]
38. Gerber, F.; de Jong, R.; Schaepman, M.E.; Schaepman-Strub, G.; Furrer, R. Predicting Missing Values in Spatio-Temporal Remote Sensing Data. *IEEE Trans. Geosci. Remote Sens.* **2018**, *56*, 2841–2853. [CrossRef]
39. Fablet, R.; Rousseau, F. Missing data super-resolution using non-local and statistical priors. In Proceedings of the 2015 IEEE International Conference on Image Processing (ICIP), Quebec City, QC, Canada, 27–30 September 2015; pp. 676–680. [CrossRef]

40. Bernard, D.; Boffetta, G.; Celani, A.; Falkovich, G. Inverse Turbulent Cascades and Conformally Invariant Curves. *Phys. Rev. Lett.* **2007**, *98*, 024501. [CrossRef]
41. Moazenzadeh, R.; Mohammadi, B.; Shamshirband, S.; Chau, K.W. Coupling a firefly algorithm with support vector regression to predict evaporation in northern Iran. *Eng. Appl. Comput. Fluid Mech.* **2018**, *12*, 584–597. [CrossRef]
42. Faizollahzadeh Ardabili, S.; Najafi, B.; Shamshirband, S.; Minaei Bidgoli, B.; Deo, R.C.; Chau, K.W. Computational intelligence approach for modeling hydrogen production: A review. *Eng. Appl. Comput. Fluid Mech.* **2018**, *12*, 438–458. [CrossRef]
43. Yaseen, Z.M.; Sulaiman, S.O.; Deo, R.C.; Chau, K.W. An enhanced extreme learning machine model for river flow forecasting: state-of-the-art, practical applications in water resource engineering area and future research direction. *J. Hydrol.* **2018**. [CrossRef]
44. Taormina, R.; Chau, K.W. Data-driven input variable selection for rainfall-runoff modeling using binary-coded particle swarm optimization and Extreme Learning Machines. *J. Hydrol.* **2015**, *529*, 1617–1632. [CrossRef]
45. Taherei Ghazvinei, P.; Hassanpour Darvishi, H.; Mosavi, A.; Yusof, K.b.W.; Alizamir, M.; Shamshirband, S.; Chau, K.w. Sugarcane growth prediction based on meteorological parameters using extreme learning machine and artificial neural network. *Eng. Appl. Comput. Fluid Mech.* **2018**, *12*, 738–749. [CrossRef]
46. Wu, C.; Chau, K. Rainfall runoff modeling using artificial neural network coupled with singular spectrum analysis. *J. Hydrol.* **2011**, *399*, 394–409. [CrossRef]
47. Muja, M.; Lowe, D.G. Scalable nearest neighbor algorithms for high dimensional data. *IEEE Trans. Pattern Anal. Mach. Intell.* **2014**, *36*, 2227–2240. [CrossRef]
48. Cleveland, W.S. Robust locally weighted regression and smoothing scatterplots. *J. Am. Stat. Assoc.* **1979**, *74*, 829–836. [CrossRef]
49. Gaultier, L.; Ubelmann, C.; Fu, L.L. The Challenge of Using Future SWOT Data for Oceanic Field Reconstruction. *J. Atmos. Ocean. Technol.* **2015**, *33*, 119–126. [CrossRef]

remote sensing

MDPI

Article

Evaluation of Heavy Precipitation Simulated by the WRF Model Using 4D-Var Data Assimilation with TRMM 3B42 and GPM IMERG over the Huaihe River Basin, China

Lu Yi [1,2], Wanchang Zhang [2,*] and Kai Wang [3]

1 State Key Laboratory of Pollution Control and Resource Reuse, School of the Environment, Nanjing University, Nanjing 210093, China; dg1225033@nju.edu.cn
2 Key Laboratory of Digital Earth Science, Institute of Remote Sensing and Digital Earth, Chinese Academy of Sciences, Beijing 100094, China
3 State Key Laboratory of Atmospheric Boundary Layer Physics and Atmospheric Chemistry, Institute of Atmospheric Physics, Chinese Academy of Sciences, Beijing 100029, China; kai.wang@mail.iap.ac.cn
* Correspondence: zhangwc@radi.ac.cn

Received: 29 March 2018; Accepted: 20 April 2018; Published: 22 April 2018

Abstract: To obtain independent, consecutive, and high-resolution precipitation data, the four-dimensional variational (4D-Var) method was applied to directly assimilate satellite precipitation products into the Weather Research and Forecasting (WRF) model. The precipitation products of the Tropical Rainfall Measuring Mission 3B42 (TRMM 3B42) and its successor, the Integrated Multi-satellitE Retrievals for Global Precipitation Measurement (GPM IMERG) were assimilated in this study. Two heavy precipitation events that occurred over the Huaihe River basin in eastern China were studied. Before assimilation, the WRF model simulations were first performed with different forcing data to select more suitable forcing data and determine the control experiments for the subsequent assimilation experiments. Then, TRMM 3B42 and GPM IMERG were separately assimilated into the WRF. The simulated precipitation results in the outer domain (D01), with a 27-km resolution, and the inner domain (D02), with a 9-km resolution, were evaluated in detail. The assessments showed that (1) 4D-Var with TRMM 3B42 or GPM IMERG could both significantly improve WRF precipitation predictions at a time interval of approximately 12 h; (2) the WRF simulated precipitation assimilated with GPM IMERG outperformed the one with TRMM 3B42; (3) for the WRF output precipitation assimilated with GPM IMERG over D02, which has spatiotemporal resolutions of 9 km and 50 s, the correlation coefficients of the studied events in August and November were 0.74 and 0.51, respectively, at the point and daily scales, and the mean Heidke skill scores for the two studied events both reached 0.31 at the grid and hourly scales. This study can provide references for the assimilation of TRMM 3B42 or GPM IMERG into the WRF model using 4D-Var, which is especially valuable for hydrological applications of GPM IMERG during the transition period from the TRMM era into the GPM era.

Keywords: precipitation; 4D-Var data assimilation; TRMM 3B42; GPM IMERG; WRF

1. Introduction

Precipitation is a basic and vital component of the global water and energy cycles [1]. A robust knowledge of precipitation processes at finer spatiotemporal resolutions has become increasingly important for hydrological modeling, flood monitoring, soil moisture estimation, and water resource management [2–6]. Currently, there are generally three mainstream methods to obtain precipitation information: traditional in situ observations, estimations from remote sensing, and numerical weather prediction (NWP) [7–9].

In situ precipitation observations are generally obtained from conventional ground rain gauge stations. This type of data is generally considered to be the most accurate measurements, and served as the grounds for true precipitation values [10,11]. However, the application of these data in the hydrological field is severely limited by poor point-to-area representativeness, incomplete opening to the public, and several well-recognized issues of the station network, such as poor spatial distribution and wind-induced deviation [12,13]. As remote sensing techniques developed, various satellite precipitation products based on visible, infrared, and microwave wavelengths have emerged during the last few decades, such as the Global Precipitation Climatology Project (GPCP) [14], the Climate Prediction Center Morphing technique (CMORPH) [15], the Tropical Rainfall Measuring Mission (TRMM) [16], and its ongoing replacement Global Precipitation Measurement project (GPM) [17]. These products not only cover a nearly global area, they also are available to the public free of charge. Nevertheless, they still have coarse spatiotemporal resolutions, which are incapable of representing consecutive precipitation process and have difficulty in detecting extreme events at high latitudes [18]. Since the NWP model is built on precise physical governing equations, it can resolve the inherent dynamics of precipitation and nearly represent the entire spatial pattern of the precipitation process [19,20]. However, when solving the equations and initialization errors, approximations due to incomplete observations often induce many uncertainties to the model outputs [21,22]. In terms of simulated area ranges, the NWP model is usually divided into general circulation models (GCMs), which cover the global or continental scales, and regional climate models (RCMs), which cover the regional scale or mesoscale. Compared with GCMs, RCMs can better simulate the exact distribution of a climatic field since they have a finer spatial resolution; this enables them to resolve finer details of land surface characteristics, such as topography, land cover, and surface vegetation [23,24]. Frequently used RCMs include the National Meteorological Center (NMC) forecast model [25–27], the next-generation Weather Research and Forecasting (WRF) model [28], the operational Japan Meteorological Agency (JMA) mesoscale model [29], and the European Center for Medium-Range Weather Forecasts (ECMWF) model [30].

Considering the merits and drawbacks of these three types of precipitation data, we attempted to integrate two of them to obtain a type of precipitation data. It is not only independent from in situ observations, it can also reflect the precipitation consecutiveness at higher spatiotemporal resolutions [31,32]. Therefore, we used the data assimilation (DA) method to integrate freely downloaded remote sensing precipitation estimations with precipitation prediction from a more physically realistic NWP model. Among the various DA algorithms, such as the polynomial interpolation method [33], optimum interpolation [34], three-dimensional variational (3D-Var) assimilation [35], four-dimensional variational (4D-Var) assimilation [36], and the Kalman filter [37,38], 4D-Var is particularly appropriate for assimilating synoptic satellite data due to its advantages regarding a definite theoretical basis, simple formulation, and no limitations on the type of assimilated data that is utilized [39].

At present, there have been extensive studies regarding integrating various precipitation data with the NWP model via the 4D-Var data assimilation method. Zupanski and Mesinger [40] first carried out a 4D-Var experiment with 24-hour accumulated precipitation data and the NMC forecast model in the United States of America (USA) and demonstrated its improvement in precipitation forecasting. Koizumi et al. [21] used the JMA mesoscale 4D-Var system to assimilate one-hour radar-based precipitation data at a spatial resolution of 20 km over the islands of Japan and demonstrated improvements in precipitation forecasts for an 18-hour forecast time. Lopez [41] assimilated the National Centers for Environmental Prediction (NCEP) stage IV gauge-corrected radar precipitation data into the ECMWF Global Integrated Forecasting System over the eastern USA, and found a substantial improvement in short-term (i.e., up to 12 h) precipitation forecasts. Lin et al. [42] assimilated NCEP stage IV rainfall data into the WRF model with the 4D-Var method in the USA, and they successfully downscaled a six-hour precipitation product with a 20-km resolution to an hourly precipitation product with a resolution of less than 10 km. These studies were mainly carried out to resolve the problems such as the highly nonlinear and discontinuous in cumulus convection parameterization [7,41], the sensitivity of the different global datasets for the initial and boundary conditions [21,41,43], and the effectiveness of applying different observational and background error covariance matrices [44,45]. We will not go into the many details of 4D-Var techniques, but rather will investigate its potential in hydrological applications.

From an application perspective, a majority of these existing studies were performed and evaluated at the mesoscale, whereas a limited number of studies focused on the basin scale evaluation, even though basin is the most commonly used unit in hydrological studies. Moreover, since the GPM was just released in 2014, there are very few studies on the feasibility and efficiency of the GPM application in 4D-Var data assimilation, and the discrepancies in assimilation effectiveness between assimilating GPM and assimilating TRMM are less investigated. Therefore, we assimilated the Integrated Multi-satellitE Retrievals for GPM (GPM IMERG) and the TRMM Multi-satellite Precipitation Analysis 3B42 (TRMM 3B42) with the 4D-Var method into the NWP model of WRF, and assessed their performances in simulating two heavy precipitation events that occurred over the Huaihe River basin (HRB) in eastern China. Before assimilation, we first drove the WRF model with two different forcing data to choose a more suitable forcing datum and determine the control experiment for the subsequent assimilation work. Then, DA experiments were carried out with different assimilation data and for different precipitation events. Finally, we evaluated the experimental precipitation results with the daily in situ observations and the hourly merged CMORPH estimations at different spatial and temporal scales in detail. The manuscript is organized as follows: Section 2 introduces the study area, study events, and data. Section 3 introduces the WRF configuration, 4D-Var methodology, experimental design, and evaluation metrics. Section 4 shows the evaluations of the simulated precipitation from the WRF and the WRF 4D-Var that is assimilated with TRMM 3B42 and GPM IMERG. Section 5 discusses the WRF sensitivity to different rainfall events, forcing data, and spatial resolutions; examines the effectiveness of WRF 4D-Var at different thresholds and time; and compares the 4D-Var performances assimilated with TRMM 3B42 and GPM IMERG. The conclusions are drawn in Section 6.

2. Data

2.1. Study Area and Events

The HRB is one of seven major river basins in China; it is located between 110°–122 °E and 31°–37 °N (Figure 1b), and covers an area of 270,000 km^2. Most of the HRB region is vast plains, except for some mountains and foothills located at the western, southern, and northeastern HRB (as shown in Figure 1b). The mountain altitudes are normally 1000–2000 meters above the sea level (a.s.l.). The HRB is in the transitional zone between the abundant rainfall area of southern China and the arid area of northern China [46], belonging to the warm temperate and semi-humid monsoon region with an average temperature of 11–16 degrees centigrade and an average annual rainfall about 910 mm. The precipitation distribution within a year is very uneven. In the flood season (June–September), the total precipitation accounts for 50–80% of annual precipitation [47]. The heavy rainfall events that occur in the summer frequently cause disaster in this area. Across the basin, there is a gradual gradient in average annual precipitation from about 1000 mm in the southeast to less than 600 mm in the northwest, while the highest precipitation occurs in the inner mountain areas; this variation is a result of the topography [48]. A robust knowledge of the precipitation processes in the HRB is vital for local flood monitoring and water resource management.

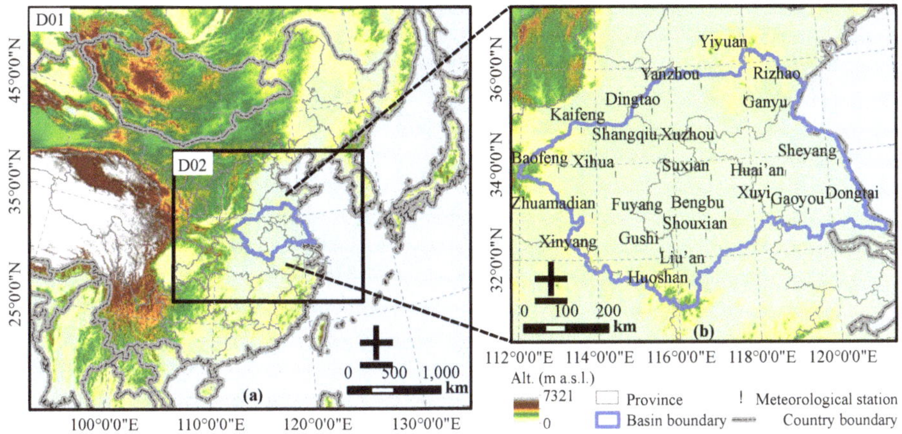

Figure 1. (**a**) Outer domain (D01, 27-km resolution) and inner domain (D02, 9-km resolution) defined in the Weather Research and Forecasting (WRF) model; (**b**) location and meteorological stations of the Huaihe River basin.

To select the studied precipitation events, daily precipitation data from 30 meteorological stations in the HRB during 2015 were obtained from the China Meteorological Administration (CMA) and analyzed (Figure 2). Based on these CMA observations, the contributions of accumulated daily precipitation to the annual precipitation amounts at each meteorological station were calculated and sorted in decreasing order. As shown in Figure 2a, all 30 CMA stations received half of the total annual precipitation within seven days (Dingtao station) or 13 days (Huoshan station). This suggests that short-term heavy precipitation events are the main contributors to the total annual precipitation amount in the HRB. Therefore, we focused our studies on short-term heavy precipitation events. The mean monthly precipitation over the HRB in 2015 was also calculated, and is illustrated in Figure 2b. The results showed that the precipitation amount during the flood season of the HRB composed 58.23% of the annual amount at all of the stations. The mean monthly precipitation in June comprised the highest percentage (21.6%), and August had the second largest percentage (18.33%). However, because the WRF forcing data of the NCEP final analyses (FNL) ds083.3 was issued in July 2015, we selected one event from August. As shown in Figure 2c, a heavy rainfall event occurred on 9 August, when the daily precipitation at many CMA stations exceeded 50 mm, and the highest exceeded 160 mm; thus, we chose the rainfall event from 0000 UTC on 9 August to 1200 UTC on 11 August (hereafter referenced as event A) as our first case study. As shown in Figure 2b, the mean monthly precipitation in November was evidently the highest during the non-flood season. Therefore, we chose the rainfall event that occurred from 0000 UTC on 5 November to 1200 UTC on 7 November (hereafter referenced as event N) as our second case study. During event N, the average daily precipitation of the basin reached 22.1 mm, and that of a single CMA station (Rizhao) reached 66.6 mm. The two case studies (event A and event N; each spanned 60 h) represent the type of convection dominant precipitation events during a flood season, and extratropical cyclone-associated precipitation events during the non-flood season, respectively.

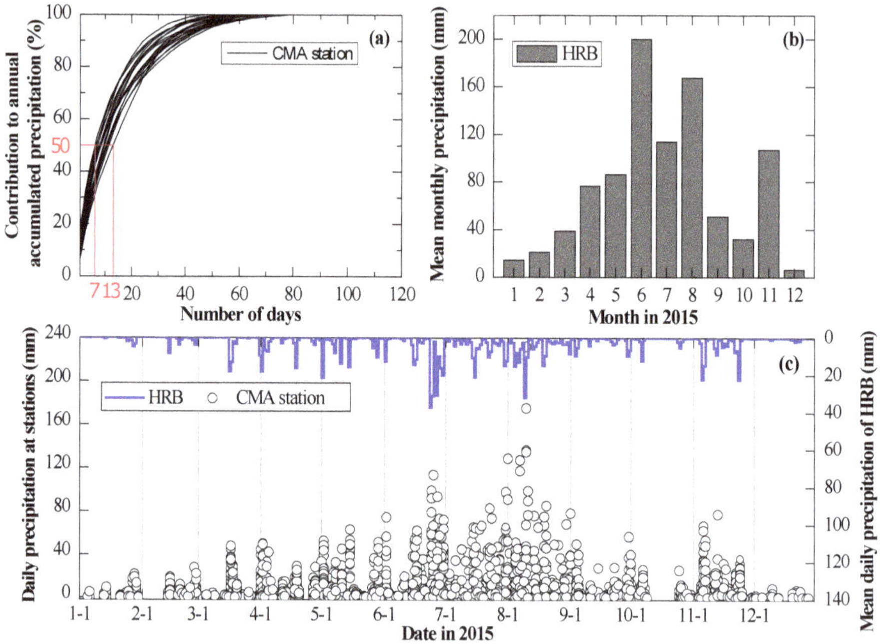

Figure 2. (**a**) Contribution (%) of accumulated daily precipitation to the total annual precipitation for 30 CMA stations across the HRB in 2015 (sorted in decreasing order); (**b**) mean monthly precipitation of the HRB in 2015; (**c**) daily precipitation at the 30 CMA stations and the mean daily precipitation of the basin (HRB).

2.2. Study Data

2.2.1. Satellite Precipitation Products for Assimilation

In this study, two satellite-based precipitation data, including the TRMM 3B42 (version 7) and the GPM IMERG (final run), were applied as observation operators. Both were processed to collect six-hour accumulated precipitation data for assimilation into the WRF model.

TRMM was jointly launched by the National Aeronautics and Space Administration (NASA) and the Japan Aerospace Exploration Agency (JAXA) in 1997. TRMM 3B42 is one product of the TRMM Multi-satellite Precipitation Analysis (TMPA); it combines remote sensing data from various microwave and infrared sensors with the monthly gauge analysis from the Global Precipitation Climatology Centre (GPCC). TRMM 3B42 covers ±50 °N/S at spatiotemporal resolutions of 0.25° and 3 h [49]. After over 17 years of productive data acquisition, the observation instruments on TRMM were turned off on 8 April 2015. However, the actual termination of TRMM was not a substantive issue for its multi-satellite products in TMPA; therefore, the TRMM 3B42 data during our study periods are still available. Tang et al. [50] compared TRMM 3B42 with the gauge observations and found that the correlation coefficient (CC) over Mainland China reached 0.42 and 0.68 at its original three-hour and daily timescales, respectively; for the region encompassing the lower reaches of the Huaihe River, the CCs reached 0.43 and 0.71 at the three-hour and daily timescales, respectively.

As the replacement of TRMM, GPM has been providing next-generation global observations of rain and snow for more than three years. The observation system on the GPM comprises one core satellite and several constellation satellites, with a dual-frequency precipitation radar and a suite of microwave radiometers. The core observatory of the GPM was designed as an extension of TRMM's highly successful rain-sensing package, which primarily focused on heavy to moderate rain over tropical and subtropical oceans [51]. A key advancement of the GPM over the TRMM is the extended capability to measure light rain (≤0.5 mm/hour) and falling snow, since the two types of precipitation account for significant proportions of precipitation at mid and high latitudes. GPM IMERG is the third-level precipitation product of GPM, which covers an area of ±60 °N/S with unprecedented resolutions of 0.1° and 30 min. Tang et al. [50] reported that when compared with the gauge observations, the CCs of GPM IMERG over Mainland China reached 0.53 and 0.71 at the hourly and daily timescales, respectively; for the region encompassing the lower reaches of the Huaihe River, the CCs reached 0.55 and 0.72 at the hourly and daily timescales, respectively.

2.2.2. Forcing Data for the WRF Model

To simulate the two heavy precipitation events over the HRB, the NCEP FNL datasets were applied to provide the WRF model with the initial states and lateral boundary conditions. Among the various NCEP FNL datasets, the commonly used 1.0° and six-hour FNL ds083.2 and the newly released 0.25° and six-hour FNL ds083.3 datasets were employed and compared. The one that performed better with the WRF model was used in the subsequent 4D-Var experiments. These two forcing data were composed of an underlying Global Forecast System (GFS) that was obtained from the Global Data Assimilation System (GDAS), which continuously collects observational data from the Global Telecommunications System (GTS) and other sources for many related analyses. The archived time series of the ds083.2 and ds083.3 datasets started on 30 July 1999 and 8 July 2015, respectively. Both of these datasets extended to a near-current date. They can be downloaded from the Research Data Archive at the National Center for Atmospheric Research (https://rda.ucar.edu/datasets/).

2.2.3. In Situ and Gauge-Corrected Data for Evaluation

To evaluate the precipitation results simulated by the WRF model and the WRF 4D-Var with TRMM 3B42 and GPM IMERG, two types of precipitation datasets were used as reference data. One was the in situ gauge observation dataset for daily precipitation; this data was obtained from the 30 meteorological stations in the HRB (Figure 1b) and provided by the CMA (hereafter referenced as CMA data). This dataset was collected as a benchmark for the point-scale evaluations. The accuracy of the CMA data was officially claimed to be approximately 100%, as before its release, it experienced a series of quality control assessments including a climate limit value check, station extreme value check, spatiotemporal consistency check, and other related checks (http://data.cma.cn/data/cdcdetail/dataCode/SURF_CLI_CHN_MUL_DAY_V3.0.html). When comparing with the CMA data, the simulated precipitation results from the WRF and the WRF 4D-Var models were changed to Beijing time, which is used in the CMA data. The other dataset was the gridded merged CMORPH data, which was used for the evaluation at the grid scale [52]. It is also released by the CMA, with spatiotemporal resolutions of 0.1° and hourly. The CMA generated the merged CMORPH data by applying the probability density function matching method and the optimal interpolation method. This dataset integrates with the following two data sources: (1) gauged hourly precipitation from more than 30,000–40,000 automatic weather stations in China after quality control; and (2) inverse precipitation products from the global CMORPH satellite, with resolutions of 30 min and 8 km provided by the Climate Prediction Center (USA). The merged CMORPH data have been opened to the public since 1 January 2008. This dataset covers mainland China, and its applied time zone was the same as that used in the WRF model, i.e., the UTC time zone. Its general error remains under 10%, and the errors in strong precipitation over sparsely-gauged areas are less than 20%. Its quality is better than that of other similar products in China (http://data.cma.cn/). When used for evaluation, the merged CMORPH data were resampled to the same spatial resolutions as those of the WRF domains (i.e., 27 km and 9 km) by the nearest neighborhood interpolation method. Moreover, to ensure evaluation quality, the coordinate system of the merged CMORPH data was transformed to be the same projected coordinate system as that used in the WRF model. It should be noted that for the grid-scale evaluation, assessments were only carried out over intersecting regions that encompassed the two WRF domains and the merged CMORPH data region since the latter only covered mainland China.

3. Methods

3.1. WRF Configuration, 4D-Var Methodology, and Experimental Design

3.1.1. WRF Configuration

To obtain precipitation at a finer resolution and reduce the computing quantity, a nested domain (Figure 1a) was applied in our WRF configuration. The outer domain (D01) was set around the HRB to be as large as possible to cover the important weather features of the HRB, which are probably caused by the Indian southwestern monsoon, the East Asian subtropical monsoon, and the Asian monsoon that originates from the Siberian high. However, D01 cannot be set to an unlimited large domain, as a large domain means more grids for calculation, which requires more computational resources. Therefore, we finally define D01 with grids of 180 * 155 and a 27-km resolution and set the inner domain (D02) to cover the whole HRB, with 226 * 175 grids and a 9-km resolution.

The physical configuration of the WRF model and its related dominant parameters were determined by obeying the rule that the configuration should incorporate the experiences obtained from comparable numerical modeling studies as much as possible, especially studies performed over the HRB. Moreover, a one-way nesting strategy was employed, since a nesting strategy was not decisive of the overall performance [53], and was not our research priority. The main configuration of the WRF model is listed in Table 1.

Table 1. The main configuration of the Weather Research and Forecasting (WRF) model.

Map and Grids	
Map projection	Lambert conformal
Center point of the domain	35.8 °N, 114 °E
Number of vertical layers	27
Horizontal grid resolution	27 km (D01), 9 km (D02)
Domain grid	180 * 155, 226 * 175
Static geographical fields time step	Standard dataset at a 30″ resolution 150 s, 50 s from the United States Geological Survey (USGS)
Physical Parameterization Schemes	
Cloud microphysics	WRF double-moment six scheme [54]
Long-wave radiation	Rapid Radiative Transfer Model (RRTM) [55]
Short-wave radiation	Dudhia scheme [56]
Land surface model	Noah land surface model (LSM) [57]
Planetary boundary layer	Yonsei University scheme [58]
Cumulus parameterization	New Grell–Devenyi 3 scheme [59] (except for the 9-km domain: no cumulus)

3.1.2. 4D-Var Methodology

The incremental 4D-Var formulation, which is commonly used in operational systems [60–63], was utilized in the WRF 4D-Var data assimilation system. The incremental approach was designed to determine an analysis increment that minimized the cost function, which was defined as a function of the analysis increment instead of the analysis itself [64]. Mathematically, the WRF 4D-Var minimizes cost function J:

$$J = J_b + J_o + J_c \tag{1}$$

which includes quadratic measures of distance to the background, observation, and balanced solution. The background cost function term J_b is:

$$J_b = \frac{1}{2}\left(x^n - x^b\right)^T B^{-1}\left(x^n - x^b\right) \tag{2}$$

where the superscripts -1 and T denote the inverse and the adjoint of a matrix or a linear operator, respectively. The final analysis of WRF 4D-Var (i.e., after the last (n) outer loop occurs) is denoted as x^n.

The background x^b has typically been a short-range forecast in previous analyses. B represents the background error covariance matrix.

The observation cost function term represents the quadratic measure of the distances between the analysis x^n and the forecast model M_k and the observation operator H_k and the observations y_k:

$$J_o = \frac{1}{2}\sum_{k=1}^{K}\{H_k[M_k(x^n)] - y_k\}^T R^{-1}\{H_k[M_k(x^n)] - y_k\} \quad (3)$$

When calculating the observation cost function J_o, a linear approximation is made, and the entire assimilation time window is split into K observation windows. The nonlinear observation operator H_k is approximately transformed into a tangent linear observation operator, and the nonlinear model M_k is also transformed into a tangent linear model. x^n denotes the guess vector, and $x^n - x^{n-1}$ denotes the analysis increment. R represents the observation error covariance matrix.

The balancing cost function J_c measures the quadratic distance between the analysis and a balanced state, and it is expressed as follows:

$$J_c = \frac{1}{2}\gamma_{df}\left[M_{\frac{N}{2}}(x^n - x^{n-1}) - \sum_{i=0}^{N} f_i M_i(x^n - x^{n-1})\right]^T C^{-1}\left[M_{\frac{N}{2}}(x^n - x^{n-1}) - \sum_{i=0}^{N} f_i M_i(x^n - x^{n-1})\right] \quad (4)$$

where N represents the total number of integration steps over the assimilation window, and γ_{df} represents the weight assigned to the J_c term. f_i is the coefficient for the digital filter [65,66]. C is a diagonal matrix containing variances of wind, temperature, and dry surface pressure, with values of $(3\ \mathrm{m/s})^2$, $(1\ \mathrm{K})^2$, and $(10\ \mathrm{hPa})^2$, respectively.

3.1.3. Experimental Design

In this study, two series of experiments were carried out over the HRB. The first series of experiments, which are labeled the CTL experiments in this manuscript, were performed with the WRF model. They were carried out to choose the most optimal forcing data for the WRF model and determine the control experiments for the subsequent 4D-Var assimilation study. According to the different applied forcing data and the different study events, the CTL experiments were labeled from CTL1 to CTL4 (Table 2). The second series of experiments utilized the 4D-Var via direct assimilation of TRMM 3B42 and GPM IMERG. They were referred to as the DA experiments (Table 2). For the DA experiments, the background error covariance matrices, B, for August and November in 2015 were separately generated with one-month ensemble forecasts every 12 h using the NMC method [67]. Meanwhile, one variable, named "eps", was set to 0.0001 to make the WRF DA convergence criterion more stringent. In addition, by considering the impacts of the WRF spin-up [61,68,69], we utilized the WRF forecast during the first 12-hour period of each CTL experiment as the first guess of the 4D-Var assimilation in each DA experiment. Thus, the simulation period for each DA experiment was reduced to 48 h. For each 24 h segment of the 48 h period, there was one WRF 4D-Var model that provided the initial and boundary conditions for the other WRF model with a forecast period of 24-hour cycles; this cycling mode was used to connect the simulations during the entire 48 h. With the application of 32 GB of memory on eight nodes and 16 CPU/nodes on the supercomputer used in our study, each WRF 4D-Var that ran over a six-hour time window took approximately 12 h. Limited by the available computational resources, 4D-Var was only employed in D01. Given the problem of the WRF spin-up, the first 12 h of the CTL experiments were not included in the following evaluations. Therefore, in this study, the evaluation periods for both the CTL experiments and DA experiments spanned from 1200 UTC on 9 August to 1200 UTC on 11 August in 2015 for event A, and from 1200 UTC on 5 November to 1200 UTC on 7 November in 2015 for event N.

Table 2. Designs of the WRF control (CTL) experiments and the WRF data assimilation (DA) experiments.

No.	Studied Event	Forcing Data	No.	Studied Event	Assimilated Data
CTL1	Event A	FNL ds083.2	DA1	Event A	TRMM 3B42
CTL2		FNL ds083.3	DA2		GPM IMERG
CTL3	Event N	FNL ds083.2	DA3	Event N	TRMM 3B42
CTL4		FNL ds083.3	DA4		GPM IMERG

3.2. Evaluation Metrics

To evaluate the simulated precipitation results of the CTL and DA experiments, we applied two categories of metrics, which are shown in Table 3. The first category of metrics contained the mean error (*ME*), the relative error (*RE*), the root mean square error (*RMSE*), and the *CC*, which were used to describe the errors, deviations, and correlations between the simulated precipitation and the reference precipitation at the point scale. The second category of metrics contained skill scores commonly used in meteorological studies, including the bias score (*BIAS*), the false alarm ratio (*FAR*), the probability of detection (*POD*), the probability of false detection (*POFD*), and the Heidke skill score (*HSS*). These skill scores were constructed based on a contingency table [70] and applied in the evaluation at the grid scale. *BIAS* is an indicator of how well the model predicts the number of occurrences of an event. FAR indicates the fraction of forecasts detected by the WRF model that turns out to be wrong. *POD* represents the ratio of correct forecasts to the number of events that occurred, which is commonly known as the hit rate. *POFD* represents the faction of false alarms to the total number of events that did not occur. *HSS* is one of the most frequently used and comprehensive skill scores for summarizing square contingency tables and combining the characteristics of hints and random detections. Considering that the detection accuracy of the CMA ground observation was 0.1 mm/day, the rain/no rain threshold was set to 0.1 mm in this study.

Table 3. Statistical metrics applied in the evaluations *.

Statistical Metrics	Equation	Perfect Value
Mean Error (*ME*; unit: mm)	$\mathrm{ME} = \frac{1}{N}\sum_{i=1}^{N}(P_{P,i} - P_{O,i})$	0
Relative Error (*RE*)	$\mathrm{RE} = \frac{\sum_{i=1}^{N}(P_{P,i} - P_{O,i})}{\sum_{i=1}^{N}P_{O,i}}$	0
Root Mean Square Error (*RMSE*; unit: mm)	$\mathrm{RMSE} = \sqrt{\frac{1}{N}\sum_{i=1}^{N}(P_{P,i} - P_{O,i})^2}$	0
Correlation Coefficient (*CC*)	$\mathrm{CC} = \frac{\sum_{i=1}^{N}(P_{P,i} - P'_{P,i})(P_{O,i} - P'_{O,i})}{\sqrt{\sum_{i=1}^{N}(P_{P,i} - P'_{P,i})^2\sum_{i=1}^{n}(P_{O,i} - P'_{O,i})^2}}$	1
Bias Score (*BIAS*)	$\mathrm{BIAS} = \frac{(A+B)}{(A+C)}$	1
False Alarm Ratio (*FAR*)	$\mathrm{FAR} = \frac{B}{(A+B)}$	0
Probability of Detection (*POD*)	$\mathrm{POD} = \frac{A}{(A+C)}$	1
Probability of False Detection (*POFD*)	$\mathrm{POFD} = \frac{B}{(B+D)}$	0
Heidke Skill Score (*HSS*)	$\mathrm{HSS} = \frac{(s - S_{Ref})}{(S_{Perf - S_{Ref}})}$ $S = \frac{(A+D)}{N}$ $S_{Ref} = \frac{[(A+B)(A+C)+(B+D)(C+D)]}{N^2}$	1

* $P_{P,i}$ and $P_{O,i}$ denote the predicted and observed precipitation values of the *i* grid, where $P'_{P,i}$ and $P'_{O,i}$ are their means, respectively. *A* represents the precipitation predicted by the WRF model and observed by the reference data; *B* represents the precipitation predicted by the WRF model but not observed by the reference data; *C* represents the precipitation not predicted by the WRF model but observed by the reference data; and *D* represents the precipitation not predicted by the WRF model and not observed by the reference data.

4. Results

4.1. Evaluation of Simulated Precipitation in the CTL Experiments

4.1.1. Evaluation of Simulated Precipitation in the CTL Experiments at the Point Scale

To evaluate the precipitation simulated by the WRF model at the point scale, the error indices of ME, RE, CC, and RMSE were used. Precipitation values in D01 and D02 were both extracted from grids located closest to the CMA stations, then accumulated into daily values and compared against the daily in situ CMA data. As delineated in Figure 3, for both event A and event N, the values of *ME* and *RE* in D01 and D02 were positive, which indicated that the WRF-predicted precipitation results were generally overestimated. Although the CC values of event N were lower than those of event A, the points that represented the relationship between the station precipitation and the predicted precipitation was evidently much closer to the 1:1 line for event N (Figure 3b) than that for event A (Figure 3a). Moreover, the values of *ME*, *RE*, and *RMSE* for event N were also lower than those for event A. This means that the predicted precipitation results for event N showed better agreement with the CMA observations than those for event A. In addition, the error metrics for the simulated precipitation from D02 were generally comparable to those from D01, which indicated that the resolution increase for precipitation from 27 km (D01) to 9 km (D02) was realized through the nested domain at a very slight accuracy reduction cost at the point and daily scales. For the variations made by different forcing data, there appeared to be some slight negative changes in the *ME*, *RE*, CC, and *RMSE* values from CTL4 to CTL3, but the *CCs* of CTL 2 were only 0.01 lower than those of CTL1, and the *ME*, *RE*, and *RMSE* values of CTL2 were evidently much lower than those of CTL1. Therefore, we conclude that the WRF performance driven by the FNL ds083.3 dataset outperformed that driven by the ds083.2 dataset at the daily and point scales.

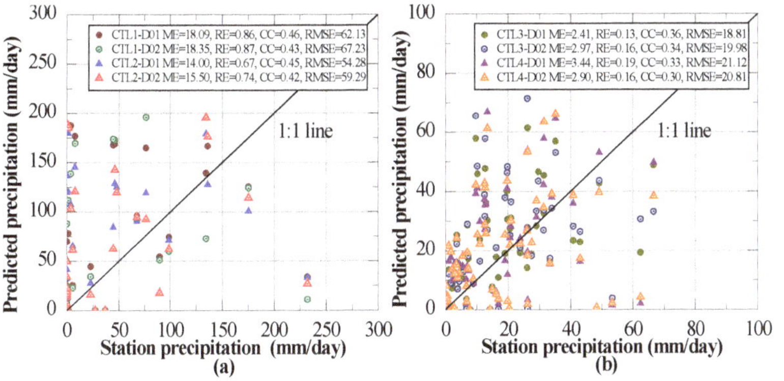

Figure 3. Scatter plots of daily precipitation (mm/day) observed by the China Meteorological Administration (CMA) meteorological stations and predicted by the WRF control experiments: CTL1 and CTL2 (**a**) and CTL3 and CTL4 (**b**).

4.1.2. Evaluation of the Simulated Precipitation in the CTL experiments at the Grid Scale

To make the evaluations of the simulated precipitation more complete, we also compared the simulated precipitation from the CTL experiments with the hourly merged CMORPH data. The skill scores of *BIAS*, *FAR*, *POD*, *POFD*, and *HSS* were used to quantify the assessment. The evaluated results are displayed in Figure 4. It is observed that the box lengths of *BIAS*, *FAR*, and *POFD* for event N were much shorter than those for event A, which indicated lower deviations and better performances for the simulation of event N. This finding was consistent with the evaluated results at the daily and point scales. Thus, we conclude that the WRF performance for event N was better than that for event A.

This was probably attributed to the over or underestimation of the WRF model for localized extreme precipitation intensities when simulating strong convective precipitation events.

By comparing the evaluation results between D01 and D02, it is found for event A that the values of *BIAS* over D02 were much closer to 1 than those for event N; the mean values of *FAR* and *POFD* over D02 were much lower, and the maximum and mean values of *HSS* over D02 were higher. For the rainfall in event N, although the comprehensive index values of *HSS* over D02 were slightly lower than those over D01, the *BIAS* values over D02 were much closer to 1 than those over D01, and the *FAR* values also decreased more over D02 than those over D01. Thus, advantages in the WRF improvement regarding spatial resolution were conveyed both in convective and non-convective rainfall events at the grid and hourly scales.

Based on the above evaluations, the forcing data of NCEP FNL ds083.3 dataset were selected to drive the following WRF 4D-Var experiments. The CTL2 experiment was utilized as the reference for DA1 and DA2 (event A), and the CTL4 experiment was determined as the reference for DA3 and DA4 (event N).

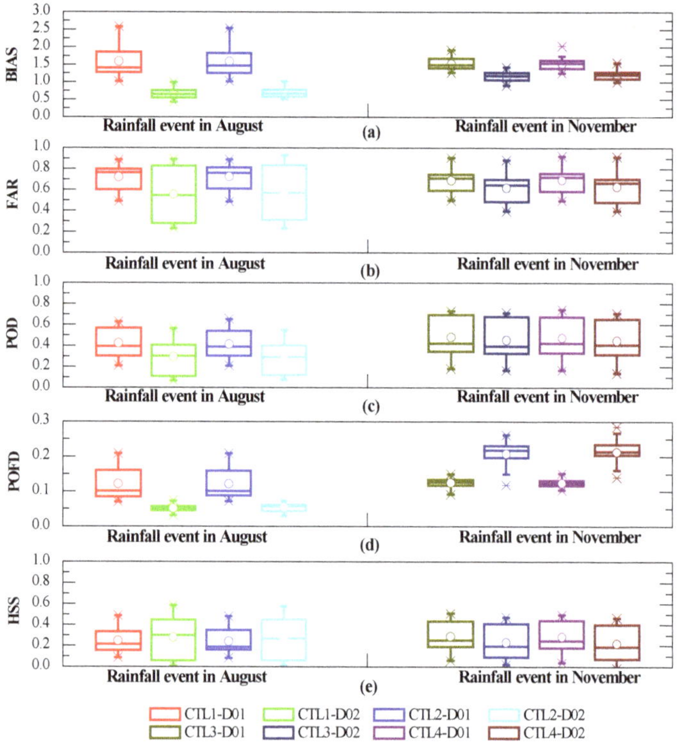

Figure 4. Box plots * of evaluation scores *BIAS* (**a**), *FAR* (**b**), *POD* (**c**), *POFD* (**d**), and *HSS* (**e**) for hourly precipitation (exceeding 0.1 mm/h) simulated by the WRF CTL experiments and estimated by the merged Climate Prediction Center Morphing technique (CMORPH) data. * The lower and upper edges of the central box represent the first and third quartiles (25% and 75%, respectively), and the band and the circle inside the box represent the 50th percentiles and the mean values, respectively. The ends of the outliers represent the minimum and maximum values of the score distributions. The asterisks represent several possible alternative values.

4.2. Evaluation of the Simulated Precipitation in the DA Experiments

The DA experiments output precipitation values with spatiotemporal resolutions of 27 km and 150 s in D01, and 9 km and 50 s in D02. Figure 5 shows the spatial patterns of the 12-hour accumulated precipitation extracted from the reference CTL experiments, the DA experiments and the merged CMORPH dataset over a subset of the D01 domain. It is clear in Figure 5 that the CTL2 experiment captures the shift in precipitation during event A over the HRB from the southwest to the northeast, and the CTL4 experiment also captures the shift in precipitation during event N over the HRB from the west to the east of the HRB. However, compared with the merged CMORPH data, which served as the true measurements at the grid scale, the spatial distributions and the amounts of simulated precipitation in CTL2 and CTL4 were much wider and larger than those reflected in the merged CMORPH dataset. For each 24-hour period, these discrepancies between the simulated precipitation and the merged CMROPH data became more evident in the second 12-hour period than those in the first 12-hour period, which may be related to the error accumulations when the WRF model was running. It can also be found in Figure 5 that the simulated precipitation after assimilation with TRMM 3B42 and GPM IMERG was more agreeable with the merged CMORPH dataset.

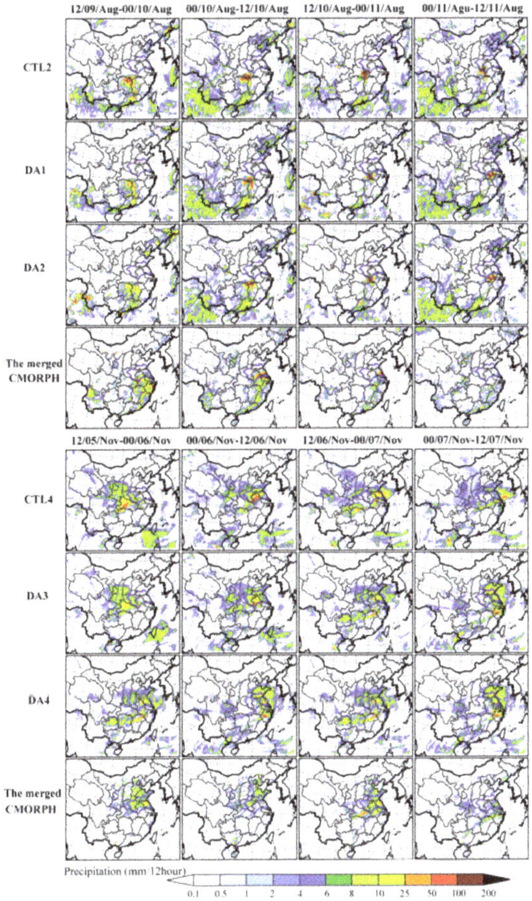

Figure 5. Spatial patterns of 12-hour accumulated precipitation obtained from experiments CTL2, CTL4, DA1-4, and the merged CMORPH estimations over a subset of D01.

4.2.1. Evaluation of Simulated Precipitation in the DA Experiments at the Point Scale

To evaluate the simulated precipitation in the DA experiments at the point scale, precipitation values were extracted from the grids closest to the CMA stations from the DA simulations, which were accumulated into daily values and compared against the daily CMA data. The evaluated error indices of ME, RE, CC, and RMSE for the control experiments and the DA experiments are displayed in Figure 6. As shown in Figure 6a, the application of 4D-Var to event A had an obvious change in simulated precipitation via CLT2 from over-forecasting (above the 1:1 line) to under-forecasting (below the 1:1 line). This variation can also be found in the numerical changes in *ME* from positive values to negative values. For event N (Figure 6b), the *ME*s changed differently. The *ME*s for DA3 were still positive (indicating over-forecasting), and the *ME*s for DA4 were negative (indicating under-forecasting). This means that after assimilating GPM IMERG, WRF precipitation outputs tend to be lower than those from the corresponding CTL experiments for both event A and event N. By assimilating with the remotely sensed precipitation products, the *RMSE*s of all of the DA experiments all significantly decreased, which indicated reduced deviations in the precipitation outputs of the WRF 4D-Var. Moreover, the CCs of the DA experiments all significantly increased. The biggest CC increase for event A was 0.39 (from CTL2-D01 to DA2-D01) and 0.22 for event N (from CTL4-D01 to DA4-D01). These positive variations manifested improvements in the precipitation simulations due to the assimilation of TRMM 3B42 and GPM IMERG onto WRF at the daily and point scales.

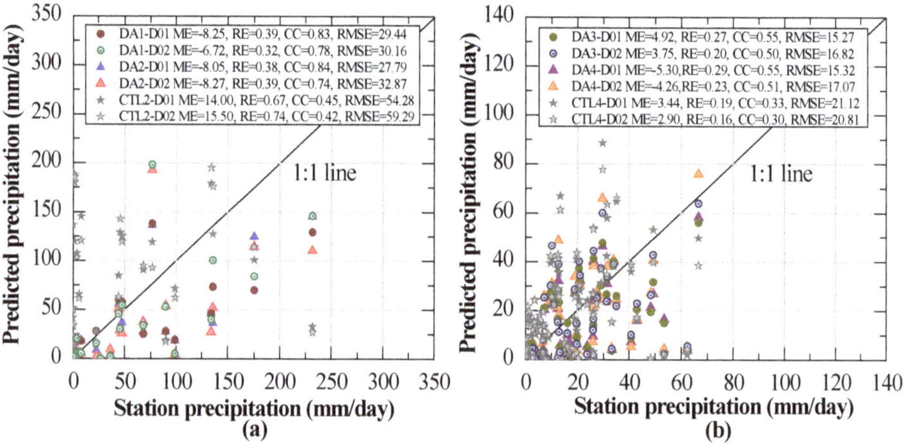

Figure 6. Scatter plots of daily precipitation (mm/day) observed by CMA meteorological stations and simulated by the CTL and DA experiments for rainfall events in August (**a**) and November (**b**) of 2015.

4.2.2. Evaluation of Simulated Precipitation in the DA experiments at the Grid Scale

The simulated precipitation results of the DA experiments in D01 and D02 were processed into hourly values and evaluated with the grid data (i.e., the hourly merged CMORPH data). The evaluations were quantified by the skill scores of *BIAS*, *FAR*, *POD*, *POFD*, and *HSS*. These skill scores for the hourly precipitation from the CTL2, CTL4, and DA experiments changed with time during the 48 h in each experiment (Figure 7). On the whole, for the reason of error accumulation during the simulations, the *FAR* values showed increasing tendency with time, the values of *POD* and *HSS* presented decreasing tendency with time, the values of BIAS and POFD fluctuated with time, and the amplitudes were weakened by the application of 4D-Var in DA experiments. Looking into the *HSS* variation, for the reason of 4D-Var data assimilation, the *HSS* values for the hourly precipitation predictions increased at first, but as time passed, they decreased because of the error accumulation during the simulations. Moreover, there appear abrupt changes between the 24th hour and the 25th hour in Figure 7a–e. These sudden variations were caused by the newly initial and boundary conditions accompany with the start of the second 24-hour forecast in each DA experiment.

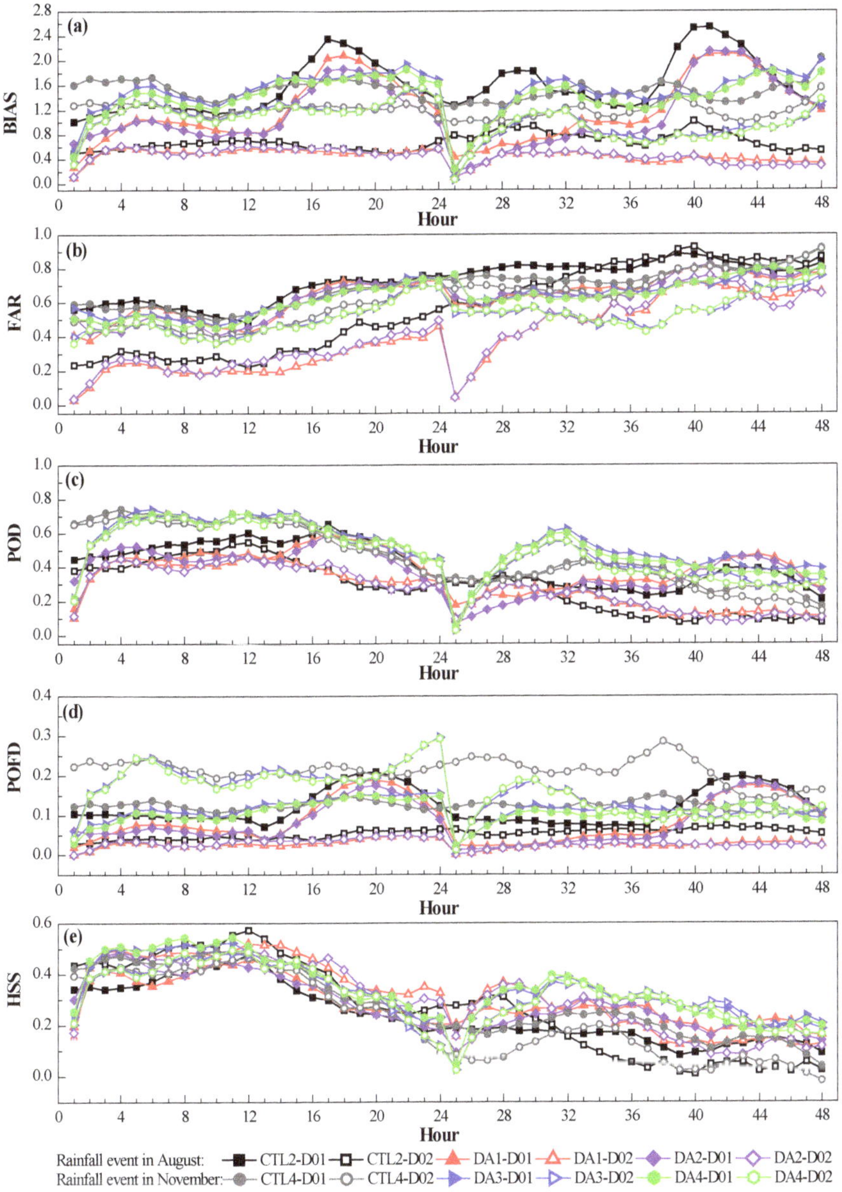

Figure 7. Skill scores of BIAS (**a**), FAR (**b**), POD (**c**), POFD (**d**), and HSS (**e**) for the hourly precipitation (48 h in all during the study period for each event) obtained from the CLT2, CTL4, and DA experiments.

Besides, it is also seen in Figure 7 that in the same domains, all of the values of *BIAS*, *FAR*, and *POFD* for the DA experiments are primarily lower than those in the CTL experiments. This suggests that the 4D-Var-assimilated TRMM 3B42 or GPM IMERG can effectively reduce several false alarms regarding rainfall occurrences. It is noteworthy that the variations in hit rates, which were indicated by the *POD*s, were different between event A and event N (Figure 7c). When comparing CTL2-D01 and CTL2-D02, the *POD* average values of DA1-D01, DA1-D02, DA2-D01, and DA2-D02 were reduced by 0.02, 0.01, 0.04, and 0.01, respectively. For event N, the *PODs* of DA3 and DA4 all remarkably improved compared with those of CTL4. These different variations in *POD* were probably attributed to the different precipitation mechanisms of event A and N and the imperfection of WRF in predicting extreme local precipitation. Regardless, the comprehensive scores of *HSS* for the DAs finally improved. In comparison with the mean hourly *HSS* values for the reference CTL experiments, the *HSS* mean values of DA1-D01, DA1-D02, DA2-D01, DA2-D02, DA3-D01, DA3-D02, DA4-D01, and DA4-D02 increased by 0.05 (0.28), 0.05 (0.32), 0.04 (0.28), 0.04 (0.31), 0.05 (0.34), 0.09 (0.31), 0.06 (0.34), and 0.09 (0.31), respectively (the *HSS* values are in parentheses). This definitely demonstrates the positive improvements in the precipitation simulations that were made by the 4D-Var assimilation with TRMM 3B42 and GPM IMERG via the WRF model at the grid and hourly scales. Moreover, it can be concluded that the improvements in WRF 4D-Var for event A mainly benefited from the corrections of the false alarms for non-occurrences, since the *POD* values for DA1 and DA2 decreased, but their *BIASs*, *FARs*, and *POFDs* also reduced, which finally improved their *HSSs*. For event N, the reductions in *BIASs*, *FARs*, and *POFDs* were minor, but the *PODs* experienced evident improvements, and the *HSSs* finally became better than those of CTL4.

Considering precipitation applications in a hydrological basin, the simulated precipitation results of the DA experiments in the HRB were specifically evaluated; they were extracted, accumulated and spatially averaged. Figure 8 shows the hourly mean precipitation values of the basin via the CTL2, CTL4, and DA experiments during the studied 48 h. As shown in Figure 8, the agreements between the DAs' simulated precipitation and the merged CMORPH estimations were better than those between the CTL simulated results and the merged CMORPH estimations. Nevertheless, it should be noted that the impact of the WRF 4D-Var was not always consistent throughout the whole forecast period. For each 24-hour simulation, the mean precipitation of the basin via the DAs had the greatest agreement with the merged CMORPH estimations at the beginning (about the first 12 h). During the next 6 h, the mean precipitation values of the basin also maintained relatively good consistency with the merged CMORPH data. However, in the following hours, the predictions showed evident deviations from the merged CMORPH estimations. The reduction in the positive impact on the precipitation simulations suggested a time interval for the efficiency of the WRF 4D-Var. This may be related to the inevitable errors that accumulate while the WRF 4D-Var DA system is running [53]. Moreover, when we applied the cycling mode, the initial conditions of the subsequent 24-hour forecast were provided by the previous 24-hour forecast; this caused error accumulation from the first 24-hour simulation, and as a result, the WRF 4D-Var performance for the subsequent 24 h became generally worse than that for the first 24 h.

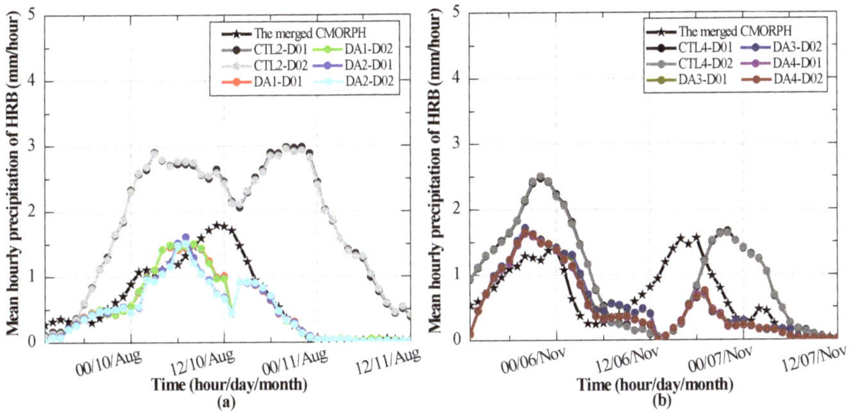

Figure 8. Mean hourly precipitation (mm/h) of the basin via the merged CMORPH estimates and the CTL2, CTL4, and DA experiments for rainfall events in August (**a**) and November (**b**).

5. Discussions

5.1. WRF Sensitivity to Different Rainfall Events, Forcing Data, and Spatial Resolutions

To quantitatively assess the WRF sensitivity to different rainfall events, different forcing data and spatial resolutions, the error indices of *ME*, *RE*, *CC*, and *RMSE* and the skill scores of *BIAS*, *FAR*, *POD*, *POFD*, and *HSS* for the CTL experiments were inter-compared and are displayed in Figure 9a,b, respectively. Both the sharp decreases in *ME*, *RE*, and *RMSE* from CTL1-2 to CTL3-4 (Figure 9a) and the evident *BIAS* value much closer to 1 (Figure 9b) demonstrated a better WRF performance for event N than that for event A. This finding is consistent with that concluded by Lin et al., [42], where they found that the WRF model was relatively harder to use when forecasting convective and dominant precipitation events than precipitation events caused by extratropical cyclones.

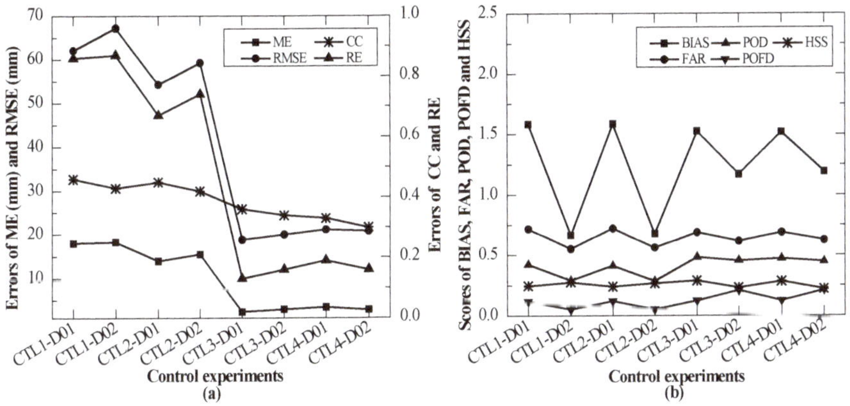

Figure 9. (**a**) Error indices for daily precipitation between the CTL experiment simulations and the CMA observations; (**b**) skill score averages of hourly precipitation between the CTL experiment simulations and the merged CMORPH estimations.

The impacts of different forcing data on the WRF precipitation simulations were well reflected in the error indices at the daily and point scales. As shown in Figure 9a, the CCs of CTL2 over D01 and D02 were both 0.01 lower than the CCs of CTL1. However, the *ME*, *RE*, and *RMSE* values of CTL2 significantly decreased by 22.61%, 22.09%, and 12.63% compared with those of CTL1_D01, respectively, and decreased by 15.53%, 14.94% and 11.81% compared with those of CTL1_D02, respectively. This indicated that significant improvements were made by using forcing data ds083.3 for event A. For event N, contrasted with CTL3, the CCs of CTL4 decreased by 0.03 in D01 and 0.04 in D02. The *ME*, *RE*, and *RMSE* values of CTL4_D01 slightly increased by 1.03 mm, 0.06 mm, and 2.31 mm respectively, compared with those of CTL3_D01. The values of CTL4-D02 increased by −0.07 mm, 0.00 mm, and 0.83 mm, respectively, compared to those of CTL3_D02. These results indicated that slightly negative impacts of ds083.3 were found when predicting non-convective precipitation. These differences in the error indices caused by different forcing data suggested that the WRF model is very sensitive to its initial and lateral boundary conditions, especially for convective precipitation. Considering the notable positive improvements caused by ds083.3 for the simulation of convective precipitation, the slightly negative impacts of ds083.3 on the simulation of non-convective precipitation were omitted from this study. Therefore, we considered forcing data ds083.3 to be better than ds083.2 to drive the WRF model.

As shown in Figure 9b, the mean hourly *BIASs* and *FARs* of the CTLs over D02 showed more evident decreases than those over D01, and these reductions of CTL1 and CTL2 were especially remarkable. The *POD* values over D02 were slightly lower than those over D01 for event A, but the mean hourly *PODs* for event N over D01 and D02 were nearly the same. For event A, the *POFDs* in D02 were visually lower than those in D01, and the comprehensive *HSS*s in D02 were higher than those in D01. For event N, the *HSS*s in D02 were lower than those in D01, while clearly lower values of the *BIAS*s and *FAR*s over D02 indicated some improvements in the reduction of false predictions. Hence, we conclude that the improvement of spatial resolution (from 27 km to 9 km) in the WRF model has an evident positive impact on precipitation predictions for strong convection-dominated rainfall events, and slightly negative impacts on the predictions of rainfall events associated with extratropical cyclones. Considering the practical need for finer spatial resolutions and the overall better performances over the D02, we recognize that the WRF performance is better over D02 than that over D01.

5.2. The Effectiveness of WRF 4D-Var at Different Thresholds and Time

To examine the WRF 4D-Var effectiveness in detail, the simulated daily precipitation from the CTL2, CLT4, and DA experiments at different thresholds ranging from 1 mm/day to 70 mm/day were evaluated with the daily merged CMORPH data. The assessments were quantified with the skill scores of BIAS, FAR, POD, POFD, and HSS. As shown in Figure 10, the performances of the WRF and WRF 4D-Var models decreased as the precipitation threshold increased. The predictions for strong precipitation events greater than 50 mm/day were severely over-forecasted in CTL2 and CTL4, as their *BIAS*s mostly surpassed 2 and even reached 12 (Figure 10c). After assimilating the satellite precipitation products, over-forecasting was controlled, as the values of *BIAS* and *FAR* for the strong precipitation event obviously decreased (Figure 10a–d). As shown in Figure 10q,s, the *HSS*s of DA1-D02 and DA3-D01 for precipitation exceeding 40 mm/day were evidently much larger than those in other experiments, which suggested that the resultant heavy precipitation event assimilated with TRMM 3B42 was more accurate than that assimilated with GPM IMERG during the first day of each simulation. For the second day, these two simulated precipitation results tended to be over-forecasted. For light rain, the simulated precipitation results of DA2 and DA4 were better than those of DA1 and DA3. The *BIASs* for both DA2 and DA4 were generally much closer to 1, and they had lower values of *FAR* and *POFD* and higher values of *POD* and *HSS*. This was mainly attributed to the key advancement of the GPM over the TRMM, because the GPM has a better capability of measuring light rain; this is because its core observatory carried the first spaceborne Ku/Ka band dual-frequency precipitation radar, which is more sensitive to light rain rates [50].

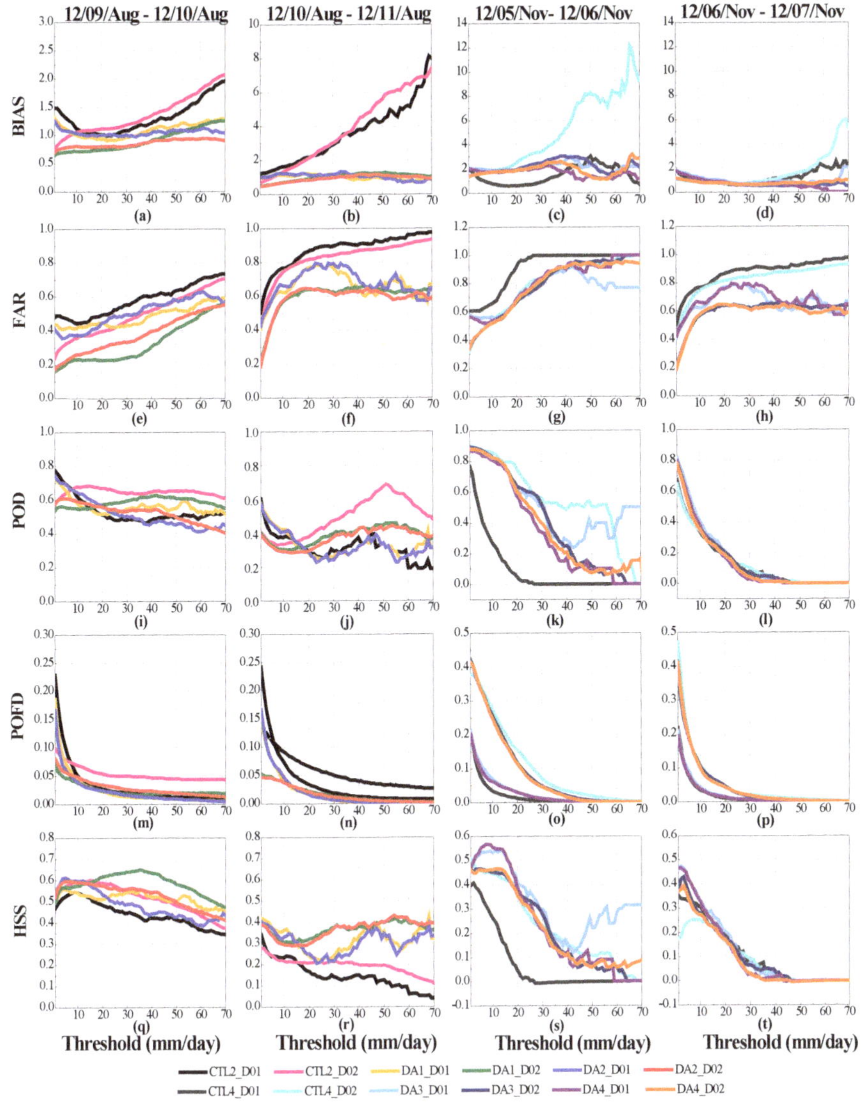

Figure 10. Skill scores of daily precipitation (mm/day) obtained from the CTL2, CTL4, and DA experiments in comparison with the merged CMORPH estimations at different thresholds ranging from 1 mm/day to 70 mm/day. (**a**-**d**), (**e**-**h**), (**i**-**l**), (**m**-**p**) and (**q**-**t**) are the skill scores of *BIAS*, *FAR*, *POD*, *POFD* and *HSS*, respectively, for the daily precipitation in the first and second days of the event A and the event N.

Looking into the time interval issues regarding the effectiveness of the 4D-Var, which was mentioned in Section 4.2, the 12-hour accumulated precipitation of the DA experiments, the CTL2 and the CTL4 were evaluated with the skill scores. The increments of the skill scores between the DA experiments and its corresponding control experiment were analyzed. As portrayed in Figure 11, the increments of *BIAS*, *FAR*, and *POFD* at the first and third 12-hour periods exhibited more extensive decreases than those during the second and fourth 12-hour periods, especially for event A, which is represented with red lines. The *POD*s for event A reduced, but those for event N increased. In Figure 11e, most of the *HSS* increments in the first and third 12-hour periods were higher than those in the second and fourth 12-hour periods. Thus, we conclude that substantial improvements via the WRF 4D-Var with TRMM 3B42 and GPM IMERG can be sustained for approximately 12 h.

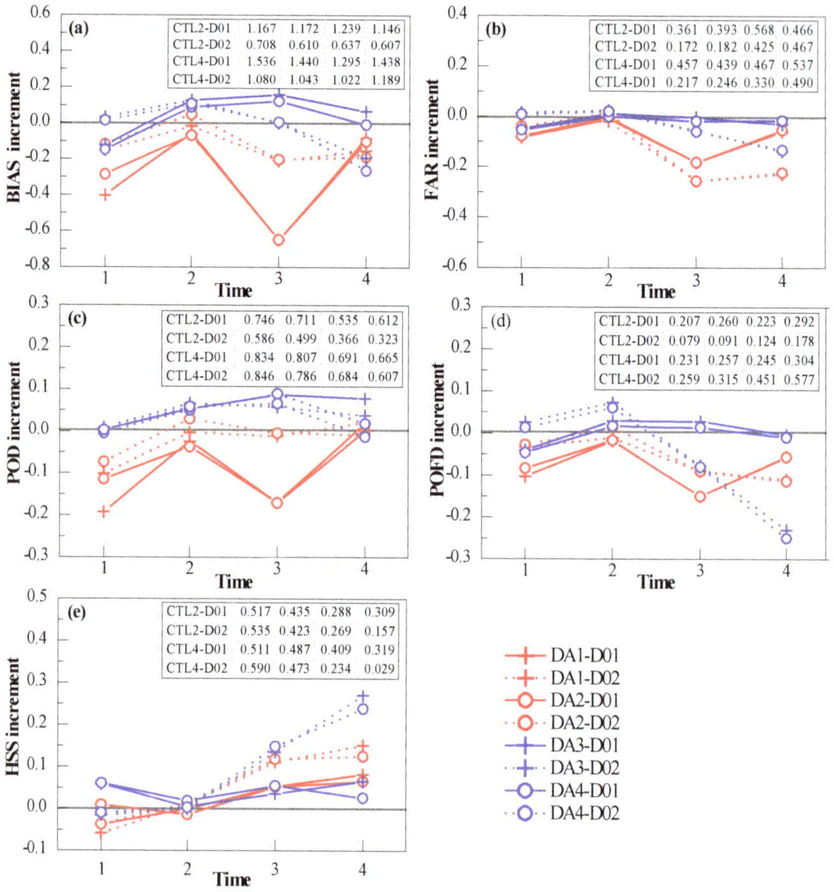

Figure 11. Increments of skill scores of *BIAS* (**a**), *FAR* (**b**), *POD* (**c**), *POFD* (**d**), and *HSS* (**e**) in the DA experiments compared to their corresponding CTL experiments. The table in each graph lists the evaluation scores of CTL2 and CTL4. The x-axis represents the first, second, third, and fourth 12-hour periods in the overall study period.

5.3. Comparison of the 4D-Var Performance Assimilated with TRMM 3B42 and GPM IMERG

To comprehensively compare the 4D-Var performance after assimilation with TRMM 3B42 and GPM IMERG, normalized Taylor diagrams [71] were used to compare the spatial patterns of daily and 48-hour (i.e., the whole study period) precipitation data between the simulations of the DA, CTL experiments, and the merged CMORPH data. The Taylor diagrams and are shown in Figure 12.

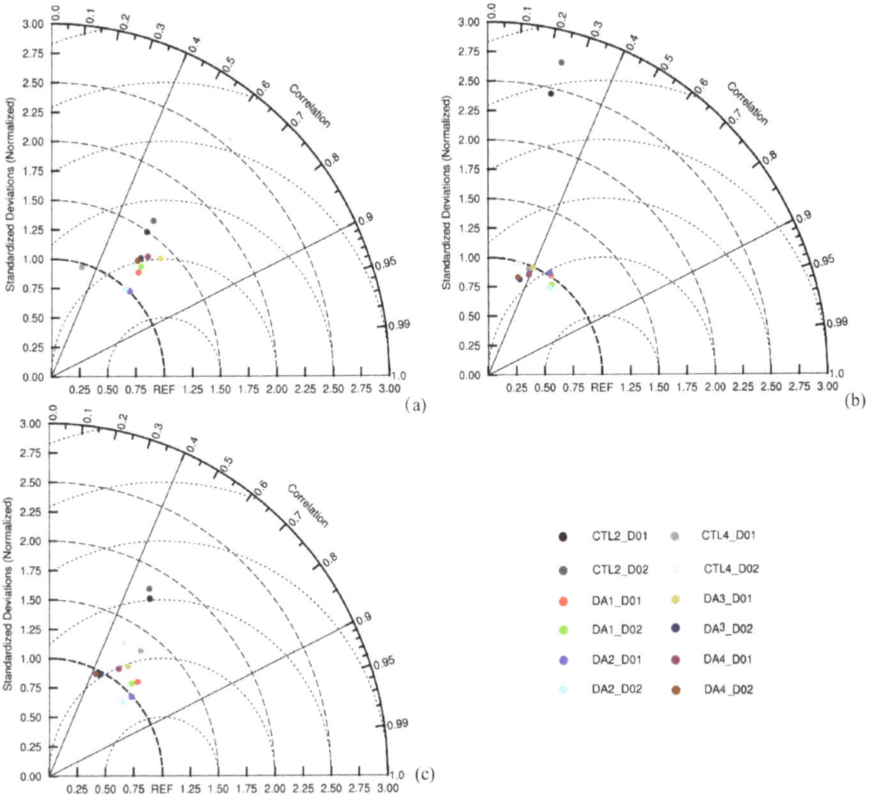

Figure 12. Normalized Taylor diagrams of the precipitation simulated by the CTL2, CTL4, and DA1-4 experiments on the first day (**a**), second day (**b**), and over the whole 48 h (**c**).

For the first day (Figure 12a), DA2 agrees the best with the merged CMORPH data, as the two points delegated DA2_D01 and DA2_D02 are the closest to the point labeled as REF in the figure, and the two points also show the highest correlation coefficients. Simultaneously, they almost have the same standard deviations as those of the merged CMORPH estimations, since they are located on the bold dashed line. For DA2-D01 and DA2-D02, their correlation coefficients are 0.692 and 0.663, respectively, and both their *RMSE*s occur at approximately 0.75 mm. DA3 and DA4 slightly differ in performance, as their representative points are very close to each other. For the performances on the second day (shown in Figure 12b), CTL2 exhibits a remarkable departure from the REF point, as simulation errors accumulated from the first day to the second day during the WRF was running These performances rank from best to worst for event A as follows: DA2-D02, DA1 D02, DA1-D01, and DA2-D01. For event N, the results of DA3 and DA4 over D01 and D02 are almost the same. It is clear in Figure 12c that DA2 evidently outperforms DA1. Thus, we conclude that the WRF 4D-Var with GPM IMERG generally outperforms the WRF 4D-Var with TRMM 3B42.

6. Conclusions

To reduce data acquisition difficulty for precipitation in hydrological studies and obtain independent, consecutive, and high-resolution precipitation data, we used a 4D-Var data assimilation method to assimilate the remotely sensed precipitation products of the TRMM 3B42 and GPM IMERG into the atmospheric WRF model. By focusing on two heavy precipitation events that occurred during the flood and non-flood seasons over the HRB in 2015, CTL experiments were first carried out to choose the best forcing data for the WRF model and determine the control experiment for the subsequent DA experiments. Then, DA experiments were carried out to investigate the feasibility and efficiency of the GPM IMERG to be assimilated into the WRF model with the 4D-Var method, and the 4D-Var performances assimilating with the GPM IMERG and the TRMM 3B42 were compared as well. All of the simulated precipitation values from the CTL experiment and the DA experiment were evaluated with in situ CMA observations and hourly merged CMORPH data.

CTL experiments were performed based on the WRF model with different forcing data and for different events. The assessment of the simulated precipitation in the CTL experiments found that when predicting heavy rainfall events over the HRB, the WRF performance for event N, which represented non-convective precipitation, outperformed the performance for event A, which represented convective precipitation. Moreover, the simulated precipitation generated by forcing data ds083.3 and the output from nested domain D02, which had a higher spatial resolution (9 km), could generally yield better agreement with the in situ CMA data and the merged CMORPH data.

DA experiments were carried out with forcing data ds083.3. The 4D-Var performances that were assimilated with TRMM 3B42 and GPM IMERG based on the WRF model were evaluated in detail. The simulated precipitation results of the DA experiments were assessed at spatial scales of D01, D02, and the HRB, and at hourly, 12-hour, daily, and 48-hour timescales. The evaluation results showed that (1) the 4D-Var with both the TRMM 3B42 and GPM IMERG based on the WRF model could significantly improve the precipitation simulations. The improvements made by GPM IMERG generally outperformed those made by TRMM 3B42, as GPM IMERG was more sensitive to light rain ($\leq$0.5 mm/hour), which accounted for significant portions of the precipitation occurrences at mid and high latitudes. (2) For event A, the enhancement of simulated precipitation was mainly attributed to the corrections of false alarms for non-occurrences. For event N, this improvement was primarily due to more accurate forecasting of these occurrences. The accuracy enhancement for event A was larger than that for event N. (3) The accuracy improvement in simulated precipitation over D01 (27 km) by 4D-Var could be effectively achieved over D02 (9 km); assimilation in D01 and downscaling to D02 with a nested domain based on the WRF model could provide an effective way to obtain finer-resolution precipitation forecasts. (4) Due to error accumulations in the WRF running, essential improvements made by the 4D-Var were maintained for approximately 12 h; it was also not recommended to use the cycling mode for error accumulations in the WRF model.

Further studies can be conducted to deepen the understanding of the 4D-Var algorithm from the following aspects: (1) investigating the performance of a longer duration precipitation simulation; (2) assimilating more remotely sensed precipitation products into other NWP models; and (3) analyzing the sensitivity of the WRF 4D-Var to background errors, which would be worthwhile for future studies.

Acknowledgments: This study was financially supported by the National Key Research and Development Program of China (Project Nos. 2016YFA0602302 and 2016YFB0502502) and the National Natural Science Foundation of China (Project No. 41175088). We are very grateful to Yan Liu and Dong Wang from the Institute of Atmospheric Physics (Chinese Academy of Sciences) for their computer resources. Great thanks should also be given to the NCAR Command Language (NCL) email list (ncl-talk@ucar.edu), which freely and substantially helped regarding data processing and plotting with the NCL.

Author Contributions: Wanchang Zhang and Lu Yi conceived this research. Lu Yi performed the experiments under computer assistance from Kai Wang. Lu Yi analyzed the results and wrote the paper. Wanchang Zhang and Kai Wang gave comments and modified the manuscript.

Conflicts of Interest: The authors declare no conflicts of interest.

References

1. Palmer, T.N.; Ralsanen, J. Quantifying the risk of extreme seasonal precipitation events in a changing climate. *Nature* **2002**, *415*, 512–514. [CrossRef] [PubMed]
2. Groisman, P.Y.; Karl, T.R.; Easterling, D.R.; Knight, R.W.; Jamason, P.F.; Hennessy, K.J.; Suppiah, R.; Page, C.M.; Wibig, J.; Fortuniak, K.; et al. Changes in the probability of heavy precipitation: Important indicators of climatic change. *Clim. Chang.* **1999**, *42*, 243–283. [CrossRef]
3. Groisman, P.Y.; Knight, R.W.; Karl, T.R. Heavy precipitation and high streamflow in the contiguous United States: Trends in the twentieth century. *Bull. Am. Meteorol. Soc.* **2001**, *82*, 219–246. [CrossRef]
4. Zhang, W.; Villarini, G. Heavy precipitation is highly sensitive to the magnitude of future warming. *Clim. Chang.* **2017**, *145*, 249–257. [CrossRef]
5. Zeng, J.Y.; Li, Z.; Chen, Q.; Bi, H.Y.; Qiu, J.X.; Zou, P.F. Evaluation of remotely sensed and reanalysis soil moisture products over the Tibetan Plateau using in-situ observations. *Remote Sens. Environ.* **2015**, *163*, 91–110. [CrossRef]
6. Zeng, J.Y.; Chen, K.S.; Bi, H.Y.; Chen, Q. A Preliminary Evaluation of the SMAP Radiometer Soil Moisture Product over United States and Europe Using Ground-Based Measurements. *IEEE Trans. Geosci. Remote Sens.* **2016**, *54*, 4929–4940. [CrossRef]
7. Pan, X.D.; Li, X.; Cheng, G.D.; Hong, Y. Effects of 4D-Var data assimilation using remote sensing precipitation products in a WRF Model over the complex terrain of an arid region river basin. *Remote Sens.* **2017**, *9*, 693. [CrossRef]
8. Alemohammad, S.H.; McLaughlin, D.B.; Entekhabi, D. Quantifying precipitation uncertainty for land data assimilation applications. *Mon. Weather Rev.* **2015**, *143*, 3276–3299. [CrossRef]
9. Ward, E.; Buytaert, W.; Peaver, L.; Wheater, H. Evaluation of precipitation products over complex mountainous terrain: A water resources perspective. *Adv. Water Resour.* **2011**, *34*, 1222–1231. [CrossRef]
10. Chen, Y.J.; Ebert, E.; Walsh, K.E.; Davidson, N. Evaluation of TRMM 3B42 precipitation estimates of tropical cyclone rainfall using PACRAIN data. *J. Geophys. Res. Atmos.* **2013**, *118*, 2184–2196. [CrossRef]
11. McCabe, M.F.; Rodell, M.; Alsdorf, D.E.; Miralles, D.G.; Uijlenhoet, R.; Wagner, W.; Lucieer, A.; Houborg, R.; Verhoest, N.E.C.; Franz, T.E.; et al. The future of earth observation in hydrology. *Hydrol. Earth Syst. Sci.* **2017**, *21*, 3879–3914. [CrossRef]
12. Steiner, M.; Smith, J.A.; Burges, S.J.; Alonso, C.V.; Darden, R.W. Effect of bias adjustment and rain gauge data quality control on radar rainfall estimation. *Water Resour. Res.* **1999**, *35*, 2487–2503. [CrossRef]
13. Lorenz, C.; Kunstmann, H. The hydrological cycle in three state-of-the-art reanalyses: Intercomparison and performance analysis. *J. Hydrometeorol.* **2012**, *13*, 1397–1420. [CrossRef]
14. Huffman, G.J.; Adler, R.F.; Arkin, P.; Chang, A.; Ferraro, R.; Gruber, A.; Janowiak, J.; McNab, A.; Rudolf, B.; Schneider, U. The Global Precipitation Climatology Project (GPCP) combined precipitation dataset. *Bull. Am. Meteorol. Soc.* **1997**, *78*, 5–20. [CrossRef]
15. Joyce, R.J.; Janowiak, J.E.; Arkin, P.A.; Xie, P.P. CMORPH: A method that produces global precipitation estimates from passive microwave and infrared data at high spatial and temporal resolution. *J. Hydrometeorol.* **2004**, *5*, 487–503. [CrossRef]
16. Garstang, M.; Kummerow, C.D. The joanne simpson special issue on the Tropical Rainfall Measuring Mission (TRMM). *J. Appl. Meteorol.* **2000**, *39*, 1961. [CrossRef]
17. Hou, A.Y.; Kakar, R.K.; Neeck, S.; Azarbarzin, A.A.; Kummerow, C.D.; Kojima, M.; Oki, R.; Nakamura, K.; Iguchi, T. The global precipitation measurement mission. *Bull. Am. Meteorol. Soc.* **2014**, *95*, 701. [CrossRef]
18. Gaona, M.F.R.; Overeem, A.; Leijnse, H.; Uijlenhoet, R. First-year evaluation of GPM rainfall over the Netherlands: IMERG Day 1 final run (VO3D). *J. Hydrometeorol.* **2016**, *17*, 2799–2814. [CrossRef]
19. Schmidli, J.; Goodess, C.M.; Frei, C.; Haylock, M.R.; Hundecha, Y.; Ribalaygua, J.; Schmith, T. Statistical and dynamical downscaling of precipitation: An evaluation and comparison of scenarios for the European Alps. *J. Geophys. Res. Atmos.* **2007**, *112*. [CrossRef]
20. Zhang, X.X.; Anagnostou, E.; Frediani, M.; Solomos, S.; Kallos, G. Using NWP simulations in satellite rainfall estimation of heavy precipitation events over mountainous areas. *J. Hydrometeorol.* **2013**, *14*, 1844–1858. [CrossRef]
21. Koizumi, K.; Ishikawa, Y.; Tsuyuki, T. Assimilation of precipitation data to the JMA mesoscale model with a four-dimensional variational method and its impact on precipitation forecasts. *Sola* **2005**, *1*, 45–48. [CrossRef]

22. Mazrooei, A.; Sinha, T.; Sankarasubramanian, A.; Kumar, S.; Peters-Lidard, C.D. Decomposition of sources of errors in seasonal streamflow forecasting over the US Sunbelt. *J. Geophys. Res. Atmos.* **2015**, *120*. [CrossRef]
23. Case, J.L.; Crosson, W.L.; Kumar, S.V.; Lapenta, W.M.; Peters-Lidard, C.D. Impacts of high-resolution land surface initialization on regional sensible weather forecasts from the WRF Model. *J. Hydrometeorol.* **2008**, *9*, 1249–1266. [CrossRef]
24. Flesch, T.K.; Reuter, G.W. WRF model simulation of two Alberta flooding events and the impact of topography. *J. Hydrometeorol.* **2012**, *13*, 695–708. [CrossRef]
25. Janjic, Z.I. The step-mountain coordinate -physical package. *Mon. Weather Rev.* **1990**, *118*, 1429–1443. [CrossRef]
26. Black, T.L. The new NMC mesoscale ETA model—Description and forecast examples. *Weather Forecast.* **1994**, *9*, 265–278. [CrossRef]
27. Mesinger, F.; Janjic, Z.I.; Nickovic, S.; Gavrilov, D.; Deaven, D.G. The step-mountain coordinate - model description and performance for cases of description and performance for cases of Alpine Lee Cyclongensis and for a case of an Appalachian redevelopment. *Mon. Weather Rev.* **1988**, *116*, 1493–1518. [CrossRef]
28. Dudhia, J.; Klemp, J.; Skamarock, W.; Dempsey, D.; Janjic, Z.; Benjamin, S.; Brown, J.; Ams, A.M.S. A collaborative effort towards a future community mesoscale model (WRF). In Proceedings of the 12th Conference on Numerical Weather Prediction, Phoenix, AZ, USA, 11–16 January 1998; pp. 242–243.
29. Saito, K.; Fujita, T.; Yamada, Y.; Ishida, J.I.; Kumagai, Y.; Aranami, K.; Ohmori, S.; Nagasawa, R.; Kumagai, S.; Muroi, C.; et al. The operational JMA nonhydrostatic mesoscale model. *Mon. Weather Rev.* **2006**, *134*, 1266–1298. [CrossRef]
30. Molteni, F.; Buizza, R.; Palmer, T.N.; Petroliagis, T. The ECMWF ensemble prediction system: Methodology and validation. *Quart. J. R. Meteorol. Soc.* **1996**, *122*, 73–119. [CrossRef]
31. Yang, B.; Qian, Y.; Lin, G.; Leung, R.; Zhang, Y. Some issues in uncertainty quantification and parameter tuning: A case study of convective parameterization scheme in the WRF regional climate model. *Atmos. Chem. Phys.* **2012**, *12*, 2409–2427. [CrossRef]
32. Angevine, W.M.; Brioude, J.; McKeen, S.; Holloway, J.S. Uncertainty in Lagrangian pollutant transport simulations due to meteorological uncertainty from a mesoscale WRF ensemble. *Geosci. Model Dev.* **2014**, *7*, 2817–2829. [CrossRef]
33. Panofsky, H.A. Objective Weather-manp analysis. *J. Meteorol.* **1949**, *6*, 386–392. [CrossRef]
34. Leneman, O.A.Z. Random sampling of random processes—Optimum linear interpolation. *J. Frankl. Inst. Eng. Appl. Math.* **1966**, *281*, 302. [CrossRef]
35. Sasaki, Y. An objective analysis based on the variational method. *J. Meteorol. Soc. Jpn.* **1958**, *36*, 77–88. [CrossRef]
36. Sasaki, Y. Some basic formalisms in numerical variational analysis. *Mon. Weather Rev.* **1970**, *98*, 875. [CrossRef]
37. Evensen, G. Sequential data assimilation with a nonlinear quasi-geostrophic model using monte-carlo methods to forecast error statistics. *J. Geophys. Res. Oceans* **1994**, *99*, 10143–10162. [CrossRef]
38. Evensen, G. Advanced data assimilation for strongly nonlinear dynamics. *Mon. Weather Rev.* **1997**, *125*, 1342–1354. [CrossRef]
39. Tsuyuki, T. Variational data assimilation in the tropics using precipitation data part I: Column model. *Meteorol. Atmos. Phys.* **1996**, *60*, 87–104. [CrossRef]
40. Zupanski, D.; Mesinger, F. 4-dimensional variational assimilation of precipitation data. *Mon. Weather Rev.* **1995**, *123*, 1112–1127. [CrossRef]
41. Lopez, P. Direct 4D-Var assimilation of NCEP stage IV radar and gauge precipitation data at ECMWF. *Mon. Weather Rev.* **2011**, *139*, 2098–2116. [CrossRef]
42. Lin, L.F.; Ebtehaj, A.M.; Bras, R.L.; Flores, A.N.; Wang, J.F. Dynamical precipitation downscaling for hydrologic applications using WRF 4D-Var data assimilation: Implications for GPM era. *J. Hydrometeorol.* **2015**, *16*, 811–829. [CrossRef]
43. Chambon, P.; Zhang, S.Q.; Hou, A.Y.; Zupanski, M.; Cheung, S. Assessing the impact of pre-GPM microwave precipitation observations in the Goddard WRF ensemble data assimilation system. *Quart. J. R. Meteorol. Soc.* **2014**, *140*, 1219–1235. [CrossRef]
44. Ballard, S.P.; Li, Z.H.; Simonin, D.; Caron, J.F. Performance of 4D-Var NWP-based nowcasting of precipitation at the Met Office for summer 2012. *Quart. J. R. Meteorol. Soc.* **2016**, *142*, 472–487. [CrossRef]

45. Verlinde, J.; Cotton, W.R. Fitting microphysical observations of nonsteady convective clouds to a numerical model: An application of the adjoint technique of data assimilation to a kinematic model. *Mon. Weather Rev.* **1993**, *121*, 2776–2793. [CrossRef]
46. Xia, J.; She, D.X.; Zhang, Y.Y.; Du, H. Spatio-temporal trend and statistical distribution of extreme precipitation events in Huaihe River Basin during 1960–2009. *J. Geogr. Sci.* **2012**, *22*, 195–208. [CrossRef]
47. Cao, Q.; Qi, Y.C. The variability of vertical structure of precipitation in Huaihe River Basin of China: Implications from long-term spaceborne observations with TRMM precipitation radar. *Water Resour. Res.* **2014**, *50*, 3690–3705. [CrossRef]
48. Zhou, Y.K.; Ma, Z.Y.; Wang, L.C. Chaotic dynamics of the flood series in the Huaihe River Basin for the last 500 years. *J. Hydrol.* **2002**, *258*, 100–110. [CrossRef]
49. Huffman, G.J.; Adler, R.F.; Bolvin, D.T.; Gu, G.J.; Nelkin, E.J.; Bowman, K.P.; Hong, Y.; Stocker, E.F.; Wolff, D.B. The TRMM Multisatellite Precipitation Analysis (TMPA): Quasi-global, multiyear, combined-sensor precipitation estimates at fine scales. *J. Hydrometeorol.* **2007**, *8*, 38–55. [CrossRef]
50. Tang, G.Q.; Ma, Y.Z.; Long, D.; Zhong, L.Z.; Hong, Y. Evaluation of GPM Day-1 IMERG and TMPA version-7 legacy products over Mainland China at multiple spatiotemporal scales. *J. Hydrol.* **2016**, *533*, 152–167. [CrossRef]
51. Skofronick-Jackson, G.; Huffman, G.; Stocker, E.; Petersen, W. Successes with the Global Precipitation Measurment (GPM) mission in 2016 Ieee International Geoscience and Remote Sensing Symposium. In Proceedings of the IEEE International Geoscience and Remote Sensing Symposium (IGARSS), Beijing, China, 10–15 July 2016; pp. 3910–3912.
52. 52. Xuan, Z.; Yali, L.; Xueliang, G. Application of a CMORPH-a WS merged hourly gridded precipitation product in analyzing charateristics of short-duration heavy rainfall over southern China. *J. Trop. Meteorol.* **2015**, *31*, 333–344.
53. Maussion, F.; Scherer, D.; Finkelnburg, R.; Richters, J.; Yang, W.; Yao, T. WRF simulation of a precipitation event over the Tibetan Plateau, China—An assessment using remote sensing and ground observations. *Hydrol. Earth Syst. Sci.* **2011**, *15*, 1795–1817. [CrossRef]
54. Lim, K.S.S.; Hong, S.Y. Development of an effective double-moment cloud microphysics scheme with prognostic Cloud Condensation Nuclei (CCN) for weather and climate models. *Mon. Weather Rev.* **2010**, *138*, 1587–1612. [CrossRef]
55. Mlawer, E.J.; Taubman, S.J.; Brown, P.D.; Iacono, M.J.; Clough, S.A. Radiative transfer for inhomogeneous atmospheres: RRTM, a validated correlated-k model for the longwave. *J. Geophys. Res. Atmos.* **1997**, *102*, 16663–16682. [CrossRef]
56. Dudhia, J. Numerical study of convection observed during the winter monsoon experiment using a mesoscale two-dimensional model. *J. Atmos. Sci.* **1989**, *46*, 3077–3107. [CrossRef]
57. Chen, F.; Dudhia, J. Coupling an advanced land surface-hydrology model with the Penn State-NCAR MM5 modeling system. *Part I: Model implementation and sensitivity. Mon. Weather Rev.* **2001**, *129*, 569–585. [CrossRef]
58. Hong, S.Y.; Noh, Y.; Dudhia, J. A new vertical diffusion package with an explicit treatment of entrainment processes. *Mon. Weather Rev.* **2006**, *134*, 2318–2341. [CrossRef]
59. Grell, G.A.; Devenyi, D. A generalized approach to parameterizing convection combining ensemble and data assimilation techniques. *Geophys. Res. Lett.* **2002**, *29*. [CrossRef]
60. Courtier, P.; Thepaut, J.N.; Hollingsworth, A. A strategy for operational implementation of 4D-Var, using an incremental approach. *Quart. J. R. Meteorol. Soc.* **1994**, *120*, 1367–1387. [CrossRef]
61. Veerse, F.; Thepaut, J.N. Multiple-truncation incremental approach for four-dimensional variational data assimilation. *Quart. J. R. Meteorol. Soc.* **1998**, *124*, 1889–1908. [CrossRef]
62. Lorenc, A.C. Modelling of error covariances by 4D-Var data assimilation. *Quart. J. R. Meteorol. Soc.* **2003**, *129*, 3167–3182. [CrossRef]
63. Barker, D.; Huang, X.Y.; Liu, Z.Q.; Auligne, T.; Zhang, X.; Rugg, S.; Ajjaji, R.; Bourgeois, A.; Bray, J.; Chen, Y.S.; et al. The weather research and forecasting model's community variational/ensemble data assimilation system WRFDA. *Bull. Am. Meteorol. Soc.* **2012**, *93*, 831–843. [CrossRef]
64. Huang, X.Y.; Xiao, Q.N.; Barker, D.M.; Zhang, X.; Michalakes, J.; Huang, W.; Henderson, T.; Bray, J.; Chen, Y.S.; Ma, Z.Z.; et al. Four-dimensional variational data assimilation for WRF: Formulation and preliminary results. *Mon. Weather Rev.* **2009**, *137*, 299–314. [CrossRef]

65. Lynch, P.; Huang, X.Y. Initialization of the hirlam model using a digital-filter. *Mon. Weather Rev.* **1992**, *120*, 1019–1034. [CrossRef]
66. Gauthier, P.; Thepaut, J.N. Impact of the digital filter as a weak constraint in the preoperational 4DVAR assimilation system of Meteo-France. *Mon. Weather Rev.* **2001**, *129*, 2089–2102. [CrossRef]
67. Parrish, D.F.; Derber, J.C. The national-meteorological-centers spectral statistical-interpolation analysis system *Mon. Weather Rev.* **1992**, *120*, 1747–1763. [CrossRef]
68. Kleczek, M.A.; Steeneveld, G.J.; Holtslag, A.A.M. Evaluation of the Weather Research and Forecasting mesoscale model for GABLS3: Impact of boundary-layer schemes, boundary conditions and spin-up. *Bound.-Layer Meteorol.* **2014**, *152*, 213–243. [CrossRef]
69. Srinivas, D.; Rao, D.V.B. Implications of vortex initialization and model spin-up in tropical cyclone prediction using Advanced Research Weather Research and Forecasting Model. *Nat. Hazards* **2014**, *73*, 1043–1062. [CrossRef]
70. Wilks, D.S. *Statistical Methods in the Atmospheric Sciences*; Academic Press: Cambridge, MA, USA, 2006; Volume 91, p. 627.
71. Taylor, K.E. Summarizing multiple aspects of model performance in a single diagram. *J. Geophys. Res. Atmos.* **2001**, *106*, 7183–7192. [CrossRef]

remote sensing

MDPI

Article

Soil Moisture Variability in India: Relationship of Land Surface–Atmosphere Fields Using Maximum Covariance Analysis

Kishore Pangaluru [1,*], Isabella Velicogna [1,2], Geruo A [1], Yara Mohajerani [1], Enrico Ciracì [1], Sravani Charakola [3], Ghouse Basha [4] and S. Vijaya Bhaskara Rao [3]

1 Department of Earth System Science, University of California, Irvine, CA 92697, USA; isabella@uci.edu (I.V.); geruoa@uci.edu (G.A.); ymohajer@uci.edu (Y.M.); enrico.ciraci@gmail.com (E.C.)
2 Jet Propulsion Laboratory, California Institute of Technology, Pasadena, CA 91109, USA
3 Department of Physics, Sri Venkateswara University, Tirupati 517502, India; sravanicpepa@gmail.com (S.C.); drsvbr.acas@gmail.com (S.V.B.R.)
4 National Atmospheric Research Laboratory, Gadanki 517112, India; mdbasha@gmail.com
* Correspondence: kishore@uci.edu; Tel.: 1-949-824-3516

Received: 19 December 2018; Accepted: 2 February 2019; Published: 8 February 2019

Abstract: This study investigates the spatial and temporal variability of the soil moisture in India using Advanced Microwave Scanning Radiometer-Earth Observing System (AMSR-E) gridded datasets from June 2002 to April 2017. Significant relationships between soil moisture and different land surface–atmosphere fields (Precipitation, surface air temperature, total cloud cover, and total water storage) were studied, using maximum covariance analysis (MCA) to extract dominant interactions that maximize the covariance between two fields. The first leading mode of MCA explained 56%, 87%, 81%, and 79% of the squared covariance function (SCF) between soil moisture with precipitation (PR), surface air temperature (TEM), total cloud count (TCC), and total water storage (TWS), respectively, with correlation coefficients of 0.65, −0.72, 0.71, and 0.62. Furthermore, the covariance analysis of total water storage showed contrasting patterns with soil moisture, especially over northwest, northeast, and west coast regions. In addition, the spatial distribution of seasonal and annual trends of soil moisture in India was estimated using a robust regression technique for the very first time. For most regions in India, significant positive trends were noticed in all seasons. Meanwhile, a small negative trend was observed over southern India. The monthly mean value of AMSR soil moisture trend revealed a significant positive trend, at about 0.0158 cm^3/cm^3 per decade during the period ranging from 2002 to 2017.

Keywords: soil moisture; precipitation; temperature; total cloud cover; GRACE; total water storage; MCA analysis

1. Introduction

Spatial and temporal changes of soil moisture (SM) are essential to the exchange of water and energy over land, monitoring of land surface conditions, and quantifying the sensitivity to global warming and human pressure. SM, precipitation, orography, and vegetation land cover are some of the basic variables that affect the hydrological cycle. Therefore, an accurate quantification of SM anomalies can result in significant changes in modeled atmospheric hydrological processes by land–atmospheric interactions [1]. In 2010, the Global Climate Observing System (GCOS) panel endorsed SM as one of the 50 Essential Climate Variables (ECVs), supporting both the work of the United Nations Framework Convention on Climate Change (UNFCCC) and the International Panel on Climate Change [2].

Several investigations have indicated the importance of SM in influencing weather and climate anomalies. SM can influence the climate system through different feedback processes [3–5].

Some relationships between SM and monthly seasonal variability in surface temperature as well as precipitation have been explored [6–8]. Shukla and Mintz [9] found an increase in precipitation amounts in a dry soil simulation experiment over Southeast Asia and the Indian region. Later, Ashraf et al. [10] reported that pre-monsoon soil moisture has a large impact on monsoon onsets, using perturbations simulations with the regional climate model. Varikoden and Revadekar [11] studied the relationship between SM and precipitation, concluding that the variability in SM influences the wetness or dryness of the monsoon season. Recently, Kantharao and Rakesh [12] reported that SM and precipitation are positively correlated in June throughout India. Raman et al. [13] used soil moisture simulations over the Indian continent to find that wet soil conditions intensify the large-scale circulation, which further enhances convective activity and precipitation. In most of these studies, a model output is used to perform numerical experiments on the role of SM in climate variability. However, a large uncertainty existed in the modeled SM and associated land surface couplings [14]. For instance, Koster et al. [15] evaluated an ensemble of 16 simulations of soil moisture from the Global Land Atmosphere Coupling Experiment (GLACE) and found a strong SM–PR coupling in the interior Peninsula in India. They inferred that the extent of couplings between land surface and the atmosphere vary significantly between models.

Nowadays, remote sensing datasets provide a useful approach in understanding the behavior of seasonal and annual variability and land-atmosphere interactions [16–18]. Satellite remote sensing enables regional-scale evaluation of SM dynamics that is previously infeasible due to the sparse coverage of in situ measurements. Furthermore, the use of satellite observations provides unique SM information that is independent from the potential bias and uncertainties common in model simulations. SM variation is influenced by multiple parameters: precipitation, evapotranspiration, vegetation, land cover, and land use [6,7]. In addition, surface temperature, total cloud cover (TCC), and total water storage (TWS) can also directly or indirectly control SM to some degree. Total cloud cover likely exerts an indirect but composite control on the soil moisture dynamics: cloud coverage often coincides with the rainfall occurrence associated with soil moisture replenishment, while in the absence of rainfall, it also controls the input radiation energy on the surface that is directly linked to the soil dry-down process. Large-scale groundwater depletion was observed in India [19]; it has yet to be investigated how the variations in deeper water storage, as presented in TWS observations, influence near-surface soil moisture dynamics. To the best of our knowledge, this is the first study to examine the relationship between such variables and soil moisture in Indian regions. In this paper, satellite remote sensing data was used to evaluate the monthly, seasonal, and annual characteristics and interannual variability of soil moisture over India, and to aim to provide a better understanding of the spatiotemporal variation of soil moisture in this region. The Maximum covariance analysis (MCA) technique was applied over India's soil moisture and land-surface fields, including precipitation, temperature, TCC, and TWS. Further, the regional teleconnection and direct and indirect effects of climate factors on SM and land surface–atmospheric fields over India were explored in the study. Such variability creates substantial uncertainty in the sign and magnitude of decadal-scale trends in regional soil moisture. We estimated the SM spatial trends in India using robust regression analysis. This paper is organized as follows: the employed datasets are described in Section 2, the methodology is given in Section 3, the results and discussion are given in Section 4, and the overall summary and conclusions drawn from this study are presented in Section 5.

2. Data

2.1. AMSR Soil Moisture

The Advanced Microwave Scanning Radiometer-Earth Observing System (AMSR-E) operates with six bands, ranging from 6.9 to 89 GHz at HH-VV polarization with the radiometer sensor on-board the Aqua satellite since 2002 [20]. AMSR-E is the first satellite radiometer sensor that includes measurements of soil moisture and vegetation/roughness correction [21]. AMSR-E brightness temperatures in the 6.9 GHz band (C-band) are reported on a 25 × 25 km^2 grid. The C-band observations are sensitive to soil moisture in the upper most ~1 cm of the Earth's surface [20]. NASA (National Aeronautics and Space Administration), JAXA (Japan Aerospace Exploration Agency), and other groups developed several algorithms (using different physical formulations, parameters, and ancillary data) to retrieve soil moisture from brightness temperature measured by AMSR-E [20] with an accuracy goal of less than 0.06 m^3/m^3. JAXA launched the AMSR2 as part of the global observation mission-water (GCOM-W) as a follow-on to AMSR-E. The AMSR-E and AMSR2 soil moisture retrievals have a near daily temporal fidelity, and are presented on a 25 km nearly equal-area grid with an effective resolution that is close to 50 km. This study uses the global Land Parameter Data Record version 2 (LPDF V2) and the AMSR-E/2 SM record derived by the University of Montana (UMT) [22,23]. The updated LPDR new algorithms provide a long-term global record spanning the observation periods of June 2002 to April 2017 from both AMSR-E and AMSR2 (hereafter referred to as AMSRSM). The observational gap between the AMSR-E and AMSR-2 records is bridged using the overlapping FY3B-MWRI (Microwave Radiation Imager) record [23]. The AMSRSM record helps to build a consistent long-term dataset for monitoring the Earth's soil moisture.

2.2. SMAP Soil Moisture Data:

The National Aeronautics and Space Administration's (NASA) Soil Moisture Active Passive (SMAP) satellite mission [24] was launched on January 2015 and provides unprecedented accuracy, resolution, and coverage [25]. SMAP generates different levels of products that are projected onto fixed ease-grid at 36 km (passive), 9 km (active/passive and enhanced), and 3 km (active) resolutions. Pan et al. (2016) compared SMAP SM with in situ measurements at point and regional scales using a one-year dataset over southeastern US. Later studies by Zeng et al. [26] and Colliander et al. [27] validated the SMAP SM product with 13 core validation sites. These studies found a very promising performance by SMAP, with the bias ranging from -0.088 m^3 m^{-3} to 0.072 m^3 m^{-3}. The correlation values of SMAP and AMSR2 with the in situ network were 0.74 and 0.65 respectively. The corresponding bias values were –0.0460 and 0.0418 m^3/m^3 for SMAP and AMSR2, respectively [28]. Here, we used Level 3 SMAP data (SMAPSM) during the period from April 2015 to April 2017.

2.3. GRACE Total Water Thickness

The Gravity Recovery and Climate Experiment (GRACE) satellites, launched in March 2002, provide accurate monthly geoid changes from which useful hydrological information can be obtained. GRACE can provide estimates of monthly changes in continental water storage [29,30]. This study uses the JPL RL05Mv2 mascon solution [31] from April 2002 to June 2017. JPL (Jet Propulsion Laboratory) mascons were obtained directly from the range-rate data on a global set of 4551 3° × 3° equal area spherical caps. The solid Earth contribution to the geoid from glacial isostatic adjustment was globally corrected using the model proposed by Geruo et al. [32]. The remaining signal is attributed to changes in total water storage (TWS). The mascon solution with the coastline resolution improvement (CRI) filter and the corresponding gain factors following Wiese et al. [33] were utilized. The CRI filter and gain factors have been shown to improve the GRACE leakage error by as much as 38–81% locally [30].

This study also utilized high-resolution (0.25 × 0.25) gridded daily precipitation (PR) data from the Indian Meteorological Department (IMD). The monthly precipitation values were estimated from daily values throughout India. Rainfall records were quality-controlled against rain gauge stations at about 6995 locations in India [34,35]. PR datasets from 2002 to 2016 were utilized. Furthermore, ERA-Interim ((https://apps.ecmwf.int/datasets/data/interim-full-daily/levtype=sfc/; [36]) monthly total cloud cover (referred to as TCC hereinafter), and soil moisture (ERASM) datasets were used from the period spanning from 2002 to 2017. Lastly, monthly surface temperature (TEM) datasets were obtained from the Climate Research Unit (CRU) for period 2002–2016.

3. Methodology

All datasets have been regridded to a 1 × 1 degree longitude–latitude grid, using bilinear interpolation for consistent regional scale comparison between the various geophysical fields. In the case of GRACE data, the interpolation was done from a 0.5 × 0.5 degree grid after the application of gain factors. Note that while this is below the intrinsic resolution of GRACE (a few hundred kilometers), this study does not intend on resolving each grid independently, but rather acknowledge the correlation between nearby grids and aim to examine the spatial distribution of the trends and the relationship between various geophysical fields across India. The hydrological variables are averaged on monthly, seasonal, and annual scales using daily datasets. In order to find the relation between variables, maximum covariance analysis (MCA) [37] was used to isolate the most coherent pairs of spatial patterns and identify a linear relationship between two different geophysical fields that are most closely related to each other. This method is commonly applied to observations of two distinct variables as well as comparisons of a single variable within two different measurements. Statistical assessments in MCA are commonly based entirely on the Monte Carlo method, by evaluating the expected rectangular covariance to that of a randomly scrambled ensemble [38]. The major advantage of MCA is verifying one data field and the corresponding modes with another field data, where the modes are represented by the variance in each field. MCA is a powerful method to investigate dominant interactions that maximize the covariance between soil moisture and land surface–atmospheric fields (PR, TEM, TWS, and TCC). To investigate linear trends in spatial moisture patterns, the study employs a robust regression analysis technique using iterative reweighted least-squares, an improvement to ordinary least-squares and less affected by outliers [39]. This method allows the reconstruction problem to be tackled in a computationally efficient manner with a large number of outliers [40]. The statistical confidence level of the trend in each grid was calculated using the non-parametric Mann–Kendall test [41,42].

4. Results and Discussion

4.1. Spatial Monthly Variability of Soil Moisture

Figure 1 shows the spatial monthly variations of AMSR soil moisture during the period from June 2002 to April 2017. North central and interior peninsula regions show the highest soil moisture over monsoon (June - September) and post-monsoon (October and November) months. The majority of months with lower soil moisture are mainly found over the northwest region of India. Note that insufficient observations are available over the Himalayan region during winter and pre-monsoon months. The spatial distribution of soil moisture shows a similar behavior in March and April as well as in November and December. Over the interior peninsula and north central regions, AMSRSM values are ~70% larger than in other regions. The large SM may largely be attributed to monsoon precipitation and advection of moisture from neighboring regions. Anusha et al. [43] mentioned the larger SM over central regions of India using one year of summer soil moisture data. Figure 1 clearly shows the seasonal variation of soil moisture, which is an input to the agricultural production of the country.

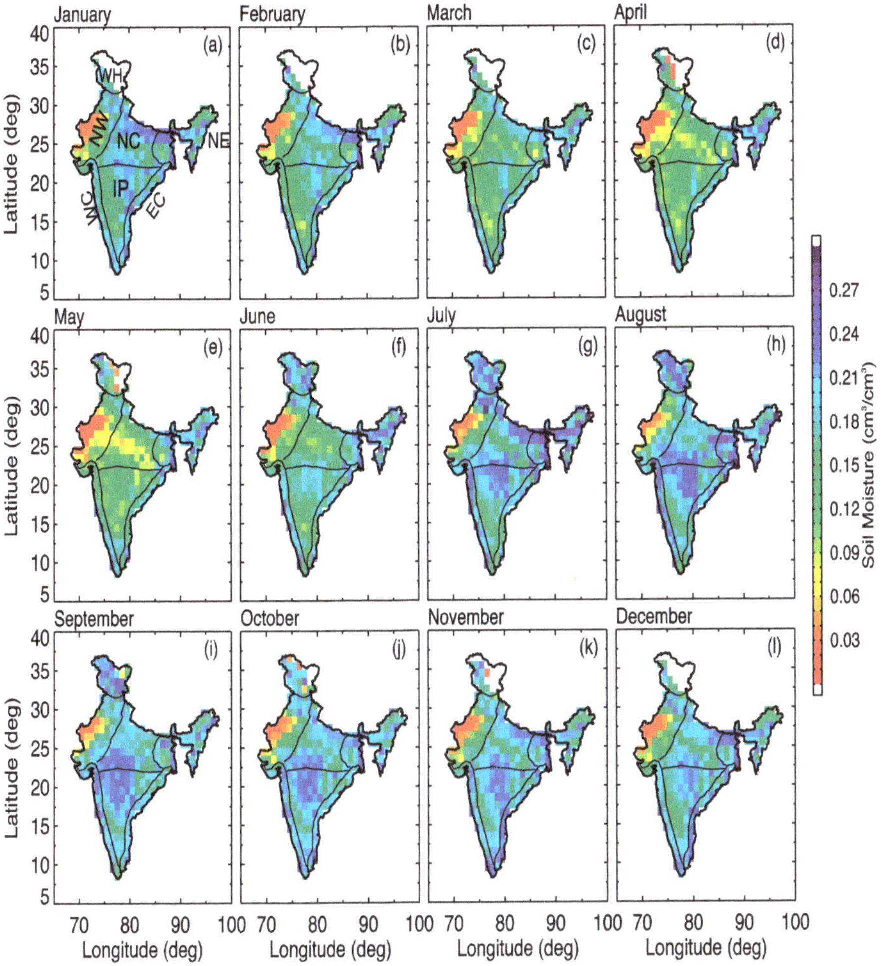

Figure 1. Spatial monthly climatology of soil moisture observed by AMSR over India for the period June 2002–April 2017. Seven different regions are indicated in the first panel: East coast (EC), Interior Peninsula (IP), North central (NC), North east (NE), North west (NW), West coast (WC), Western Himalayas (WH).

4.2. Seasonal and Annual Gridded Soil Moisture

The spatial distribution pattern of annual and seasonal soil moisture, and relative difference of soil moisture over India are discussed in this section. The seasonal and annual soil moisture variations of SMAP (top panel) and AMSR (bottom panel) are shown in Figure 2. For an equal comparison, SMAPSM and AMSRSM data were used only within the period of March 2015–April 2017. The seasonal means refer to the following months: winter (December, January, and February), pre-monsoon (March–May), monsoon (June-September), and post-monsoon (October and November). In all seasons, the minimum soil moisture (<0.02 cm^3/cm^3) values are in the northwestern (NW) region. SMAP shows relatively drier conditions than AMSRSM in winter, especially over northwest, north central, and interior peninsula regions. In the spring season, both SMAP and AMSR report drier soil moisture conditions over the northwest, extending further to the north central region. During monsoon and fall seasons, there is relatively more soil moisture over north central and Interior peninsula regions in both satellite measurements. The maximum values of soil moisture over central and Interior Peninsula regions are observed during monsoon season. The monthly soil moisture distribution in these regions tends to increase from June to November and then decrease gradually from December to March, with the minimum being in May. Unnikrishnan et al. [44] also observed maximum soil moisture during monsoon season over central India using United Kingdom meteorological Office (UKMO) datasets. The annual mean soil moisture is comparable in both datasets, with the exception of the Western Himalayas region. These seasonal trends are important because an increase (decrease) in food grain yield is associated with an increase (decrease) in soil moisture.

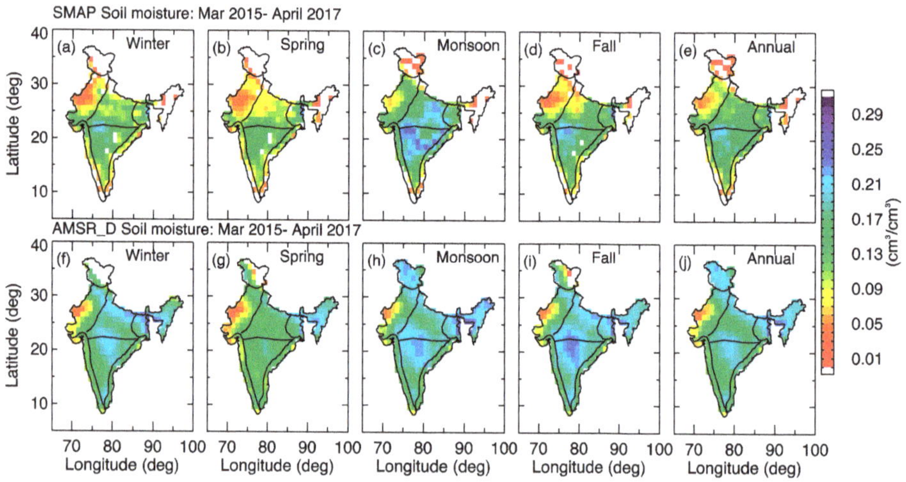

Figure 2. Seasonal and annual climatological soil moisture from SMAP (top panels) and AMSR (bottom panels) in India from March 2015 to April 2017.

The annual and seasonal behavior of the AMSR and SMAP soil moisture are shown in Figure 2. AMSR shows higher soil moisture values than SMAP over northwest and north central regions of India in winter, spring, and fall, as well as the annual period. In the monsoon season, AMSRSR is lower (drier) than SMAP in all regions over India except for the Himalayas, since AMSRSR is more sensitive to the very top layer. During the monsoon and fall seasons, AMSRSM is higher over North central and Interior peninsula regions, whereas SMAP shows maximum soil moisture values in the monsoon season but not in the fall season. It is evident from Table 1 that in the Northeast (NE) the mean monsoon soil moisture is highest in AMSR followed by that in SMAP. Furthermore, SMAP and AMSR display the strongest intra-seasonal variation of soil moisture over central India. In the monsoon season, the SMAP maximum and minimum seasonal mean value is observed in the Interior peninsula (IP) and Western Himalaya (WH), respectively. Meanwhile, AMSR has a minimum value of 0.10 cm^3/cm^3 in the Northwest (NW) during spring and a maximum value of 0.22 cm^3/cm^3 in the Northeast (NE) during the monsoon season. In general, the annual regional means of soil moisture from AMSR and SMAP show general agreement, with the mean values ranging from 0.13 to 0.20 cm^3/cm^3 and 0.10 to 0.16 cm^3/cm^3, respectively.

Table 1. Seasonal and annual mean soil moisture (cm^3/cm^3) and their standard deviations observed by AMSR and SMAP datasets over seven homogeneous regions in India.

Region	Winter (DJF)	Spring (MAM)	Monsoon (JJAS)	Fall (SO)	Annual
AMSR (March 2015–April 2017)					
East coast (EC)	0.188 (0.025)	0.154 (0.069)	0.206 (0.039)	0.155 (0.076)	0.201 (0.039)
Interior Peninsula (IP)	0.178 (0.032)	0.148 (0.029)	0.196 (0.027)	0.203 (0.027)	0.178 (0.026)
North Central (NC)	0.193 (0.031)	0.151 (0.032)	0.192 (0.025)	0.202 (0.025)	0.181 (0.026)
North east (NE)	0.202 (0.045)	0.198 (0.037)	0.218 (0.038)	0.205 (0.030)	0.206 (0.033)
North west (NW)	0.125 (0.032)	0.101 (0.035)	0.141 (0.042)	0.135 (0.047)	0.126 (0.061)
West coast (WC)	0.180 (0.045)	0.166 (0.050)	0.202 (0.049)	0.200 (0.047)	0.185 (0.046)
Western Himalaya (WH)	0.188 (0.025)	0.154 (0.069)	0.206 (0.039)	0.155 (0.076)	0.201 (0.039)
SMAP (March 2015–April 2017)					
East coast (EC)	0.119 (0.065)	0.125 (0.063)	0.168 (0.105)	0.137 (0.085)	0.133 (0.078)
Interior Peninsula (IP)	0.148 (0.052)	0.138 (0.044)	0.198 (0.078)	0.173 (0.057)	0.160 (0.059)
North central (NC)	0.135 (0.067)	0.107 (0.041)	0.194 (0.062)	0.139 (0.064)	0.147 (0.049)
North east (NE)	0.139 (0.092)	0.143 (0.103)	0.148 (0.093)	0.169 (0.103)	0.141 (0.091)
North west (NW)	0.083 (0.032)	0.072 (0.041)	0.127 (0.048)	0.091 (0.050)	0.095 (0.051)
West coast (WC)	0.126 (0.074)	0.108 (0.060)	0.164 (0.094)	0.144 (0.084)	0.133 (0.077)
Western Himalaya (WH)	0.088 (0.064)	0.069 (0.03)	0.052 (0.034)	0.059 (0.050)	0.050 (0.036)

4.3. *Interannual Variability of Soil Moisture*

The year-to-year variations of soil moisture, precipitation, and TWS in India are shown in Figure 3. Normalized values were estimated by subtracting monthly soil moisture values from the long-term climatology calculated from averaging the monthly values over India, and by dividing by the monthly standard deviation. The AMSR soil moisture time series is shown by a gray color with open circles, but it also contains locally estimated scatterplot smoothing (LOESS) curves (smoothing parameter α=0.75) also plotted in Figure 3. The LOESS smoothing technique based on locally weighted regression smoothing [45]. The time series exhibit consistent year-to-year and seasonal variability, and the nature of the intra-seasonal variability looks similar during periods of major droughts or major floods [46]. On the other hand, two dominant intra-seasonal oscillatory modes and large-scale standing patterns are observed using outgoing longwave radiation (OLR) datasets [47]: one at the equatorial Pacific and the other over the equatorial Indian Ocean. Spatial and temporal variations of rainfall influence the spatiotemporal distribution of runoff, soil moisture, and groundwater reserves, which in turn affect the frequency of droughts and floods. Ultimately, these drought and floods affect the patterns of vegetation productivity in India.

The normalized plot (Figure 3) shows variations in soil moisture during the period of 2002 to 2016. Both SMAP and AMSR records show consistent variability during the overlapping period. Before the SMAP record started, there were the three major droughts shown by the vertical gray bars in Figure 3, in 2002, 2009, and 2014. The year 2009 and 2002 had the third and fourth largest major droughts in the past 100 years, after 1918 and 1972. In contrast to the AMSR record, the ERA-Interim reanalysis soil moisture estimate did not capture the year 2009 drought. Neena et al. [48] also mentioned the failure of global models in the 2009 severe drought, when a seasonal rainfall deficit of 21.5% in 2002 and ~24% was recorded [49]. In addition, the AMSR SM record also successfully captured the extreme flooding events from 2005–2006 [49].

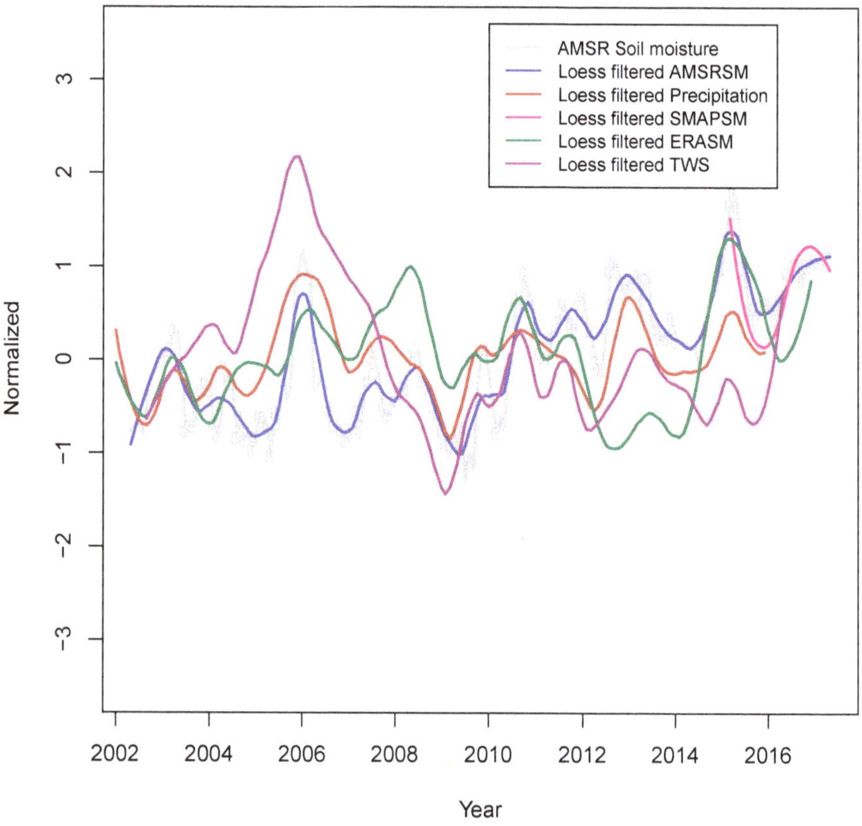

Figure 3. Time series (normalized units) of soil moisture (AMSR, SMAP, and ERA), precipitation (IMD), and total water storage (GRACE) in India. The vertical gray bars indicate drought periods and the light blue bar represents a period of excess of water.

4.4. Soil Moisture–Precipitation (SM–PR) MCA Results

Here, Maximum Covariance Analysis (MCA) was used to investigate the spatiotemporal relationship of the covariance between SM and precipitation datasets. Figure 4a,b show the first mode of maximum covariance patterns of precipitation and soil moisture for 2002–2016. Here, the monthly spatial anomalies of precipitation and soil moisture data were utilized. The first mode shows similar general patterns in the northwest, north central, and northern parts of the Interior Peninsula. Over the Northeast, precipitation patterns resemble those of soil moisture. Both variables exhibit the opposite sign in the Western Himalayas compared to the northeast, which is the main driving force of SM change. Note that the precipitation datasets have less coverage over the Western Himalayas. A squared covariance factor (SCF) of 56% (see Figure 4c) is observed between two fields, indicating the large-scale coupling between soil moisture and precipitation monthly anomalies. In the temporal domain, these time series datasets are strongly correlated ($r = 0.62$), which indicates a strong association between SM and PR, both of which exhibiting a slight downward trend since 2010 (Figure 4c). It can be seen in Figure 4c that soil moisture and precipitation demonstrate a strong seasonal cycle and coupling in India, as most rainfall occurs during the monsoon season in central India. These findings are consistent with previous observations and model simulations. For instance, Jung et al. [50] observed a significant negative trend over the global tropics (28°S–38°N) using soil moisture and evapotranspiration data. Douville et al. [51] studied model-derived soil moisture over Asia and Africa, concluding that surface wetness in India contributes to increased rainfall. Later, Orlowsky and Senevirante [52] found remarkable coupling strength between soil moisture and precipitation for specific regions.

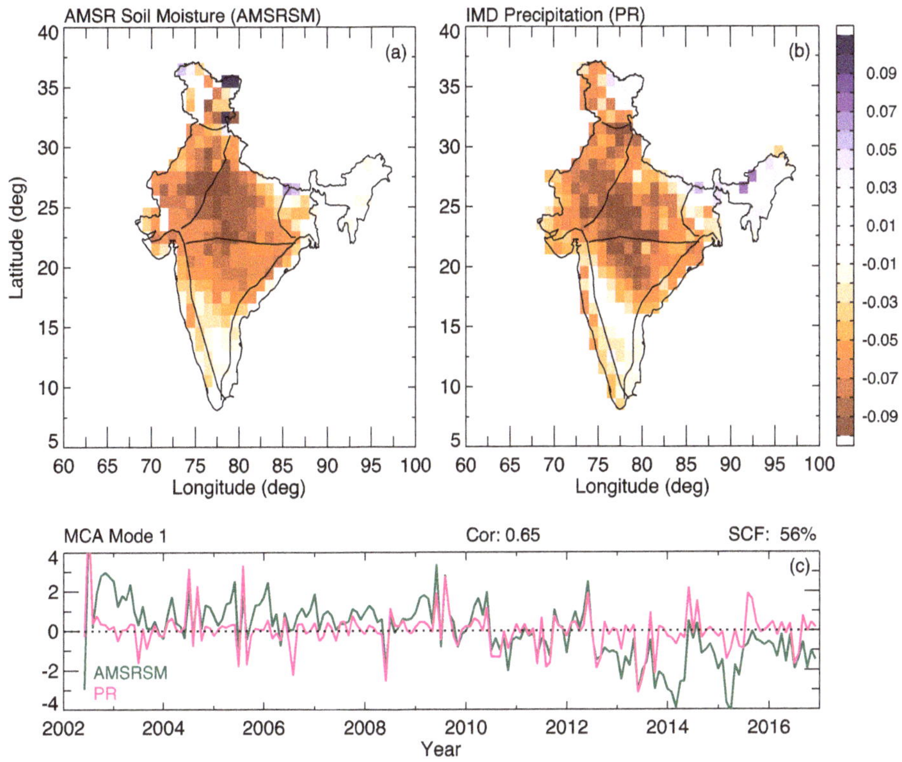

Figure 4. Spatial patterns of the first MCA mode associated with (**a**) AMSR soil moisture (AMSRSM) and (**b**) IMD precipitation (PR) during the period 2006–2016. (**c**) The corresponding covariance coefficients associated with AMSR soil moisture (green curve) and precipitation (pink curve). The squared coefficient factor (SCF) and correlation coefficients are indicated on the top of the plot.

4.5. Local Temperature Impact on Soil Moisture (SM–TEM)

To investigate how the spatial pattern of the surface temperature in India is affected by soil moisture, we once again utilized MCA during the period 2006 to 2016. The covariance analysis was applied after removing the long-term trends from both the datasets, the results of which are shown in Figure 5. The soil moisture pattern reveals more variability over the northeastern central and northeastern Interior Peninsula regions, which shows strong anti-correlation with temperature. A positive temperature anomaly is observed over the eastern part of India, concurrent with negative soil moisture anomalies in the same region. Higher temperatures can lead to a greater potential evaporation over land, resulting in a declining soil moisture. Furthermore, warm temperatures frequently accompany the shortage of rainfall, leading to agricultural droughts that are normally associated with soil moisture deficiency. Recent studies with CRU temperature and CMIP5 simulations also indicate that central and northeastern regions show the highest warming compared with the other regions [53]. The first mode of MCA explores the largest variability between soil moisture and TEM. The total amount of variance of the first mode reaches 28% for soil moisture and 42% for temperature. The squared covariance factor of 87% between the two fields is illustrated in Figure 5c.

The correlation between the two time series is about −0.72, which is significant at the 95% level. MCA analysis indicates an inverse relationship between SM and TEM variance; the higher the TEM variance, the lower the SM variance. Note that the time series show similar patterns, with a decreasing trend from 2008 to 2014. It indicates that temperature controls the evapotranspiration and consequently affects the soil moisture. Our results for SM couplings with PR and TEM suggest that temperature has a larger contribution than precipitation to soil moisture variation in India, consistent with previous study in a similar humid basin environment [54]. The inverse relationship indicates that a decrease in soil moisture may be responsible for an increase in temperature. This conclusion is generally true in most regions, with some exceptions that exhibit complicated soil moisture–climate interaction and feedback.

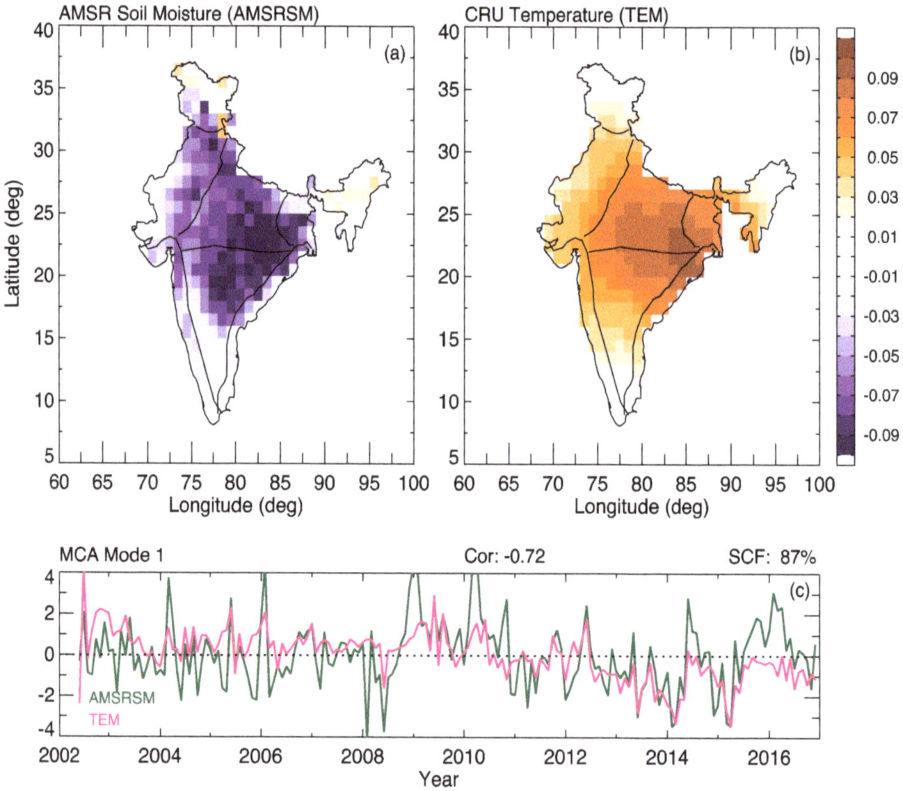

Figure 5. Spatial patterns of the first MCA mode associated with (**a**) AMSR soil moisture and (**b**) CRU temperature during the period 2006–2016. (**c**) The corresponding covariance coefficients associated with AMSR soil moisture (green curve) and temperature (pink curve). The squared coefficient factor (SCF) and correlation coefficients are indicated on the top of the plot.

4.6. Relation Between TCC, TWS and SM

Figures 6 and 7 show the first leading modes of covariance between SM and TCC, and SM and TWS for the period of 2006–2016. In Figure 6, the SM pattern tends to be zonally elongated from the Northwest to the Interior Peninsula. The SM covariance anomalies match those of TCC. The maximum variance of SM concentrated over central India is closely associated with total cloud events in that region (Figure 6a,b), which shows a strong similarity between the two variables. It suggests that a large increase in cloud cover can reduce the surface air temperature and increase the soil moisture. The SCF between SM and TCC is about 81%. The time series of AMSRSM and TCC show clear decreasing trends.

The decreasing trends may be linked to ENSO (El Nino Southern Oscillation) variation. However, TCC feedback remains a source of uncertainly in global climate change Intergovernmental Panel on Climate Change (IPCC 2007 [55]). Rao et al. [56] observed decreasing trends in total cloud cover in many regions throughout India from 1951 to 2000. TCC and TWS variability has a statistically significant impact on SM over several regions in India. It is clear that the higher correlations, the bigger the influence of TCC on evapotranspiration; indeed, the highest values can be found in monsoon and post-monsoon followed by the winter and summer. About 60% of the world is covered by clouds, and the influence of clouds on both the water balance and global radiation budget, even small variations, can alter the climate response [57,58]. The spatial patterns of variance for SM and TWS are consistent, with the exception of the northwest, northeast, and west coast regions, where TWS shows negative variances but where SM still has positive variances. The first mode of MCA yields a squared covariance factor of about 79% between the SM and TWS. The time series of SM and TWS show significant increasing trends with a positive correlation of about 0.62, suggesting that the soil moisture variability is the main driver of the regional change in the water balance. TWS changes reflect the balance or imbalance of water fluxes (precipitation, evapotranspiration, and runoff) and are strongly affected by regional climate conditions. Furthermore, the interannual and decadal climate variability caused by large-scale ocean–atmosphere interactions might influence moisture advection, rainfall, and eventually TWS changes at long-term timescales [59].

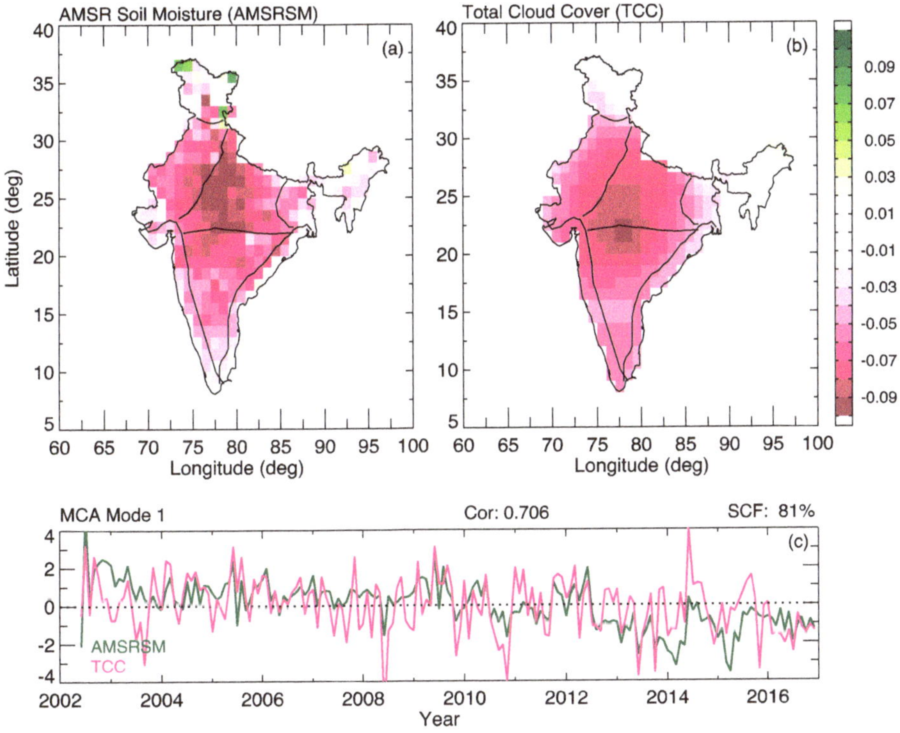

Figure 6. Spatial patterns of the first MCA mode associated with (**a**) AMSR soil moisture and (**b**) total cloud cover (TCC) during the period 2006–2016. (**c**) The corresponding covariance coefficients associated with AMSR soil moisture (green curve) and TCC (pink curve). The squared coefficient factor (SCF) and correlation coefficients are indicated on the top of the plot.

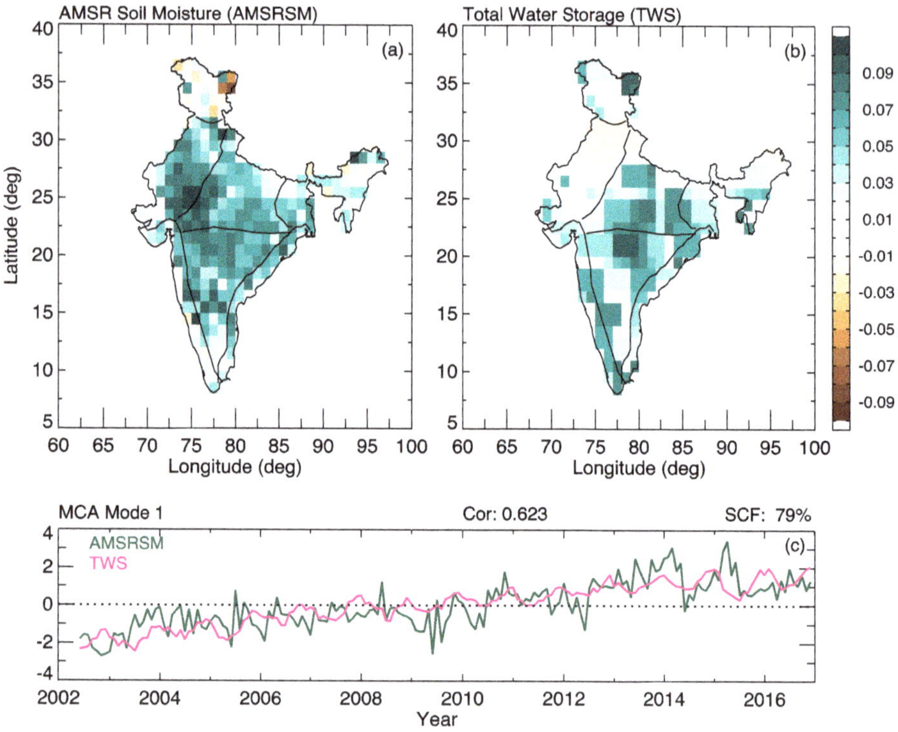

Figure 7. Spatial patterns of the first MCA mode associated with (**a**) AMSR soil moisture (AMSRSM) and (**b**) total water storage (TWS) during the period 2006–2016. (**c**) The corresponding covariance coefficients associated with AMSR soil moisture (green curve) and TWS (pink curve). The squared coefficient factor (SCF) and correlation coefficients are indicated on the top of the plot.

The cumulative SCF across MCA modes from SM and PR, TEM, TCC, and TWS as well as the correlation between the time series obtained from each mode are shown in Figure 8. The first two leading modes account for most of the variation, which account for 56%, 10% (PR), 87%, 5% (TEM), 81%, 10% (TCC), and 79%, 9% (TWS) of the SCF, respectively. Significance levels were estimated at each mode using a moving block bootstrap procedure as described by Wilks [57]. The significant modes are shaded in red in Figure 8. The SCF explained by the first four modes exceeds 95% significance.

The four leading modes explain about 97%, 97%, 78%, and 95% of the SCF for SM–TCC, SM–TEM, SM–PR, and SM–TWS, respectively. The rest of the MCA modes only explain approximately 5% of the squared covariance, with the exception of SM–PR modes. The first four leading modes represent the major characteristics of SM and other fields (PR, TEM, TCC, and TWS) in India. The correlation coefficients between AMSRSM and the variables are 0.65 (PR), −0.72 (TEM), 0.71 (TCC), and 0.62 (TWS) for the first leading mode. The correlation coefficients decrease as the mode increase, which indicates a teleconnection between SM–PR, SM–TWS and SM–TCC. There is an anti-correlation with surface temperature. SM is indirectly related to the surface temperature, because increasing surface temperatures leads to a decreasing soil moisture due to evaporation.

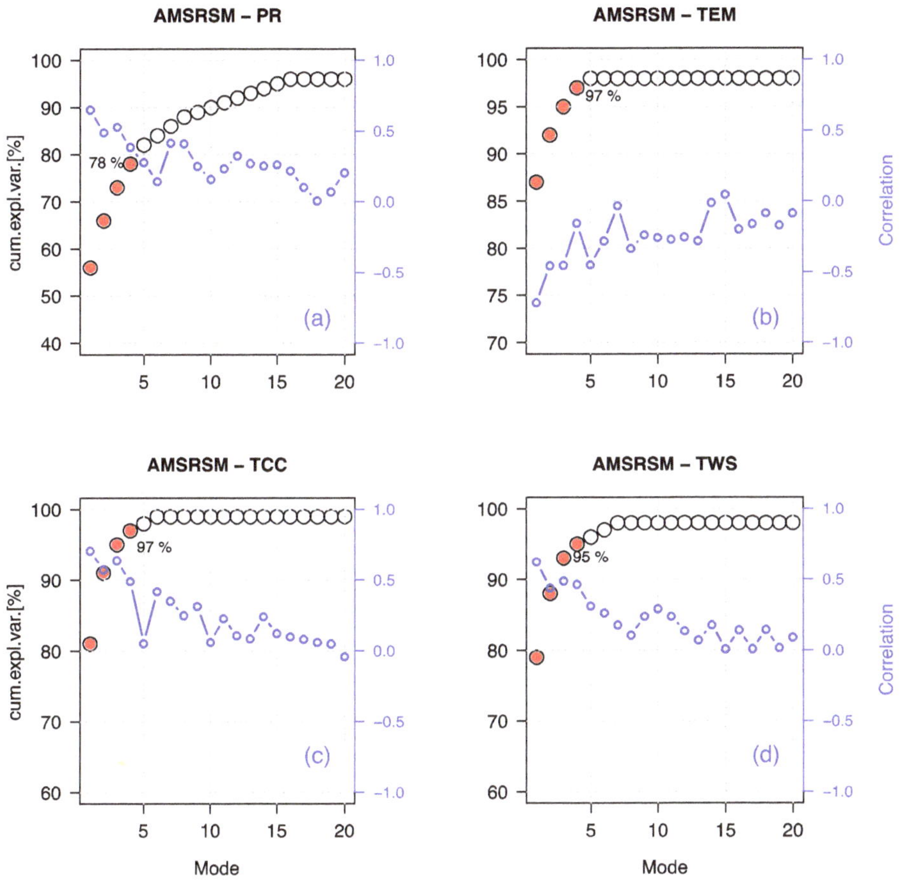

Figure 8. The cumulative squared covariance coefficient (SCF) percentage and correlation coefficients as a function of MCA mode between (**a**) AMSR soil moisture (AMSRSM) and precipitation (PR), (**b**) AMSR soil moisture (AMSRSM) and temperature (TEM), (**c**) AMSR soil moisture (AMSRSM) and total cloud cover (TCC), and (**d**) AMSR soil moisture (AMSRSM) and total water storage (TWS).

4.7. Spatial Soil Moisture Trends

Next, seasonal (a–d) and annual (e) SM trends were analyzed, with the results shown in Figure 9. Robust regression analysis was utilized to obtain the seasonal and annual SM trends for each grid cell throughout India during the period from June 2002 to April 2017. The statistical significance of the trends was calculated by using the Monte Carlo test as described in Allen and Smith [58] and marked as asterisks where the trends are statistically significant at the 95% confidence level. In all seasons, there is an increasing SM trend in most regions. Seasonal SM exhibits a positive trend, which is particularly pronounced in the northwestern region, and a moderate trend in several other regions. However, decreasing trends (-0.01 to -0.025 cm^3/cm^3 per decade) appear in the Western Himalayas and southern India. Therefore, there is clear evidence that there are large precipitation variations, especially in north central and interior peninsula regions due to monsoon rainfall. The decreasing trends may be due to the southeastern monsoon. The rainfall is high over the eastern part and decreases northwestward across the zone and near the northwestern boundary. As the monsoon continues, the air moisture content is reduced and the monsoon precipitation gradually decreases from east to west [60]. The central part of India is characterized by the presence of a large area with positive SM trends that are statistically significant (95%) over winter, spring, and fall seasons. The positive trends vary between 0.03 to 0.05 cm^3/cm^3 per decade.

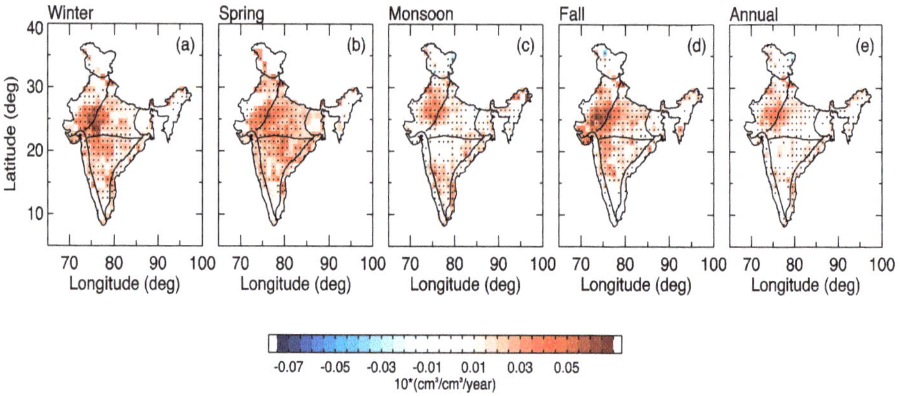

Figure 9. Spatial distribution of AMSR soil moisture trends for (**a**) winter, (**b**) spring, (**c**) monsoon, (**d**) fall, and (**e**) annual from June 2002 to April 2017. Solid black marks indicate regions where the trends are significant at the 95% confidence level based on the Mann–Kendall test.

The fact that increasing or decreasing soil moisture trends may have an adverse effect on food grain yield points towards the existence of a potentially linear or nonlinear relationship between SM and better food grain yield. Rajeevan and Nayak [59] have reported increasing trends in annual mean soil moisture in 15 out of 27 stations in India during the period of 1991–2013. There were significantly increasing trends in north, central, and northeast India as well as on the west coast. They stated that in situ SM datasets are costly, labor intensive, and not extensively quality-controlled. Nevertheless, relative surface atmospheric moisture studies in India conducted by Jaswal and Koppar [61] have also reported significantly increasing trends of specific humidity during summer and winter as well as annual time frames, especially over northwest, central, and southeast regions of India. Rajeevan and Nayak [59] found positive soil moisture trends over 15 different meteorological stations in India during the period 1991–2012.

Figure 10 shows the monthly variations and the linear trend of the AMSR and SMAP soil moisture in Indian for the periods June 2002–April 2017 and Mar 2015–April 2017. The monthly means in the figure clearly show the interannual variability of AMSR soil moisture. It is noteworthy that the monthly SM time series of AMSR and SMAP are correlated with each other, although they exhibit some small differences in their year-to-year variability. The monthly mean values across all of India have a moderate standard deviation ranging from -0.085 to 0.20 cm^3/cm^3 towards the end of the observation. The blue-shaded region indicates the internal regional variability of the soil moisture. The maximum standard deviation is seen August and September, indicating a high degree of uncertainty, which may be due to the year to year variation of summer monsoon. The results show a positive trend of 0.016 cm^3/cm^3 per decade for the entire period of AMSR observations for all of India. The more positive trends are found in the northwest and north central regions.

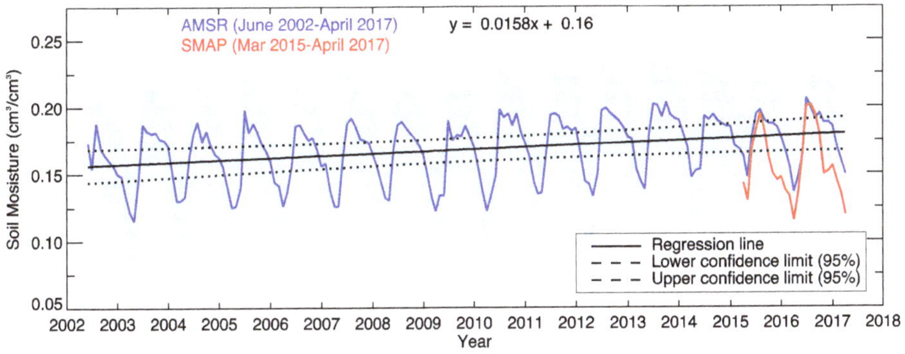

Figure 10. Monthly time series of AMSR soil moisture from June 2002 to April 2017. SMAP soil moisture soil moisture was also plotted for the period Mar 2015 to April 2017. Robust regression analysis was performed at the 95% confidence level to obtain the trend, which is shown by the dotted line. Shaded areas represent ±2 standard deviations associated with the mean monthly soil moisture.

5. Conclusions

This study explored the spatial distribution, characteristics, and temporal variability of soil moisture on monthly, seasonal, and annual temporal scales using AMSR and SMAP satellite observations. The spatiotemporal coherency between soil moisture and geophysical fields (PR, TEM, TCC, and TWS) were examined using the powerful tool of Maximum Covariance Analysis (MCA). In addition, trends were estimated using robust regression with a 95% confidence level for annual and seasonal (winter, spring, monsoon, and fall) time frames. The trends indicate the increase or decrease of soil moisture in India on regional scales, which are essential for a better understanding of agricultural yields. The most important findings of our analysis are summarized as follows:

Firstly, the characteristics of satellite-observed soil moisture in India were presented using 15 years of data. The mean seasonal soil moisture from AMSR shows a maximum value of about 0.22 cm^3/cm^3 over the northeast region followed by the East Coast (0.21 cm^3/cm^3). The monthly mean values of AMSR and SMAP soil moisture show a reasonably good agreement. However, the spatial distributions of annual and seasonal soil moisture show some discrepancies between AMSR and SMAP in some regions (West Coast and Western Himalayas). The relative SM difference over the seven regions in India shows that AMSR is higher than SMAP over the east coast in winter (20.41%), but lower over the Interior Peninsula in the monsoon season (−7.34%).

The normalized time series of different observations (SM, TWS, and PR) all clearly show drought years in 2002, 2009, and 2014, as well as an excess of water in the year 2006. The MCA method is a powerful technique that identifies spatial patterns of maximum covariance between two fields. These results indicate that the spatial and temporal patterns of SM and surface–atmosphere fields (PR, TEM, TCC, and TWS) agree well. The first MCA mode explains about 56%, 87%, 81%, and 79% of the squared covariance between SM–PR, SM–TEM, SM–TCC, and SM–TWS, respectively, with the corresponding time series having correlation coefficients of 0.65, −0.72, 0.71, and 0.62. These results indicate strong teleconnection patterns associated with soil moisture and land surface–atmosphere parameters.

The annual and seasonal trends were computed at each grid cell (1° × 1°) for the period June 2002–April 2017. It is worth mentioning here that the positive trends are obtained for most regions on annual and seasonal scales (for all seasons) throughout India. The hilly regions, both in the northeast and the west coast of India, all exhibit small negative trends, but these are only in small regions in northern and southern parts of India. The regional moisture budget shows that changes in precipitable water and precipitation efficiency vary temporally and spatially in India. The maximum positive trend occurs in the northwest and north central regions of India in all seasons. Moderate positive trends for the annual scale are observed over northwest and north central regions. In addition, the trends were estimated using monthly means of soil moisture throughout India, revealing an increasing trend of about 0.0158 cm^3/cm^3 per dec from 2002 to 2017. The outcomes acquired in this trend analysis provide evidence of significant overall increase in SM content throughout India. Positive trends in SM indicate favorability for irrigation. However, some regions show small SM trends, which are more useful for single-cropping. The two crop-growing seasons, the Kharif (May–October) and the Rabi (October–April) were identified to have a high importance for agricultural production in India. In the Rabi period, increased temperatures may have profound implications for crop yields as well as fewer requirements for irrigation, especially in northwestern India [8]. This research is directly applicable to seasonal water management for accounting the water budgets (recharge, runoff, evapotranspiration). Understanding the changes and variability in runoff, evapotranspiration, precipitation, and soil moisture under retrospective and projected climate in the key crop-growing seasons is of utmost importance.

These satellite measurements provide valuable insights into the spatial and temporal characteristics of soil moisture in India, given the parse nature of ground-based measurements. Long-term observations of regional characteristics are beneficial for a better understanding of crop cultivation and agricultural yield, and ultimately boosting the economy of the country. High-resolution datasets of soil moisture are crucial for a better understanding of the future across various regions of India.

Author Contributions: Data curation, G.A., E.C., S.C., G.B. and S.V.B.R.; Funding acquisition, I.V.; Writing—original draft, and analysis, K.P.; Writing—review & editing, Y.M.

Acknowledgments: We acknowledge all groups who contributed the datasets that supported this work. The standard AMSR-2 soil moisture was obtained from NASA (Goddard Earth Sciences Data Information Services Center) data archive. Authors also wish to thank the ECMWF, SMAP, IMD, and CRU people for providing the datasets. We also thank the JPL team for providing their GRACE datasets.

Conflicts of Interest: The authors declare no conflict of interest.

References

1. Seneviratne, S.I.; Corti, T.; Davin, E.L.; Hirschi, M.; Jaeger, E.B.; Lehner, I.; Orlowsky, B.; Teuling, A.J. Investigating soil moisture climate interactions in a changing climate: A review. *Earth Sci. Rev.* **2010**, *99*, 125–161. [CrossRef]
2. GCOS-138. *Implementation Plan for the Global Observing System for Climate in support of the UNFCCC-2010 Update*; World Meteorological Organization: Geneva, Switzerland, 2010.
3. Koster, R.D.; Suarez, M.J. The relative contributions of land and ocean processes to precipitation variability. *J. Geophys. Res.* **1995**, *100*, 13775–13790. [CrossRef]
4. Liu, Y. Prediction of monthly-seasonal precipitation using coupled SVD patterns between soil moisture and subsequent precipitation. *Geophys. Res. Lett.* **2003**, *30*, 1827. [CrossRef]
5. Jiang, Y.; Fu, P.; Weng, Q. Assessing the impacts of Urbanization-associated land use/cover change on land surface temperature and surface moisture. *Remote Sens.* **2015**, *7*, 4880–4898. [CrossRef]
6. Huang, J.; van den Dool, H.M.; Georgakakos, P.K. Analysis of model calculated soil moisture over the United States (1991–1993) and applications to long-rangetemperature forecasts. *J. Clim.* **1996**, *9*, 1350–1362. [CrossRef]
7. Wang, W.Q.; Kumar, A. A GCM assessment of atmospheric seasonal predictability associated with soil moisture anomalies over North America. *J. Geophys. Res.* **1998**, *103*, 28637–28646. [CrossRef]
8. Mishra, V.; Shah, R.; Thrasher, B. Soil moisture droughts under the retrospective and projected climate in India. *J. Hydrometeorol.* **2014**. [CrossRef]
9. Shukla, J.; Mintz, Y. Influence of land-surface evapotranspiration on the Earth's climate. *Science* **2015**, *215*, 1498–1501. [CrossRef]
10. Asharaf, S.; Dobler, A.; Ahrens, B. Soil moisture-precipitation feedback processes in the Indian summer monsoon season. *J. Hydrometoerology* **2012**. [CrossRef]

11. Verikoden, H.; Revadekar, J.V. Relation between the rainfall and soil moisture during different phases of Indian monsoon. *Pure Appl. Geophys.* **2018**, *175*, 1187–1196. [CrossRef]
12. Kantharao, B.; Rakesh, V. Observational evidence for the relationship between spring soil moisture and June rainfall over the Indian region. *Theor. Appl. Climatol.* **2018**, *132*, 835–849. [CrossRef]
13. Raman, S.; Mohanty, U.C.; Reddy, N.C.; Alpaty, K.; Madala, R.V. Numerical simulation of the sensitivity of summer monsoon circulation and rainfall over India to land-surface processes. *Pure Appl. Geophys.* **1998**, *152*, 781–809. [CrossRef]
14. Dirmeyer, P.A.; Dolman, J.; Sato, N. The pilot phase of the Global Soil Witness Project. *Bull. Am. Meteorol. Soc.* **1999**, *80*, 851–878. [CrossRef]
15. Koster, R.D.; Dirmeyer, P.A.; Guo, Z.; Bonan, G.; Chan, E.; Cox, P.; Gordon, C.T.; Kanae, S.; Kowalczyk, E.; Lawrence, D.; et al. Regions of strong coupling between soil moisture and precipitation. *Science* **2004**, *305*, 1138–1140. [CrossRef] [PubMed]
16. Dorigo, W.; de Jeu, R.; Chung, D.; Parinussa, R.; Liu, Y.; Wagner, W.; Fernandez-Prieto, D. Evaluating global trends (1998–2010) in harmonized multi-satellite surface soil moisture. *Geophys. Res. Lett.* **2012**, *39*, L18405. [CrossRef]
17. Pariussa, R.M.; de Jeu, R.A.M.; van der Schalie, R.; Crow, W.T.; Lei, F.; Holmes, T.R.H. A Quasi-Global approach to improve day-time satellite surface soil moisture anomalies through the land surface temperature input. *Climate* **2016**, *4*, 50. [CrossRef]
18. Liu, Y.; Zhao, W.; Wang, L.; Zhang, X.; Daryanto, S.; Fang, X. Spatial variations of soil moisture under Caragana Korshinskii Korn from different precipitation zones: Field based analysis in the Loess Plateau, China. *Forests* **2016**, *7*, 31. [CrossRef]
19. Tiwari, V.M.; Wahr, J.; Swenson, J. Dwindling groundwater resources in northern India, from satellite gravity observations. *Geo. Phys. Res. Lett.* **2009**, *36*, L18401. [CrossRef]
20. Njoku, E.G.; Chan, S.K. Vegetation and surface roughness effects on AMSR-E land observations. *Rem. Sen. Environ.* **2006**, *100*, 190–199. [CrossRef]
21. Owe, M.; De Jeu, R.A.M.; Holmes, T.R.H. Multi-sensor historical climatology of satellite-derived global land surface moisture. *J. Geophys. Res.* **2008**, *113*, R01002. [CrossRef]
22. Jones, L.A. Synthesis of Satellite Microwave Observations for Monitoring Global Land-Atmosphere CO_2 Exchange. Ph.D. Thesis, College of Forestry and Conservation, Missoula, MT, USA, 2016.
23. Du, J.; Kimball, J.S.; Jones, L.A.; Kim, Y.; Glassy, J.; Watts, J.D. A global satellite environmental data record derived from AMSR-E and AMSR2 microwave Earth observations. *Earth Syst. Sci.* **2017**, *9*, 791–808. [CrossRef]
24. Entekhabi, D.; Njoku, E.G.; O'Neill, P.E.; Kelogg, K.H.; Crow, W.T.; Edelstein, W.N.; Entin, J.K.; Goodman, S.D.; Jackson, T.J.; Johnson, J.; et al. The Soil Moisture Active Passive (SMAP) mission. *Proc. IEEE* **2010**, *98*, 704–716. [CrossRef]
25. Pan, M.; Cai, X.; Chaney, N.W.; Wood, E.F. An initial assessment of SMAP soil moisture retrievals using high-resolution model simulations and in situ observations. *Geophys. Res. Lett.* **2016**. [CrossRef]
26. Zeng, J.; Chen, K.S.; Chen, Q. A preliminary evaluation of the SMAP radiometer soil moisture product over United States and Europe using ground-based measurements. *IEEE Trans. Geosci. Remote Sens.* **2016**, *54*, 4929–4940. [CrossRef]
27. Colliander, A.; Jackson, T.J.; Bindlish, R.; Chan, S.; Das, N.; Kim, S.B.; Cosh, M.H.; Dunbar, R.S.; Dang, L.; Pashaian, L.; et al. Validation of SMAP surface soil moisture products with core validation sites. *Remote Sens. Environ.* **2017**, *191*, 215–231. [CrossRef]
28. Kim, H.; Parinussa, R.; Konings, A.G.; Wagner, W.; Cosh, M.H.; Lakshmi, V.; Zohaib, M.; Choi, M. Global-scale assessment and combination os SMAP with ASCAT (active) and AMSR2 (passive) soil moisture products. *Remote Sens. Environ.* **2018**, *204*, 260–275. [CrossRef]
29. Wahr, J.; Molenaar, M.; Bryan, F. Time-variability of the Earth's gravity field: Hydrological and oceanic effects and their possible detection using GRACE. *J. Geophys. Res.* **1998**, *103*, 30205–30230. [CrossRef]
30. Swenson, S.; Wahr, J. Methods for inferring regional surface-mass anomalies from Gravity Recovery and Climate Experiment (GRACE) measurements of time-variable gravity. *J. Geophys. Res.* **2002**, *107*, 2193. [CrossRef]
31. Rodell, M.; Famiglietti, J.S. An analysis of terrestrial water storage variations in Illinois with implications for the Gravity Recovery and Climate Experiment A(GRACE). *Wat. Resour. Res.* **2001**, *37*, 1327–1339. [CrossRef]

32. Wahr, J.; Zhong, S. Computations of the viscoelastic response of a 3-D compressible Earth to surface loading: An application to glacial isostatic adjustment in Antarctica and Canada. *Geophys. J. Int.* **2013**, *192*, 557–572.
33. Wiese, D.N.; Landerer, F.W.; Watkins, M.M. Quantifying and reducing leakage errors in the JPL RL05M GRACE mascon solution. *J. Geophys. Res.* **2016**. [CrossRef]
34. Rajeevan, M.; Bhate, J.; Kale, J.D.; Lal, B. High resolution daily gridded rainfall data for the Indian region: Analysis of break and active monsoon spells. *Curr. Sci.* **2006**, *91*, 296–306.
35. Kishore, P.; Jyothi, S.; Basha, G.; Rao, S.V.B.; Rajeevan, M.; Velicogna, I.; Sutterley, T.C. Precipitation climatology over India: Validation with observations and reanalysis datasets and spatial trends. *Clim Dyn.* **2015**. [CrossRef]
36. Dee, D.P.; Uppala, S.M.; Simmons, A.J.; Berrisford, P.; Poli, P.; Kobayashi, S.; Andrae, U.; Balmaseda, M.A.; Balsamo, G.; Bauer, D.P.; et al. The ERA-Interim reanalysis: Configuration and performance of the data assimilation system. *Quart. J. Roy. Meteor. Soc.* **2011**, *137*, 553–597. [CrossRef]
37. Von Storch, H.; Zwiers, F.W. *Statistical analysis in Climate Research*; Cambridge University Press: Cambridge, UK, 1999; p. 494.
38. Wallace, J.M.; Smith, C.; Bretherton, C.S. Singular value decomposition of wintertime sea surface temperature and 500 mb height anomalies. *J. Clim.* **1992**, *5*, 561–576. [CrossRef]
39. Holland, P.E.; Welsch, R.E. Robust regression using iteratively reweighted least-squares. *Commun. Stat.* **1997**, *A6*, 813–827. [CrossRef]
40. Miosso, C.J.; von Borries, R.; Argaez, M.; Velazquez, L.; Quintero, C.; Potes, C.M. Compressive sensing reconstruction with prior information by iteratively reweighted least-squares. *IEEE Trans. Signal Process.* **2009**, *57*, 2424. [CrossRef]
41. Mann, H.B. Nonparametric tests against trend. *Econometrica* **1945**, *13*, 245–259. [CrossRef]
42. Kendall, M. *Rank Correlation Measure*; Charles Griffin: London, UK, 1975.
43. Anusha, S.; Anandakumar, K.; Manish, R.; Thara, P. Evaluation of soil moisture data products over Indian region and analysis of spatiotemporal characteristics with respect to monsoon rainfall. *J. Hydrol.* **2016**, *542*, 47–62.
44. Unnikrishnan, C.K.; Rajeevan, M.; Vijayabhaskara Rao, S.; Manoj, K. Development of a high resolution land surface dataset for the south Asian monsoon region. *Curr. Sci.* **2013**, *105*, 1235–1246.
45. Cleveland, W.S.; Devlin, S.J. Locally weighted regression: An approach to regression analysis by local fitting. *J. Am. Stat. Assoc.* **1988**, *83*, 596–610. [CrossRef]
46. Krishnamurthy, V.; Shukla, J. Intra-seasonal and inter-annual variability of rainfall over India. *J. Clim.* **2000**, *13*, 4366–4377. [CrossRef]
47. Krishnamurthy, V.; Shukla, J. Seasonal persistence and propagation of intraseasonal patterns over the Indian summer monsoon region. *Clim. Dyn.* **2008**, *30*, 353–369. [CrossRef]
48. Neena, J.M.; Suhas, E.; Goswami, B.N. Leading role of internal dynamics in the 2009 Indian summer monsoon drought. *J. Geophys. Res.* **2011**, *116*. [CrossRef]
49. Bhat, G.S. The Indian drought of 2002- a sub-seasonal phenomenon? *Q. J. R. Meteorol. Soc.* **2006**, *132*, 2583–2602. [CrossRef]
50. Jung, M.; Reichstein, M.; Ciais, P.; Seneviratne, S.I.; Sheffield, J.; Goulden, M.L.; Bonan, G.; Cescatti, A.; Chen, J.; De Jeu, R.; et al. Recent decline in the global land evapotranspiration trend due to limited moisture supply. *Nature* **2010**, *467*, 951–954. [CrossRef] [PubMed]
51. Douville, H.; Chauvin, F.; Brooqua, H. Influence of soil moisture on the Asia and African monsoons. Part 1: Mean monsoon and daily precipitation. *J. Clim.* **2001**, *14*, 2381–2404. [CrossRef]
52. Orlowsky, B.; Seneviratne, S.I. Statistical analysis of land-atmosphere feedbacks nd their possible pitfall. *J. Clim.* **2010**, *23*, 3918–3932. [CrossRef]
53. Basha, G.; Kishore, P.; Venkat Ratnam, M.; Jayaraman, A.; Amir, A.K.; Taha, B.; Ouarda, M.J.; Velicogna, I. Historical and projected surface temperature over India during the 20th and 21st century. *Nat. Sci. Rep.* **2017**, *7*, 2987. [CrossRef]
54. Feng, H.; Liu, Y. Combined effects of precipitation and air temperature on soil moisture in different land covers in a humid basins. *J. Hydrol.* **2015**, *531*, 1129–1140. [CrossRef]
55. IPCC-AR4, 2007. *Climate Change 2007, The Scientific Basis, Contribution of Workshop Group-I to the Fourth Assessment Report of Intergovernmental Panel on Climate Change (IPCC)*; Cambridge University Press: Cambridge, UK, 2007.

56. Rao, G.S.P.; Jaswal, A.K.; Kumar, M.S. Effects of urbanization on meteorological parameters. *Mausam* **2004**, *55*, 429–440.
57. Wilks, D.S. Resampling hypothesis tests for autocorrelated fileds. *J. Clim.* **1997**, *10*, 65–82. [CrossRef]
58. Allen, M.R.; Smith, L.A. Monte Carlo SSA: Detecting irregular oscillations in the presence of colored noise. *J. Clim.* **1996**, *9*, 3373–3403. [CrossRef]
59. Rajeevan, M.N.; Nayak, S. *Observed Climate Variability and Change Over the Indian Region*; Springer Geology: Singapore, 2017.
60. He, L.; Chen, J.M.; Liu, J.; Belair, S.; Luo, X. Assessment of SMAP soil moisture for global simulations of gross primary production. *J. Geophys. Res.* **2017**. [CrossRef]
61. Jaswal, A.K.; Koppar, A.L. Recent climatology and trends in surface humidity over India for 1969–2007. *Mausam* **2011**, *62*, 145–162.

remote sensing

MDPI

Article

Assessing Hydrological Modelling Driven by Different Precipitation Datasets via the SMAP Soil Moisture Product and Gauged Streamflow Data

Lu Yi [1,2], Wanchang Zhang [2,*] and Xiangyang Li [3]

1 State Key Laboratory of Pollution Control and Resource Reuse, School of the Environment, Nanjing University, Nanjing 210093, China; dg1225033@smail.nju.edu.cn
2 Key Laboratory of Digital Earth Science, Institute of Remote Sensing and Digital Earth, Chinese Academy of Sciences, Beijing 100094, China
3 Yellow River Conservancy Commission of the Ministry of Water Resources, Zhengzhou 210046, China; Hwlixy@sina.com
* Correspondence: zhangwc@radi.ac.cn

Received: 18 October 2018; Accepted: 21 November 2018; Published: 23 November 2018

Abstract: To compare the effectivenesses of different precipitation datasets on hydrological modelling, five precipitation datasets derived from various approaches were used to simulate a two-week runoff process after a heavy rainfall event in the Wangjiaba (WJB) watershed, which covers an area of 30,000 km^2 in eastern China. The five precipitation datasets contained one traditional in situ observation, two satellite products, and two predictions obtained from the Numerical Weather Prediction (NWP) models. They were the station observations collected from the China Meteorological Administration (CMA), the Integrated Multi-satellite Retrievals for Global Precipitation Measurement (GPM IMERG), the merged data of the Climate Prediction Center Morphing (merged CMORPH), and the outputs of the Weather Research and Forecasting (WRF) model and the WRF four-dimensional variational (4D-Var) data assimilation system, respectively. Apart from the outlet discharge, the simulated soil moisture was also assessed via the Soil Moisture Active Passive (SMAP) product. These investigations suggested that (1) all the five precipitation datasets could yield reasonable simulations of the studied rainfall-runoff process. The Nash-Sutcliffe coefficients reached the highest value (0.658) with the in situ CMA precipitation and the lowest value (0.464) with the WRF-predicted precipitation. (2) The traditional in situ observation were still the most reliable precipitation data to simulate the study case, whereas the two NWP-predicted precipitation datasets performed the worst. Nevertheless, the NWP-predicted precipitation is irreplaceable in hydrological modelling because of its fine spatiotemporal resolutions and ability to forecast precipitation in the future. (3) Gauge correction and 4D-Var data assimilation had positive impacts on improving the accuracies of the merged CMORPH and the WRF 4D-Var prediction, respectively, but the effectiveness of the latter on the rainfall-runoff simulation was mainly weakened by the poor quality of the GPM IMERG used in the study case. This study provides a reference for the applications of different precipitation datasets, including in situ observations, remote sensing estimations and NWP simulations, in hydrological modelling.

Keywords: precipitation; rainfall-runoff simulation; GPM IMERG; merged CMORPH; WRF; 4D-Var; SMAP

1. Introduction

Numerical simulation of the rainfall-runoff process is an important way to research water cycle, flood monitoring, water resource management and environmental conservation [1–5]. The performance of a rainfall-runoff simulation differs from the applied hydrological model, calibrated parameters and input data [6–9]. Because of different theories of runoff generation and routing, different considerations for spatial variation of the underlying surface, and different scheme combinations, diverse hydrological models lead to different simulation results [10–12]. For a specific study area, when the applied hydrological model, required parameters, and other input data are fixed, precipitation is the most important datum that affects hydrological modelling, because its accuracy will dominantly influence the subsequent calculation qualities of net precipitation, soil moisture variation, runoff generation, and runoff routing, then finally determine the success of the rainfall-runoff simulation [13–17].

Currently, there are three mainstream ways to obtain precipitation data. One method is the use of traditional in situ observation [18]. This type of precipitation data is measured using the equipments set in precipitation stations, meteorological stations, or automatic weather stations, which are managed by different departments. The in situ observed precipitation is generally recognized as the most accurate datum that represents the true value [19,20]. However, its application in hydrology is limited by its poor point-to-area representativeness, incomplete opening and sparse station network in developing areas [21,22]. The second method is remote sensing estimation. This category of precipitation data emerged as the remote sensing techniques of visible, infrared and microwaves were developed. Commonly used satellite precipitation products contain the Global Precipitation Climatology Project (GPCP) [23], the Climate Prediction Center Morphing Technique (CMORPH) [24], the Tropical Rainfall Measuring Mission (TRMM) [25] and its successor, the Global Precipitation Measurement (GPM) [26]. These products not only cover a nearly global area but also are freely available to the public, and their spatiotemporal resolutions are becoming finer. Moreover, there are corresponding upgraded products after post-processing [27]. Nevertheless, the remotely sensed precipitation still falls short in terms of showing the consecutiveness of precipitation and detecting extreme events at high latitudes [28]. The third method is obtaining precipitation from a numerical weather prediction (NWP) model. Because this atmospheric model is built on precise physical governing equations, an NWP model can describe the inherent dynamics of precipitation, thus present nearly the entire precipitation process with specific atmospheric reanalysis data [29–31]. However, due to the incompleteness of initial and boundary conditions provided by reanalysis data, uncertainties are unavoidably included in the NWP outputs when solving the physical equations with approximations [32,33]. When applied in hydrological modelling, the poor precipitation prediction of an NWP model and the scale mismatch between it and a hydrological model are two primary problems [34–36]. To improve the accuracy of the NWP-predicted precipitation, various methods of data assimilation have been used to enhance the initial and lateral boundary conditions of an NWP model [37–41]. The generally used NWP models include the National Meteorological Center (NMC) forecast model [42], the next-generation Weather Research and Forecasting (WRF) model [43], the operational Japan Meteorological Agency (JMA) mesoscale model [44] and the European Centre for Medium-Range Weather Forecasts (ECMWF) [45].

These three methods of collecting precipitation data have been widely used in hydrological modelling [1,46–55]. Many studies have been conducted to investigate the utilities of remote sensing estimations and the NWP simulations in hydrological modelling. For example, Wu et al. [48] evaluated the performances of nine existing precipitation products, including TRMM, CMORPH, and NLDAS-2 (Phase 2 of the North American Land Data Assimilation System), in the DRIVE model system for simulating a series of flood in Iowa. Essou et al. [1] applied global atmospheric reanalysis data to drive the lumped conceptual hydrological model HSAMI over 370 American watersheds. Liechti et al. [52] used TRMM 3B42 to drive the hydraulic-hydrologic model over the African Zambezi basin. Rasmussen et al. [49] investigated the spatial-scale characteristics of the precipitation predicted by the WRF model and its applications in hydrological modelling; they concluded that the RCM predictions had larger predictive certainty at a larger scale than at a smaller scale. Lin et al. [51] used the precipitation data generated by the Canadian atmospheric Mesoscale Compressible Community Model (MC2) to drive the Chinese Xin'anjiang hydrological model and simulated a series of flood events in the Huaihe River basin at a 5-km resolution. All of these investigations demonstrated the potential of the indirectly measured precipitation to be used in hydrological modelling. Nevertheless, there are still limited studies concerning the different effectivenesses of different precipitation datasets in hydrological modelling. Therefore, five types of precipitation datasets, incorporating one traditional in situ station observation, two satellite precipitation products, and two NWP predictions, were collected, evaluated and applied in hydrological modelling in this study. Moreover, to avoid the uncertainties caused by scale mismatch between the NWP and hydrological models [49], a 1-km resolution, which the applied NWP and hydrological models could realize, was employed in the simulations. Furthermore, as soil moisture is a crucial intermediate variable in hydrological modelling that affects water exchange, evapotranspiration estimation, runoff generation and model simulation [56–61], hydrologists have extended their attention to assess and improve the accuracy of soil moisture simulation to ensure the reasonability of hydrological simulation [62–69]. Therefore, we assessed the performances of the hydrological modelling driven by different precipitation datasets via not only outlet discharge but also soil moisture.

This manuscript is structured as follows: Section 2 introduces the study area, study period and study data. Section 3 introduces the experimental design and evaluation metrics. Section 4 shows 1-km grid data obtained from different precipitation datasets, the simulated soil moisture and outlet discharges. Section 5 presents the evaluations of different precipitation datasets, the simulated soil moisture and outlet discharges. Finally, the conclusions are drawn in Section 6.

2. Data

2.1. Study Area and Study Period

2.1.1. Study Area

As the NWP predictions generally have larger predictive certainty at a larger scale than at a smaller scale [35,49], a large-scale watershed covering an area of 30,630 km^2, namely, Wangjiaba (WJB) was selected as the study area. The WJB watershed is located between 113.3°–115.8°E and 31.5°–33.4°N (Figure 1b); it is a sub-basin of the Huaihe River basin (HRB), which is one of the seven major river basins in China and lays between the Yellow and Yangtze rivers. This region has important political and economic functions because it has the highest population density in China and has 17% of the country's cultivated land [51]. The WJB watershed belongs to the warm temperate and semi-humid monsoon climate. Its northern region is characterized by hot and wet summers and cold and dry winters because it is primarily controlled by the monsoon climate at mid-latitudes, while its southern part has hot and wet summers and mild and dry winters because it is dominated by a subtropical monsoon climate [70]. The regional elevation in this watershed decreases from the west to the east. In the west, there are foothills and mountains, and the highest altitude is 1130 m above sea level (m.a.s.l.). The middle and eastern parts are vast plains. In this watershed, rapid flood drainage occurs with heavy precipitation because of the high regional topographic relief. Such rapid flood drainages cause the flat midstream and downstream reaches of the HRB difficult to drain, which finally result in floods. Therefore, the rainfall-runoff process in the upstream reach of the HRB, i.e., the WJB watershed, deserves further investigation.

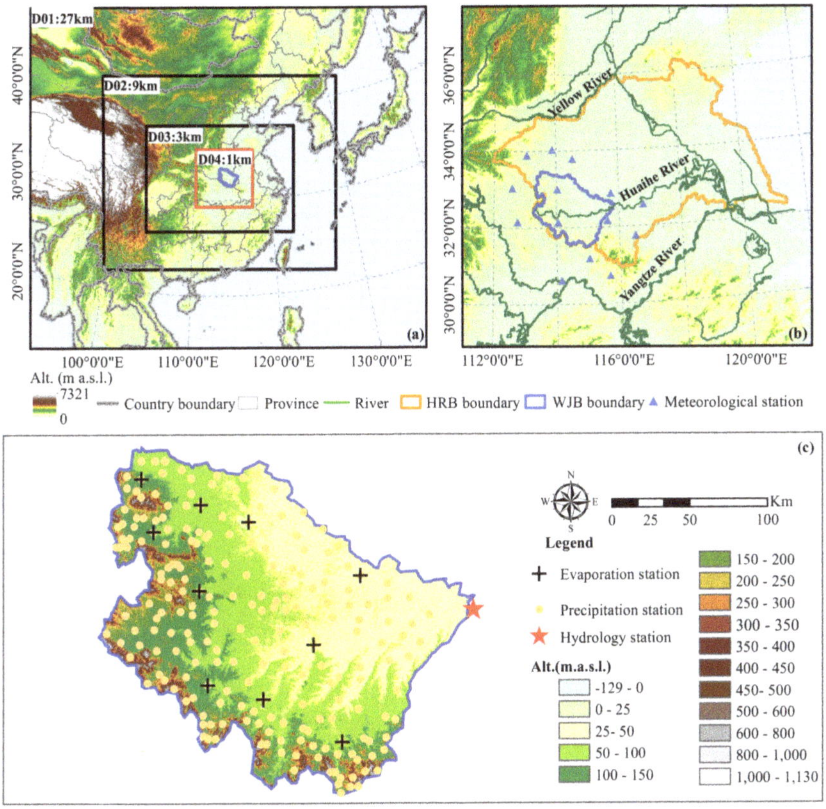

Figure 1. (**a**) The first, second, third and fourth domain sets in the WRF model: D01: 27-km resolution; D02: 9-km resolution; D03: 3-km resolution; D04: 1-km resolution; (**b**) locations of the Huaihe River basin (HRB), the Wangjiaba (WJB) watershed and its 14 meteorological stations; (**c**) altitude (Alt.) of the WJB watershed and its inner 215 precipitation stations, 10 evaporation stations and the hydrology station.

2.1.2. Study Period

It is noteworthy that the temporal spans of the hydrological modelling driven by different types of precipitation datasets are different. The hydrological modelling driven by the in situ observed and remotely sensed precipitation often span periods as long as several decades once the data are available [71,72]. In contrast, the time span of the hydrological modelling driven by the NWP-predicted precipitation is much shorter, as an NWP model commonly demands vast computational resources and computing time, particularly when it applies data assimilation and runs at a very fine grid spacing of 1 km [73–76]. Thus, to compare the effectivenesses of these different precipitation datasets on hydrological modelling, we focused our study on one short-term rainfall-runoff process and used it as a case study over the WJB watershed.

To select the study period, the rainfall events that occurred in the WJB watershed in 2015 were analysed based on the in situ daily precipitation data, which were collected from the 215 precipitation stations of the China Ministry of Water Resources (CMWR). The contributions of the accumulated daily precipitation to the annual precipitation were summed and sorted in decreasing order for each precipitation station (Figure 2a). All the 215 precipitation stations received half of the annual amount of precipitation within a minimum of 4 days (LiJi station) and a maximum of 14 days (Sanliping station). This suggests that a single heavy precipitation event is the main contributor to the amount of annual precipitation in the WJB watershed [77,78]. Because the forcing data of the applied NWP model, i.e., the final analysis (FNL) data ds083.3 were just released on 8 July 2015, we selected the specific heavy precipitation event from August, which is also in the flood season (i.e., June–September) in the WJB watershed. As shown in Figure 2b, there were two days of continuous heavy precipitation on 18 and 19 August in the WJB watershed, and the mean daily precipitation of the 215 precipitation stations reached the highest monthly value (27.4 mm) on 18 August. Thus, we chose the rainfall-runoff process caused by this heavy rainfall event as our study case. Moreover, considering the spin-up problem [79–81] of the NWP model and the natural rainfall-runoff process, we extended the study period to include the periods before and after the heavy rainfall event. Finally, a two-week period from 17 to 30 in August 2015 was selected as the study period.

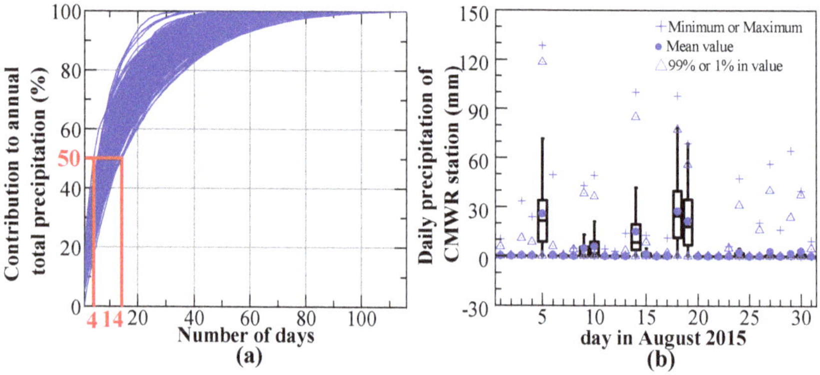

Figure 2. (**a**) Contribution (%) of the accumulated daily precipitation to the annual total precipitation for each precipitation station in the WJB watershed in 2015; the data were provided by the China Ministry of Water Resources (CMWR); (**b**) box plot * of daily precipitation from the 215 CMWR precipitation stations. * The lower and upper edges of the central box represent the first and third quartiles (25% and 75%, respectively), and the band inside the box represents 50%.

2.2. Study Data

2.2.1. In Situ Observed Precipitation

In this study, two types of in situ observed precipitation were used. One dataset was the precipitation measurements from the 14 meteorological stations within and near the WJB watershed (Figure 1b), which were provided by the China Meteorological Administration (CMA) (http://data.cma.cn). The daily CMA data are free to the public; they were downloaded and used in the calibrations and validations of the applied hydrological model. To improve the temporal resolution of the hydrological simulations, the hourly CMA data during the study period were also collected and used in the hydrological modelling. Another dataset was the precipitation measurements from the 215 precipitation stations in the WJB watershed (Figure 1c). This dataset was reported in the book of Annual Hydrological Report for the P.R. of China published by the CMWR (http://www.mwr.gov.cn). Although the CMWR data were only available in daily values, their observation network was much denser than that of the CMA data. Therefore, the CMWR data were not used for the hydrological modelling but rather applied in the selection of study case and the evaluation of the five precipitation datasets at a daily scale.

2.2.2. Remotely Sensed Precipitation

Generally, finer satellite precipitation resolutions result in higher accuracies [82–86], thus the recently released Integrated Multi-satellite Retrievals for GPM (GPM IMERG) with spatiotemporal resolutions of 0.1° and 30 min [82,87,88], and the merged CMORPH data with spatiotemporal resolutions of and 0.1° and one hour, were selected and employed in this study. The GPM IMERG was the third-level precipitation product of GPM, which covers an area of ±60°N/S. Tang et al. [70] concluded that, when compared with gauged observations, the Pearson's correlation coefficient (CC) values of the GPM IMERG over mainland China reached 0.53 and 0.71 at the hourly and daily timescales, respectively. The merged CMORPH was released by the CMA; it was produced by taking two algorithms of probability density function matching and optimal interpolation to merge the following two datasets: (1) the remote sensing precipitation product of the CMORPH data with spatiotemporal resolutions of 8 km and 30 min, which was released by the U.S. Climate Prediction Center [24]; and (2) the hourly in situ gauged precipitation data from more than 30,000–40,000 automatic weather stations in China after quality control.

2.2.3. NWP-Predicted Precipitation

In this study, the WRF model (version 3.7.1) and its WRF 4D-Var system were used to generate two types of the NWP-predicted precipitation data. The WRF model is a limited-area, non-hydrostatic, primitive-equation model with multiple options for various physical parameterization schemes. Since its release in May 2004, the WRF model has been widely used in atmospheric research and operational NWP user communities due to its advantages in terms of efficiency [89]. Generally, higher grid resolution in the WRF model can capture more local characteristic of precipitation, thus reduce the prediction biases [90–92]. Moreover, to avoid a scale mismatch between the WRF and the applied hydrological models, the nesting domain technique was used in the WRF model to dynamically downscale its input reanalysis data to a final resolution of 1 km. The nested domains were set around the WJB watershed (Figure 1a). The dominant parameters and physical configuration of the WRF model were set (Table 1), obeying the rule that the configuration should incorporate the experiences obtained from comparable atmospheric modelling studies as much as possible, especially the studies undertaken in the WJB watershed. To specify the initial state and the lateral boundary condition of the WRF model, the National Center for Environment Prediction (NCEP) FNL ds083.3 dataset (http://rda.ucar.edu/) was applied as forcing data. The spatiotemporal resolutions of the FNL data are 0.25° and 6 h; they are available from 8 July 2015 to a near-current date. This dataset is made with the same model which NCEP uses in the Global Forecast System (GFS) and obtained from the Global Data Assimilation System (GDAS), which continuously collects observational data from the Global Telecommunications System (GTS) and other sources for related analyses.

Table 1. The main configuration of the WRF model.

Map and Grids	
Map projection	Lambert conformal
Centre point of domain	35.8°N, 114°E
Number of vertical layers	27
Horizontal grid spacing	27 km, 9 km, 3 km, 1 km
Grids	180 × 155, 322 × 271, 604 × 433, 700 × 700
Static geographical fields	Standard dataset at 30″ resolution from the United States Geological Survey (USGS)
Time step	150 s, 50 s, 17 s, 6 s
Physical Parameterization Schemes	
Cloud microphysics	WRF double-moment 6 scheme [93]
Long-wave radiation	Rapid radiative transfer model (RRTM) [94]
Short-wave radiation	Dudhia scheme [95]

Table 1. *Cont.*

Land-surface model	Noah land-surface model (LSM) [96]
Planetary boundary layer	Yonsei University scheme [97]
Cumulus parameterization	New Grell-Devenyi 3 scheme [98] (only in D01)

Furthermore, considering the lower accuracy of the WRF precipitation simulation, the WRF 4D-Var data assimilation system [99–102] was also applied to improve the initial and boundary conditions to improve the accuracy of the precipitation prediction. The GPM IMERG was selected as the observation operator because of its recent release, wide coverage, free download and high accuracy [70]. Moreover, the feasibility of assimilating the GPM IMERG into the WRF model has been demonstrated in Yi et al. [18]. The GPM IMERG was accumulated into 6 h values and assimilated into the WRF 4D-Var system to improve the initial condition at every day 00 UTC. With the improved condition, a subsequent 24-h forecast was then adopted. This means that there were 14 independent WRF 4D-Var simulations during the entire 14 days of the study period. The main configuration of the WRF 4D-Var system was the same as that in the WRF model. To make the WRF 4D-Var convergence criterion more stringent, an EPS variable of 0.0001 was used. The key background error covariance matrix for the 4D-Var data assimilation was domain-specific; it was generated based on the 1-month-long (August) ensemble simulations, which were performed every 12 h using the National Meteorological Center (NMC) method [103].

2.2.4. Soil Moisture and Outlet Discharge

The measurement network for soil moisture was sparse in the WJB watershed and the gauged data were unavailable, so we used a remote sensing soil moisture product to assess the soil moisture simulated during the hydrological modelling [62,63]. The satellite-based soil moisture was chosen obeying the following rules: choose the soil moisture data with finer resolution which are generally considered to have higher accuracy [86]. Therefore, we selected the Soil Moisture Active Passive (SMAP) product with spatiotemporal resolutions of 9 km and 3 h as benchmark in the soil moisture evaluations. The SMAP mission was launched by the National Aeronautics and Space Administration (NASA); it takes advantage of the relative strengths of both active (radar) and passive (radiometer) microwave remote sensing to obtain an intermediate level of accuracy and resolution for soil moisture mapping. Among its 15 data products with different levels of data processing, the SMAP Level-4 Surface and Root-Zone Soil Moisture (L4_SM) data (version 3) were employed. L4_SM is generated by the NASA catchment land surface model, and it mainly assimilates the SMAP 9-km active-passive (AP) soil moisture product L2_SM_AP, which combines radar and radiometer measurements. It is gridded using an Earth-fixed, global, cylindrical equal-area scalable Earth grid, and version 2.0 (EASE-Grid 2.0). Reichle et al. [104] assessed the accuracy of the L4_SM product and concluded that it met the soil moisture accuracy requirements specified as an unbiased RMSE of 0.04 m^3 m^{-3} or better. The L4_SM data are available from 31 March 2015 to the present (within 3 days from real-time) and provide estimates of the surface (0–5 cm) and root-zone (0–100 cm) soil moisture values. Hereafter, the SMAP soil moisture mentioned in the manuscript refers to the data from the SMAP L4_SM product at the root zone. The hourly discharges observed at the watershed outlet in the study period were provided by the China Institute of Water Resources and Hydropower Research (IWHR).

3. Methods

3.1. Hydrological Model

To reflect the impacts of spatial characteristics of rainfall on hydrological modelling, the semi-distributed hydrological model TOPX was used in this study. The TOPX model is constructed on the basis of the topographic index (TOP) and the water balance concept of the Xin'anjiang model (X). It applies the improved simple TOPMODEL (topography-based hydrological model)-based runoff parameterization (SIMTOP) [105], and the methods of empirical unit hydrograph, linear reservoir equation, and Muskingum for its routings of overland flow, base flow, and channel flow, respectively [106,107]. Apart from some physical parameters, the input data of the TOPX model including precipitation, potential evaporation, topographic index (TI) are required as grid-based data, so as to reflect of the spatial variations of precipitation, evaporation, and topography, thus facilitate reflecting their subsequent impacts on water exchange, soil moisture variation, runoff generation, runoff routing and outlet discharge simulation.

When applying the TOPX model, the 1-km grid data of precipitation and potential evaporation were obtained from the 14 CMA meteorological stations and the 10 CMWR evaporation stations (Figure 1c), respectively. The data of TI were calculated with the method posed by Yi et al. [106] which considered the impacts of both topography and soil properties on hydrological processes. The TI was computed based on the digital elevation model (DEM) that was downloaded from the official website of the United States Geological Survey (USGS) (https://www.usgs.gov/) and the Harmonized World Soil Database (HWSD) obtained from the Cold and Arid Regions Sciences Data Center at Lanzhou (http://westdc.westgis.ac.cn/). The results of the used 1-km grid TI data and the soil type classification from the HWSD in the WJB watershed are shown in Figure 3.

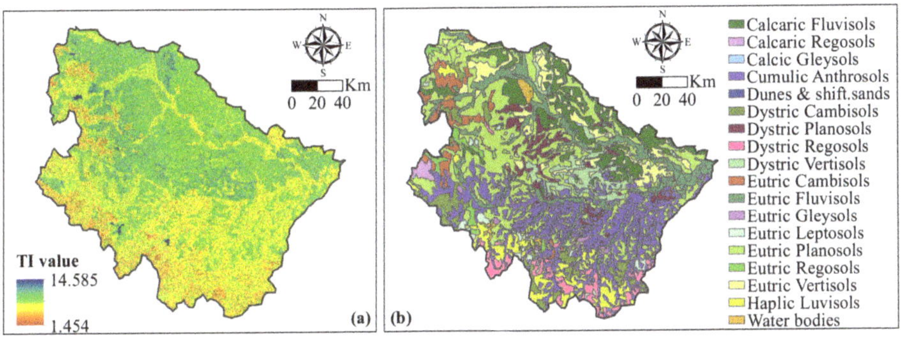

Figure 3. (**a**) The 1-km grid data of the topographic index (TI) of the WJB watershed; and (**b**) the applied soil type classification of the Harmonized World Soil Database (HWSD) used in the calculation of TI.

To achieve better compatibilities of the TOPX model in the WJB watershed, the model was calibrated and validated with the available daily data using the trial-and-error method. During calibration and validation, both the long-term rainfall-runoff process and the short-term flood process were simulated, as the latter can revise the parameters that are determined based on the former [108–110]. The long-term rainfall-runoff processes covered the periods from 2001 to 2005 and from 2014 to 2015; the 7 short-term flood events were selected from the long-term rainfall-runoff processes. As shown in Figure 4, the simulated recession curves of several selected short-term flood events decay faster than those of the observations. These fast-decay recession curves and their subsequent shorter flood durations were mainly related to the lack of consideration of the reservoir storage capacity in the TOPX model. Because we focused on the differences of model performance caused by the different precipitation inputs, such limitations of the TOPX model were not considered in the investigation. The statistical results showed that the Nash-Sutcliffe coefficient (*NS*) [111] of the TOPX model in the WJB watershed was as low as 0.700 (Table 2).

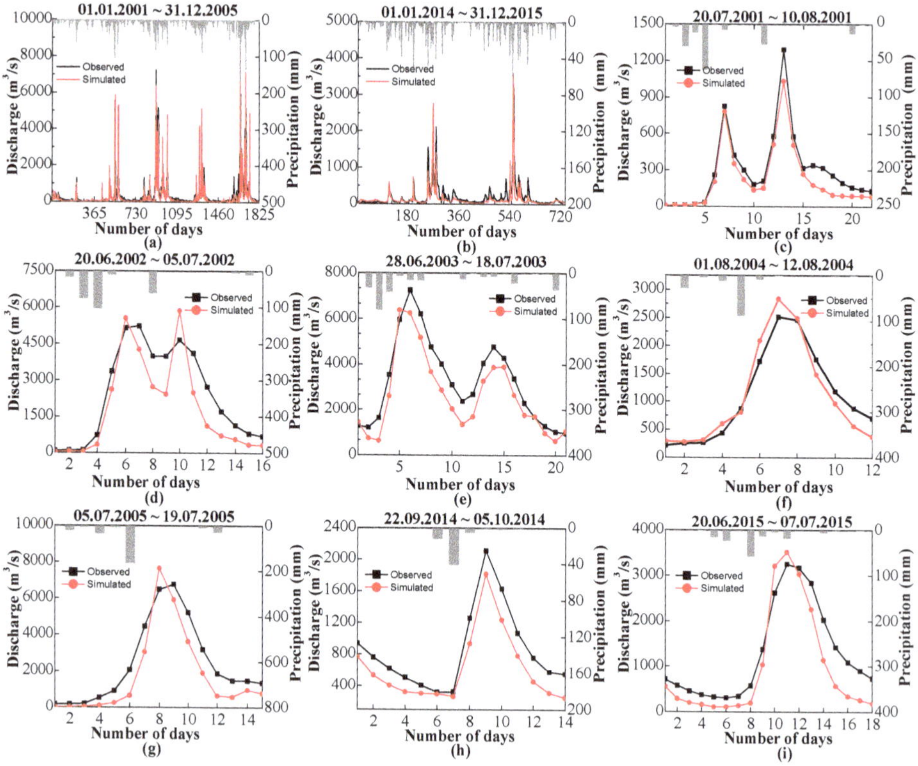

Figure 4. Long-term calibration (**a**) and validation (**b**) of the TOPX model in the WJB watershed and its short-term calibration (**c**–**g**) and validation (**h**,**i**). Discharge (mm^3/s) in the figure indicates discharge at the WJB hydrology station; precipitation (mm) denotes the average precipitation in the WJB watershed that was calculated based on the CMA observations.

Table 2. Statistical results of the simulated daily discharges in the WJB watershed during the calibrations and validations of the hydrological model TOPX *.

		Simulation Time	*NS*	*CC*	*RE*
Long-term	Calibration	1 January 2001~31 December 2005	0.787	0.898	−0.292
	Validation	1 January 2014~31 December 2015	0.700	0.857	−0.248
Short-term	Calibration	20 July 2001~10 August 2001	0.894	0.982	−0.241
		20 June 2002~05 July 2002	0.747	0.921	−0.236
		28 June 2003~18 July 2003	0.815	0.967	−0.194
		01 August 2004~12 August 2004	0.919	0.968	−0.012
		05 July 2005~19 July 2005	0.806	0.954	−0.281
	Validation	22 September 2014~05 October 2014	0.755	0.985	−0.266
		20 June 2015~07 July 2015	0.772	0.953	−0.251

* The indices of *NS*, *CC* and *RE* denote Nash-Sutcliffe coefficient, Pearson's correlation coefficient and relative error, respectively.

3.2. Experimental Design

When assessing the effectivenesses of the different precipitation datasets, the studied two-week long rainfall-runoff process was simulated with the TOPX model at an hourly scale. Therefore, a slight tuning of the model parameters calibrated at a daily scale was done to adapt the hourly simulations. The final parameters of the TOPX model used in the simulations are shown in Table 3.

Table 3. Main parameters in the hydrological model TOPX.

Parameter	Value
Decay factor (f, m^{-1})	160
Evapotranspiration coefficient (E)	0.95
Initial soil moisture content (W, mm)	90
Impact factor of vegetation root (C)	0.12
Maximum subsurface runoff ($R_{sb,max}$, mm)	80

To simulate the study case at spatiotemporal resolutions of 1 km and one hour, the hourly CMA observations were interpolated to 1-km grid data using the inverse distance weighted (IDW) algorithm, since the IDW method can furthest reflect the impact of each station observation on the interpolated point through distance weighting [112]. The 30-min GPM IMERG data were firstly accumulated to hourly grid data, then resampled from 0.1° to a 1-km resolution with the operationally used algorithm of bilinear interpolation [113]. The hourly merged CMORPH data were also resampled from 0.1° to a 1-km resolution with the bilinear interpolation algorithm. The 1-km grid data of precipitation obtained from the WRF model and the WRF 4D-Var system data were accumulated from 6 s to hourly values. Moreover, to unify the time zones of these precipitation datasets and make them consistent with the observed outlet discharges, all the employed precipitation data were adjusted to the same time zone as that of the observed outlet discharge data, i.e., Beijing time. According to the different precipitation inputs, five experiments of rainfall-runoff simulation were performed and labelled as P_CMA, P_GPM, P_CMOR PH, P_WRF and P_4D-Var, respectively (Table 4).

Table 4. Experimental design for the rainfall-runoff simulations with different precipitation datasets.

No.	Experiment Label	Precipitation Category	Precipitation Data	Resolution of Precipitation Data
1	P_CMA	In situ observation	CMA data	Point scale, hourly
2	P_GPM	Remote sensing estimation	GPM IMERG	0.1°, 30-min
3	P_CMORPH		Merged CMORPH	0.1°, hourly
4	P_WRF	NWP simulation	WRF output	1 km, 6-s
5	P_4D-Var		WRF 4D-Var output	1 km, 6-s

3.3. Evaluation of Precipitation, Soil Moisture and Outlet Discharge

Before the five hydrological modelling experiments, the precipitation input of the TOPX model, i.e., the 1-km grid data obtained from the five different precipitation datasets, were evaluated with the CMWR in situ data. Firstly, the hourly precipitation values were extracted from the grid points nearest to the 215 CMWR precipitation stations and accumulated to daily values, then compared to the daily CMWR station observations. For this point-scale evaluation, we used the error scores of the mean error (ME), relative error (RE), root mean square error ($RMSE$) and CC (Table 5), which describe the errors, deviations and correlation between the simulated data and the reference data, respectively. Secondly, as the studied rainfall-runoff process was triggered by the heavy precipitation on 18 and 19 August, the accuracies of the different heavy precipitation accumulated in these two days were also evaluated. The daily 1-km grid data of the CMWR data processed with the IDW method were accumulated in the two days. The 1-km grid data of the CMA, GPM IMERG, merged CMORPH, WRF model and WRF 4D-Var were accumulated in those two days as well and compared to the accumulated CMWR data. For this field-scale evaluation, the skill scores of the bias score ($BIAS$), false alarm ratio (FAR), probability of detection (POD) and threat score (TS) (Table 5) were used. These skill scores were constructed based on the "contingency table" [114]. $BIAS$ is an indicator of how well the estimation covers the number of occurrences of an event. FAR is the fraction of "yes" estimation that turns out to be wrong. POD is the ratio of correct estimations to the number of times the event occurred; it is commonly known as the hit rate. TS is one of the most frequently used and comprehensive skill scores for summarizing square contingency tables, and this metric combines the characteristics of hints and random detections. The equations and the perfect values of these evaluation indices are listed in Table 5.

Table 5. Statistical metrics applied in the evaluations *.

Statistical Metric	Equation	Perfect Value
Mean error (ME, mm)	$ME = \frac{1}{N}\sum_{i=1}^{N}(P_{P,i} - P_{O,i})$	0
Relative error (RE)	$RE = \frac{\sum_{i=1}^{N}(P_{P,i} - P_{O,i})}{\sum_{i=1}^{N} P_{O,i}}$	0
Root mean square error ($RMSE$, mm)	$RMSE = \sqrt{\frac{1}{N}\sum_{i=1}^{N}(P_{P,i} - P_{O,i})^2}$	0
Correlation coefficient (CC)	$CC = \frac{\sum_{i=1}^{N}(P_{P,i}-\overline{P}_{P,i})(P_{O,i}-\overline{P}_{O,i})}{\sqrt{\sum_{i=1}^{N}(P_{P,i}-\overline{P}_{P,i})^2 \sum_{i=1}^{n}(P_{O,i}-\overline{P}_{O,i})^2}}$	1
Bias score ($BIAS$)	$BIAS = \frac{(A+B)}{(A+C)}$	1
False alarm ratio (FAR)	$FAR = \frac{B}{(A+B)}$	0
Probability of detection (POD)	$POD = \frac{A}{(A+C)}$	1
Threat score (TS)	$TS = \frac{A}{(A+B+C)}$	1
Nash-Sutcliffe coefficient (NS)	$NS = 1 - \frac{\sum_{i=1}^{N}(Q_{obs,i}-Q_{sim,i})^2}{\sum_{i=1}^{N}(Q_{obs,i}-\overline{Q}_{obs})^2}$	1

* $P_{P,i}$ and $P_{O,i}$ denote the simulated and observed values, respectively, of the i grid, and $\overline{P}_{P,i}$ and $\overline{P}_{O,i}$ are their respective means. A represents the precipitation predicted by the WRF and observed by the reference data; B represents the precipitation predicted by the WRF but not observed by the reference data; C represents the precipitation not predicted by the WRF but observed by the reference data; D represents the precipitation not predicted by WRF and not observed by the reference data. $Q_{obs,i}$ and $Q_{sim,i}$ are the observed and simulated values at time i; $\overline{Q}_{obs}$ is the average value of the observations during the simulation period.

Because the soil area defined in the TOPX model extends to the root zone, we used the data in the root zone of L4_SM for the soil moisture evaluation. To compare with the SMAP soil moisture, the hourly soil moisture simulated by the P_CMA, P_GPM, P_CMORPH, P_WRF and P_4D-Var experiments were accumulated every 3 h and interpolated with kriging method to a 9-km resolution, which were the same with the spatiotemporal resolutions of the SMAP. The units of the soil moisture obtained from the TOPX model and the SMAP product are different, the former applies the depth of the water column (mm), and the latter employs the water volume content (mm^3/mm^3). Therefore, we indirectly evaluated the simulated soil moisture with the SMAP data from the perspective of relativity, which is a frequently used method in the evaluation of soil moisture [115,116]. The CC and standard deviation were employed to investigate the relationship between the simulated soil moisture and the SMAP soil moisture, and t-test was used to determine the statistical significance (p) of the correlation. In this study, we defined the CC as significant when its corresponding p was less than 0.05 [117]. The NS, CC and RE were applied to evaluate the simulated outlet discharge.

4. Results

4.1. 1-km Grid Data of Different Precipitation Datasets

The 1-km grid of the hourly data obtained from the CMA, GPM IMERG, merged CMOPRH, WRF model and WRF 4D-Var system were accumulated to the amount of total precipitation during the entire study period, the daily CMWR data were also accumulated in this way. As portrayed in Figure 5, the spatial heterogeneity of the CMWR data was very obvious (Figure 5a) because its observation network was the highest. In contrast, because of much fewer observation stations, the spatial variation of the CMA data was relatively homogeneous (Figure 5b), which was also reflected in its lowest standard deviation (9.0). The downscaled results of the GPM IMERG (Figure 5c) and the merged CMORPH (Figure 5d) still showed evident grid characteristics because of their coarser resolution of 0.1°. The watershed mean values of the latter were obviously higher than that of the former. Having the finest spatial resolution (1 km), the distributions of the precipitation simulated by the WRF model and the WRF 4D-Var system were the most continuous. Figure 5e shows that the WRF-predicted precipitation is focused in the middle and south of the WJB watershed, and its maximum (185.9 mm) and average (68.2 mm) watershed values are the highest among the studied precipitation datasets. After data assimilation with the GPM IMERG, the heavy precipitation predicted by the WRF 4D-Var system moved from the middle to the south (Figure 5f), and the average precipitation amount in the watershed decreased from 68.2 mm to 32.2 mm.

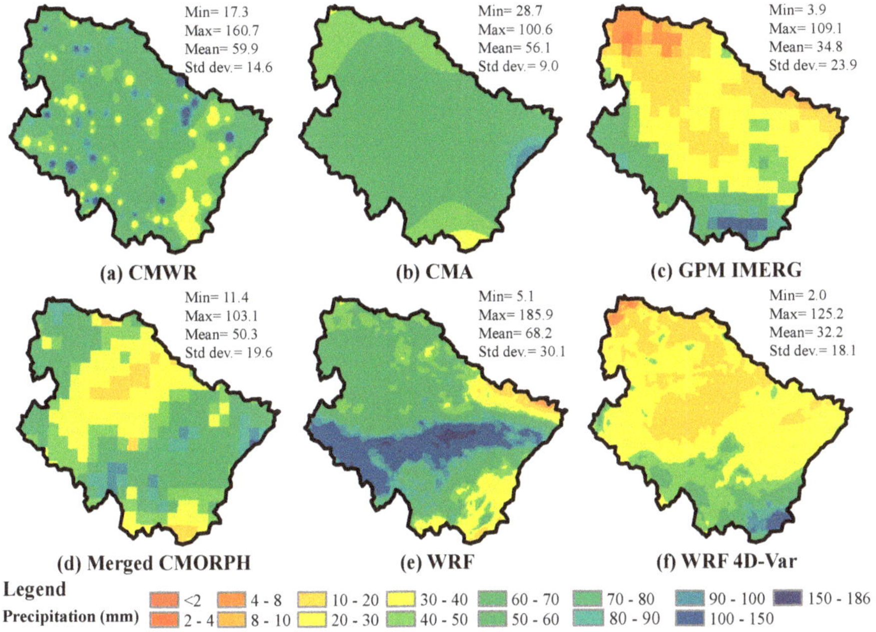

Figure 5. 1-km grid data of the total precipitation (mm) obtained from the precipitation datasets of the CMWR (**a**), CMA (**b**), GPM IMERG (**c**), merged CMORPH (**d**), WRF model (**e**) and WRF 4D-Var system (**f**). The minimum (Min), maximum (Max), mean values and the standard deviation (Std dev.) of the total precipitation are listed in the upper right corner of each graph.

The watershed average values of the different hourly precipitation datasets are portrayed in Figure 6a. They generally showed similar rainfall tendencies. Their main differences existed in terms of the times that the peaks occurred and the magnitudes of the peak flow. Figure 6b clearly shows that the watershed mean values of the daily precipitation data from the CMA, merged CMORPH, WRF model and WRF 4D-Var system present concentrated rainfall on 19 August, but the datasets of the CMWR and the GPM IMERG present the rainfall event on 18 and 19 August. In terms of the heavy rainfall, the watershed average values of the daily GPM IMERG showed evident underestimations. In contrast, the daily WRF simulations showed obvious overestimations. The WRF 4D-Var-predicted daily precipitation was more sharply reduced than the WRF-predicted precipitation after assimilating the GPM IMERG.

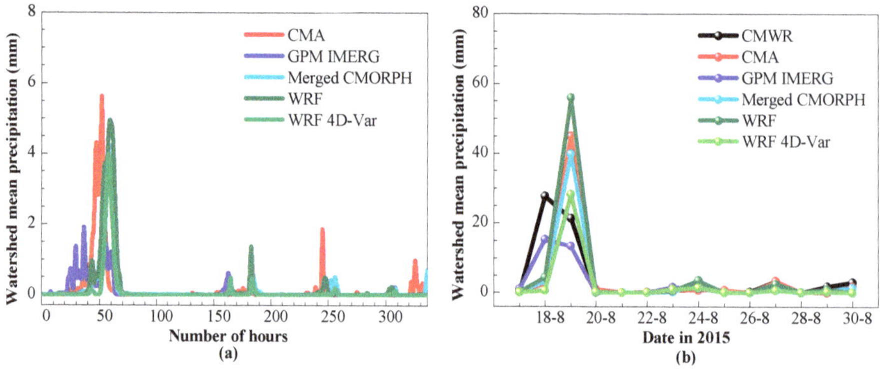

Figure 6. Averages of the hourly (**a**) and daily (**b**) precipitation (mm) in the WJB watershed obtained from different precipitation datasets during the study period.

4.2. The Simulated Soil Moisture

The 3-h averaged soil moisture simulated by the TOPX model with different precipitation datasets is shown in Figure 7. It was clear that the spatial distributions of these soil moisture all kept accordance with their corresponding precipitation fields (Figure 5). This suggested that the spatial variation of precipitation dominantly influenced the spatial distribution of soil moisture. The soil moisture generated in the P_CMORPH experiment was generally higher than that generated in the P_GPM experiment (Figure 7c,d), which was consistent with the results that the watershed average precipitation from the CMORPH was higher than that from the GPM IMERG (Figure 6). Because of the reduction of the WRF 4D-Var-predicted precipitation after assimilating the GPM IMERG, its simulated soil moisture decreased as well compared to that of the P_WRF experiment (Figure 7e,f). The WJB watershed mean values of the 3-h soil moisture simulated by the P_CMA, P_GPM, P_CMORPH, P_WRF and P_4D-Var experiments were 95.927 mm, 87.110 mm, 92.321 mm, 94.513 mm and 86.461 mm, respectively, and the watershed mean value of the SMAP soil moisture was 0.286 mm^3/mm^3.

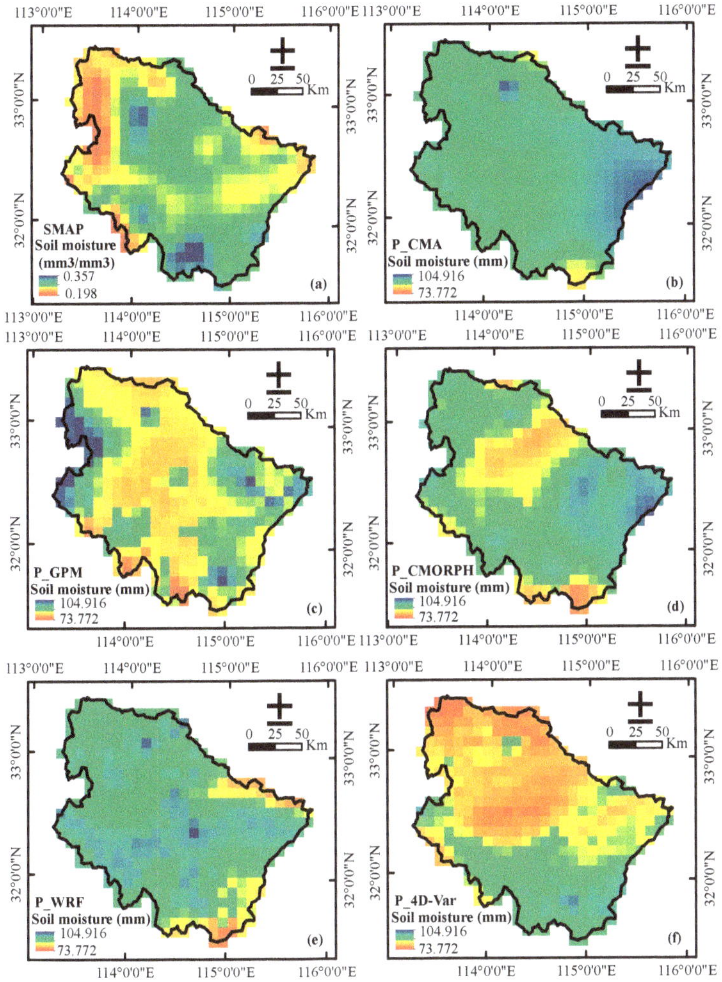

Figure 7. 9-km grid data of the average 3-h soil moisture in the WJB watershed extracted from the SMAP product (**a**) and the P_CMA (**b**), P_GPM (**c**), P_CMORPH (**d**), P_WRF (**e**), and P_4D-Var (**f**) experiments during the study period.

The watershed average values of the 3-h and daily soil moisture were calculated based on the hourly simulations of the five experiments. Figure 8 clearly shows that the SMAP soil moisture reflects a good response to the CMA-recorded precipitation process. After two days of continuous rainfall on 18 and 19 August, the watershed mean value of the SMAP soil moisture started to increase, reached its maximum on 20 August and subsequently decreased as the rainfall ceased. Compared to the SMAP soil moisture, the simulated soil moisture generally showed similar variation tendencies. Because the different precipitation peaked at different times, the different simulated soil moisture reached the highest values at different hours. The watershed mean soil moisture simulated by the P_CMA, P_CMORPH and P_WRF experiments was very similar. For the overestimation of the WRF-predicted precipitation, the soil moisture simulated by the P_WRF experiment showed the highest peak value. The soil moisture simulated by the P_GPM experiment exhibited very low values because the GPM IMERG precipitation was generally underestimated. With the assimilation of the GPM IMERG, the soil moisture simulated by the P_4D-Var experiment was evidently lower than that simulated by the P_WRF experiment, and became much closer to the simulated results of P_GPM.

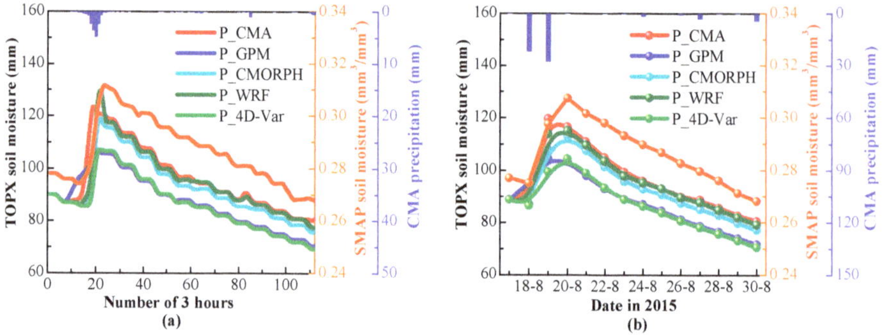

Figure 8. The 3-h (**a**) and daily (**b**) mean values of the WJB watershed soil moisture obtained from the SMAP (mm^3/mm^3) and the different rainfall-runoff experiments based on the TOPX model (mm).

4.3. The Simulaed Outlet Discharge

The simulated hourly discharges and the accumulated daily discharges at the WJB watershed outlet from the five experiments suggested similar tendencies with the observed discharges (Figure 9a,b). The peaks of the hourly simulated discharges appeared at different time due to the different occurrences of the heavy precipitation. The discharges simulated by the P_WRF experiment were evidently larger than other observed and simulated results, as the WRF-predicted precipitation was overestimated. In contrast, the discharges simulated by the GPM IMERG were generally lower than the observed discharges for its underestimation. After assimilating the GPM IMERG, the P_4D-Var experiment yielded better simulations than the P_WRF experiment. The accumulated discharges of the five experiments are shown in Figure 9c,d. It was shown that the P_CMORPH experiment generated an obviously better simulation of the outlet discharges than the P_GPM experiment. As time passed, the accumulated discharges from the P_4D-Var experiment were closer to the accumulated observations than that from the P_CMORPH experiment; this may be caused by that the overestimations of the P_4D-Var experiment after the rainfall supplemented its underestimations before the rainfall. The discharges extracted from the P_CMA experiment were the closest to the observations because of better precipitation. For the hourly accumulated results, the total simulated discharges during the entire study period for the P_CMA, P_GPM, P_CMORPH and P_4D-Var experiments were 0.328%, 13.680%, 5.667%, and 0.973% lower than the total observed discharge, respectively, and 4.849% higher for the P_WRF.

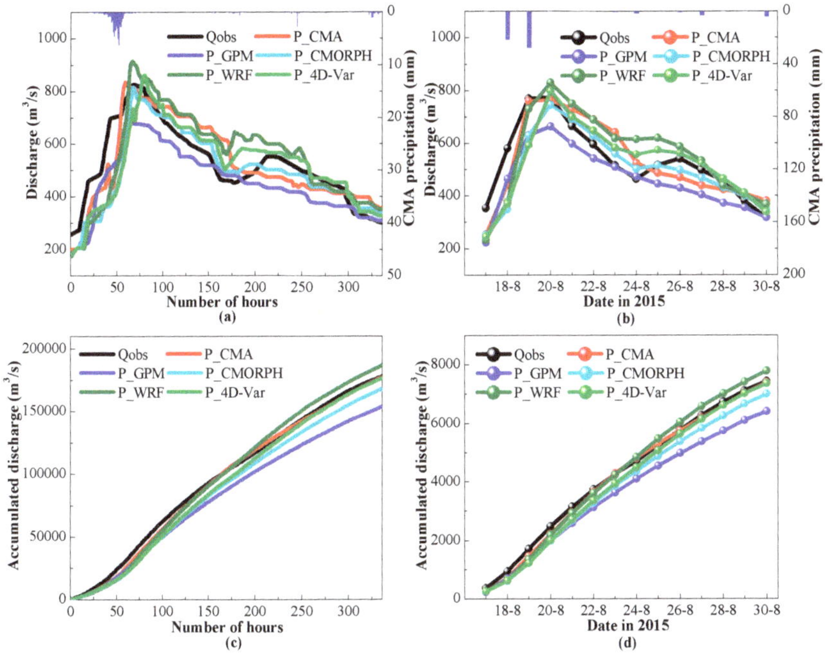

Figure 9. The results of the simulated hourly discharges (**a**), daily discharges (**b**), accumulated hourly discharges (**c**) and accumulated daily discharges (**d**) at the WJB watershed outlet.

5. Discussion

5.1. Evaluation of the Different Precipitation Datasets

The results of the point-scale evaluation between the daily data from the five different precipitation datasets and the daily CMWR in situ observations are shown in Table 6. It is clear that the *ME* and *RE* of the WRF precipitation predictions are 0.159 mm and 0.037, respectively, while the *MEs* and *REs* for the other precipitation obtained from the CMA, GPM IMERG, merged CMORPH and WRF 4D-Var are all negative. This indicated that the daily precipitation predicted by the WRF model was overestimated. The WRF predicted daily precipitation also had the highest *RMSE* (13.411), which presented the highest deviation from the CMWR data. The *CC* value of the WRF daily precipitation is only 0.343. The daily GPM IMERG has the best *CC* value (0.493). The *CC* values of the daily precipitation from the CMA, merged CMORPH and WRF 4D-Var system all passed the level of 0.4.

Table 6. Error scores of mean error (*ME*), relative error (*RE*), root mean square error (*RMSE*) and Pearson's correlation coefficient (*CC*) for the point-scale precipitation evaluations between the daily CMWR data and the daily grid data with 1-km resolution from different precipitation data.

Error Score	Precipitation Data				
	CMA	GPM IMERG	Merged CMORPH	WRF Output	WRF 4D-Var Output
ME	−0.424	−1.489	−0.805	0.159	−1.868
RE	−0.099	−0.347	−0.188	0.037	−0.436
RMSE	11.719	10.112	11.577	13.411	11.946
CC	0.459	0.493	0.433	0.343	0.444

The results of the field-scale evaluation between the two-day accumulated heavy precipitation of the five different precipitation and that of the CMWR data are shown in Figure 10. It clearly shows that as the precipitation threshold increases, the skill scores deviate further from their perfect values. This indicated that all the five applied datasets had great difficulty in estimating heavy precipitation. The CMA data had equivalent values with the CMWR data as the *BIAS* was near 1 before the threshold of 40 mm; however, it then started to decrease which indicated that the CMA data underestimated the heavy rain. For the GPM IMERG data, before the threshold of 70 mm, the rainfall estimations were always lower than those of the CMWR data; however, after the threshold of 70 mm, the rainfall estimations began to be overestimated, and then the bias began to decrease after the threshold surpassed 95 mm. This reflected that the GPM IMERG data were underestimated for most of the thresholds. The merged CMORPH data had the lowest deviation from the CMWR data; the *BIAS* values were generally lower and stayed nearest to 1 before the threshold of 90 mm. The underestimations of the heavy rainfall from the GPM IMERG and the merged CMORPH were related to the weak ability of detector to measure heavy precipitation under complicated atmospheric conditions, and the homogenization of heavy rainfall by coarse grids [82,118]. The *POD*, *FAR* and *TS* values of the merged CMORPH were generally higher than the GPM IMERG; this showed better estimation of the heavy precipitation, possibly related to its gauge correction. For the WRF prediction, its *BIAS* values showed a significant deviation from the reference value of 1, and an obvious precipitation overestimation can be found above the threshold of 50 mm. However, after assimilating the GPM IMERG, the overestimations of the WRF model for the heavy rain were well controlled, and its *BIAS*s were obviously reduced. The skill scores of the WRF 4D-Var-predicted precipitation were closer to the scores of the GPM IMERG data. This indicated that the assimilated data were the key factor to affect the final assimilation results. The generally underestimations of the heavy rain in the GPM IMERG resulted in the weaker detection of heavy rain in the WRF 4D-Var system. At the threshold of 50 mm, the *TS* values for the heavy precipitation obtained from the CMA, GPM IMERG, merged CMORPH, WRF model and WRF 4D-Var system were 0.267, 0.096, 0.216, 0.364, and 0.032, respectively.

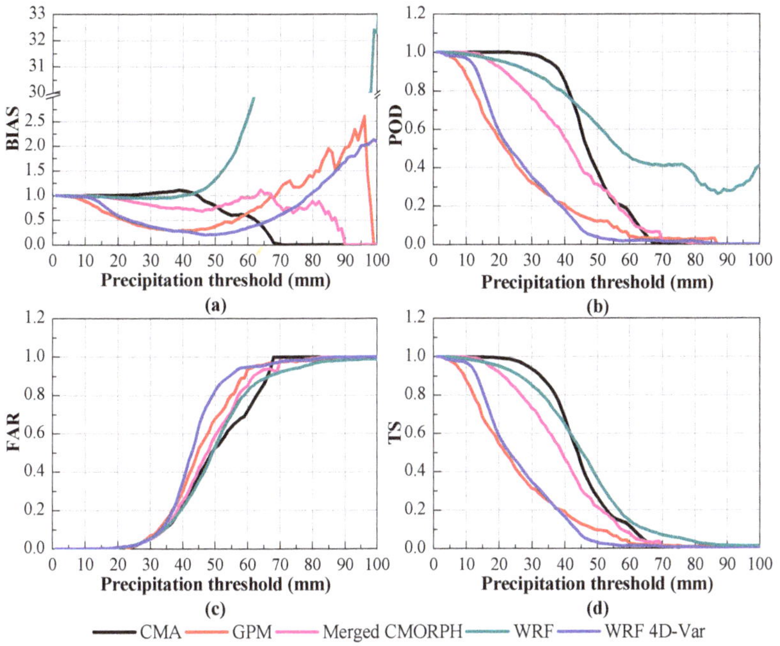

Figure 10. Skill scores of *BIAS* (**a**), *POD* (**b**), *FAR* (**c**) and *TS* (**d**) for the grid-scale evaluation of the accumulated heavy precipitation between the CMWR data and the other data from the CMA, GPM IMERG, merged CMORPH, WRF model and WRF 4D-Var system.

Based on the point-scale and field scale evaluation of the five precipitation datasets, it was concluded that although having coarser spatial resolution, the CMA data comprehensively had the best accuracy since they were in situ gauged observations. Because of gauge correction, the accuracy of the hourly merged CMORPH was better than the 30-min GPM IMERG. Although having the highest spatiotemporal resolutions, the accuracy of the WRF-predicted precipitation was worst because of uncertainties primarily introduced from its incomplete forcing data. The 4D-Var data assimilation could effectively improve the accuracy of the WRF prediction. However, the WRF 4D-Var-predicted precipitation was still not good, because the quality of the assimilated GPM IMERG was poor during the entire study period in the WJB watershed.

5.2. Evaluation of the Simulated Soil Moisture

The evaluated results of the 3-h averaged simulated soil moisture for the five experiments are shown in Figure 11. It is shown that most *CC* values between the simulated soil moisture and the SMAP soil moisture surpassed 0.6. The lower *CC* values of each experiment were mainly concentrated in the middle part of the WJB watershed, which was near the outlet, and this possibly resulted from the convergence scheme of the TOPX model. There were 412 grids with a 9-km resolution joined the correlation statistics in the WJB watershed. Except for the P_GPM and P_4D-Var experiments, 100% of the total grids for the other three experiments had *CC* values that were statistically significant at the 0.05 level. The 12 grids with non-significant *CC* values in the P_GPM experiment (Figure 11b) may be caused by the precipitation underestimation of the GPM IMERG.

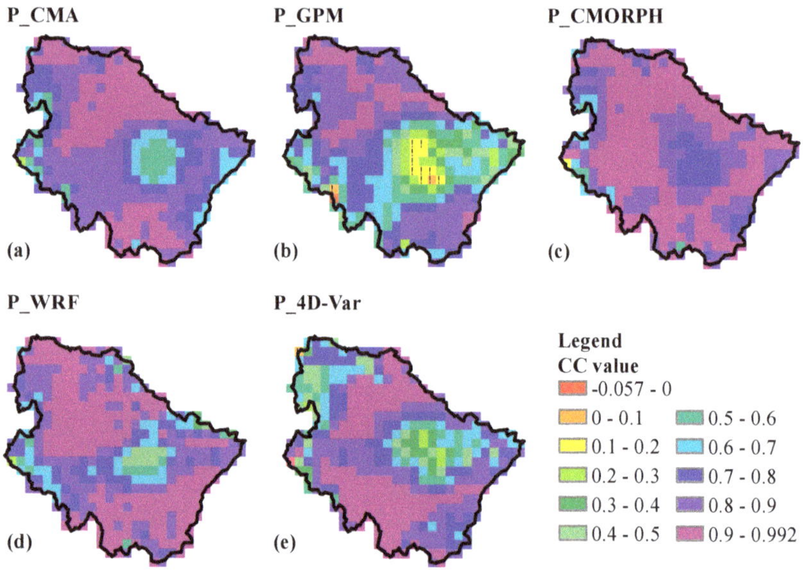

Figure 11. Pearson's correlation coefficient (*CC*) between the SMAP soil moisture and the soil moisture simulated by the rainfall-runoff experiments of P_CMA (**a**), P_GPM (**b**), P_CMORPH (**c**), P_WRF (**d**) and P_4D-Var (**e**) at 3-h intervals. The point denotes the *CC* value, which is not statistically significant at the level of 0.05.

The statistics of the *CC* values of the 3-h and daily mean soil moisture for the five experiments are listed in Table 7. It is shown that at the 3-h timescale, the P_CMORPH experiment had the highest mean value of 0.885 and the highest maximum value of 0.992, while the *CC* of the P_GPM showed the lowest value of 0.695. This indicated that the merged CMORPH precipitation data could yield better soil moisture simulations than could the GPM IMERG, because of the accuracy improvement of the merged CMORPH for merging gauged data. The negative values of the minimum *CC* appeared in the P_GPM and P_4D-Var experiments, and their mean *CC* values were also lower than those of the other experiments. These differences were caused by the obvious underestimation of the GPM IMERG. At the daily scale, the minimum values of the *CC* were all significantly improved except for that of the P_WRF, and the maximum and mean values of the *CC* in all five experiments showed the same variation as that at the 3-h scale. The good correlations between the simulated soil moisture and the SMAP soil moisture ensured the rationalities of the subsequent simulations of runoff and outlet discharge.

Table 7. Statistics of Pearson's correlation coefficient between the SMAP soil moisture and the soil moisture simulated by the rainfall-runoff experiments at 3-h and daily time scales.

No.	Experiment	Minimum 3 h/Daily	Maximum 3 h/Daily	Mean Value 3 h/Daily	Standard Deviation 3 h/Daily
1	P_CMA	0.256/0.504	0.963/0.986	0.825/0.858	0.111/0.114
2	P_GPM	−0.020/−0.096	0.967/0.986	0.695/0.705	0.208/0.235
3	P_CMORPH	0.197/0.477	0.992/0.996	0.885/0.891	0.091/0.093
4	P_WRF	0.259/0.186	0.986/0.992	0.841/0.852	0.132/0.136
5	P_4D-Var	−0.057/0.067	0.983/0.993	0.774/0.780	0.183/0.194

5.3. Evaluation of the Simulated Outlet Discharge

The evaluated results of the simulated hourly and the accumulated daily outlet discharge from the five rainfall-runoff experiments are shown in Table 8. Compared to the evident differences among the five precipitation datasets, the discrepancies among the five hydrological modelling experiments driven by them were narrowed. This may be influenced by the watershed size, because larger watershed generally had a lower magnitude-of-difference compared to the smaller watershed [119–122]. Table 8 shows that the *CC* values between the simulated discharges and the observed discharges are all above 0.787 and 0.796 at the hourly and daily timescales, respectively. Except for the *RE* of the P_WRF experiment, the *REs* of the other experiments were all negative, which means that the simulated outlet discharges were underestimated overall. Considering the comprehensive evaluation index *NS*, the P_CMA, P_GPM, P_CMORPH, P_WRF and P_4D-Var experiments reached 0.658, 0.576, 0.596, 0.464 and 0.547, respectively, on an hourly scale. These *NS* values indicated that for the rainfall-runoff simulations in the WJB watershed, the in situ CMA precipitation data had the best effectiveness in driving the TOPX model. As for the general underestimations of the heavy precipitation from the GPM IMERG and the merged CMORPH, their simulated discharges were lower than the observed values, there *RE* values were both negative. Because of better accuracy in estimating precipitation, especially for heavy precipitation, the P_CMORPH experiment performed better than the P_GPM experiment. The *NS* of the P_4D-Var experiment was 0.083 (hourly) and 0.061 (daily) higher than that of the P_WRF experiment, and it was clear that the 4D-Var data assimilation with the GPM IMERG improved the hydrological modelling performance through enhancing the accuracy of the WRF 4D-Var-predicted precipitation. However, the performance of the P_4D-Var experiment was only higher than the P_WRF experiment; this mainly resulted from the poor quality of the GPM IMERG over the study period in the WJB watershed. It was concluded that the model performance was affected by the precipitation accuracy to a large extent; the higher the precipitation accuracy, the better the model performed.

Table 8. Evaluation results of the hourly and daily outlet discharges simulated by the different rainfall-runoff experiments *.

No.	Experiment	Hourly			Daily		
		NS	*CC*	*RE*	*NS*	*CC*	*RE*
1	P_CMA	0.658	0.855	−0.003	0.694	0.876	−0.003
2	P_GPM	0.576	0.919	−0.137	0.582	0.935	−0.137
3	P_CMORPH	0.596	0.823	−0.057	0.621	0.837	−0.057
4	P_WRF	0.464	0.816	0.048	0.499	0.832	0.048
5	P_4D-Var	0.547	0.787	−0.010	0.560	0.796	−0.010

* The indices *NS*, *CC* and *RE* denote Nash-Sutcliffe coefficient, Pearson's correlation coefficient and relative error, respectively.

6. Conclusions

Precipitation is a very important component in water cycle. In order to investigate the effectivenesses of different precipitation datasets on hydrological modelling, five different precipitation datasets were used to simulate a two-week runoff process after a heavy rainfall event in the WJB watershed (30,000 km^2). The five precipitation datasets contained one traditional in situ observation, i.e., the CMA data, two satellite precipitation products, i.e., the GPM IMERG and the merged CMORPH, and two NWP-predicted precipitation data, i.e., predictions from the WRF model and the WRF 4D-Var system. According to the requirement of the applied TOPX model, the five precipitation datasets were processed to 1-km grid data and evaluated with the daily CMWR data at point and field scales. The evaluated results suggested that the accuracies of the precipitation datasets from the CMA data, merged CMORPH, GPM IMERG, WRF 4D-Var system and WRF model generally decreased in sequence. The methods of gauge correction and 4D-Var data assimilation could improve the accuracies of the merged CMORPH and the WRF 4D-Var-predicted precipitation.

The soil moisture generated in the hourly rainfall-runoff simulations was evaluated to guarantee the rationalities of the hydrological modelling. In comparisons with the SMAP soil moisture, the watershed average *CC* values of the 3-h mean soil moisture from the P_CMA, P_GPM, P_CMORPH, P_WRF and P_4D-Var experiments reached 0.825, 0.695, 0.885, 0.841 and 0.774, respectively. The evaluations suggested that the spatiotemporal variations of the soil moisture were closely related to the variations of precipitation. Finally, the hourly simulated and daily accumulated outlet discharges were assessed. The *NS* values for the hourly simulated outlet discharges from the P_CMA, P_GPM, P_CMORPH, P_WRF and P_4D-Var experiments were 0.658, 0.576, 0.596, 0.464 and 0.547, respectively. These investigations demonstrated that the accuracy of precipitation data was a crucial factor to influence the performance of hydrological modelling. For this study case, all five precipitation datasets could yield reasonable hydrological simulations. The traditional in situ-observed precipitation was still the optimum dataset to simulate the studied rainfall-runoff process. The remotely sensed precipitation products of the GPM IMERG and the merged CMORPH were the secondary options. As the accuracy of the merged CMORPH was better than the GPM IMERG, the P_CMORPH experiment performed better than the P_GPM experiment. The performances driven by the NWP-predicted precipitation were the worst. As the WRF 4D-Var-predicted precipitation was obviously improved by the 4D-Var data assimilation method, the performance of the P_4D-Var experiment outperformed the P_WRF experiment, but because of the poor quality of the assimilated GPM IMERG over the study period in the WJB watershed, the performance of the P_4D-Var experiment was still not good. Despite lower effectivenesses in hydrological modelling, the precipitation datasets of the remotely sensed and the NWP-predicted are undoubtedly valuable and deserve further research, as the accuracies of the two datasets have been improved with the development of remote sensing, data merging, numerical simulation and data assimilation technologies. Moreover, the remotely sensed precipitation data are particularly indispensable in un-gauged areas, and the NWP model can forecast precipitation in the future, thus realize flood warning and other related water resource management in a real time.

For future studies, other data assimilation methods that are not as time-consuming as the 4D-Var can be used to simulate long-term rainfall-runoff processes or more short-term flood events. Moreover, many other remote sensing precipitation products can be assimilated into the WRF model because the GPM IMERG generally underestimates precipitation. In addition, other hydrological models constructed on the different bases of runoff generation and routing can be applied in related studies.

Author Contributions: W.Z. and L.Y. conceived this research. L.Y. performed the experiments under the guidance of W.Z., L.Y. and W.Z. analysed the results and wrote the paper. W.Z. and X.L. provided comments and modified the manuscript.

Funding: This research was funded by the National Key Research and Development Program of China (grant numbers: 2016YFA0602302 and 2016YFB0502502) and the National Natural Science Foundation of China (grant number: 41175088).

Acknowledgments: We are very grateful to K.W., Y.L. and D.W. from the Institute of Atmospheric Physics (Chinese Academy of Sciences) for their support regarding computer resources and academic communications. Great thanks should also be given to the NCAR Command Language (NCL) email list (ncl-talk@ucar.edu), which freely and substantially helped us with data processing and plotting with the NCL. We highly appreciate the IWHR and the CMA for providing the hourly discharge and hourly precipitation data for this study case.

Conflicts of Interest: The authors declare no conflicts of interest.

References

1. Essou, G.R.C.; Sabarly, F.; Lucas-Picher, P.; Brissette, F.; Poulin, A. Can precipitation and temperature from meteorological reanalyses be used for hydrological modeling? *J. Hydrometeorol.* **2016**, *17*, 1929–1950. [CrossRef]
2. Valeriano, O.C.S.; Koike, T.; Yang, K.; Graf, T.; Li, X.; Wang, L.; Han, X.J. Decision support for dam release during floods using a distributed biosphere hydrological model driven by quantitative precipitation forecasts. *Water Resour. Res.* **2010**, *46*. [CrossRef]

3. Duethmann, D.; Zimmer, J.; Gafurov, A.; Guntner, A.; Kriegel, D.; Merz, B.; Vorogushyn, S. Evaluation of areal precipitation estimates based on downscaled reanalysis and station data by hydrological modelling. *Hydrol. Earth Syst. Sci.* **2013**, *17*, 2415–2434. [CrossRef]
4. Yan, D.H.; Liu, S.H.; Qin, T.L.; Weng, B.S.; Li, C.Z.; Lu, Y.J.; Liu, J.J. Evaluation of TRMM precipitation and its application to distributed hydrological model in Naqu River Basin of the Tibetan Plateau. *Hydrol. Res.* **2017**, *48*, 822–839. [CrossRef]
5. Delpla, I.; Baures, E.; Jung, A.V.; Thomas, O. Impacts of rainfall events on runoff water quality in an agricultural environment in temperate areas. *Sci. Total Environ.* **2011**, *409*, 1683–1688. [CrossRef] [PubMed]
6. Mei, Y.W.; Nikolopoulos, E.I.; Anagnostou, E.N.; Zoccatelli, D.; Borga, M. Error analysis of satellite precipitation-driven modeling of flood events in complex Alpine terrain. *Remote Sens.* **2016**, *8*. [CrossRef]
7. Shah, H.L.; Mishra, V. Uncertainty and bias in satellite-based precipitation estimates over Indian subcontinental basins: Implications for real-time streamflow simulation and flood prediction. *J. Hydrometeorol.* **2016**, *17*, 615–636. [CrossRef]
8. Gebregiorgis, A.S.; Hossain, F. Understanding the dependence of satellite rainfall uncertainty on topography and climate for hydrologic model simulation. *IEEE Trans. Geosci. Remote Sens.* **2013**, *51*, 704–718. [CrossRef]
9. Nourani, V.; Fard, A.F.; Gupta, H.V.; Goodrich, D.C.; Niazi, F. Hydrological model parameterization using NDVI values to account for the effects of land cover change on the rainfall-runoff response. *Hydrol. Res.* **2017**, *48*, 1455–1473. [CrossRef]
10. Refsgaard, J.C.; Knudsen, J. Operational validation and intercomparison of different types of hydrological models. *Water Resour. Res.* **1996**, *32*, 2189–2202. [CrossRef]
11. Yang, D.; Herath, S.; Musiake, K. Comparison of different distributed hydrological models for characterization of catchment spatial variability. *Hydrol. Process.* **2000**, *14*, 403–416. [CrossRef]
12. Demirel, M.C.; Booij, M.J.; Hoekstra, A.Y. The skill of seasonal ensemble low-flow forecasts in the Moselle River for three different hydrological models. *Hydrol. Earth Syst. Sci.* **2015**, *19*, 275–291. [CrossRef]
13. Chen, J.; Brissette, F.P. Hydrological modelling using proxies for gauged precipitation and temperature. *Hydrol. Process.* **2017**, *31*, 3881–3897. [CrossRef]
14. Baymani-Nezhad, M.; Han, D. Hydrological modeling using Effective Rainfall routed by the Muskingum method (ERM). *J. Hydroinform.* **2013**, *15*, 1437–1455. [CrossRef]
15. Nourani, V.; Baghanam, A.H.; Adamowski, J.; Gebremichael, M. Using self-organizing maps and wavelet transforms for space-time pre-processing of satellite precipitation and runoff data in neural network based rainfall-runoff modeling. *J. Hydrol.* **2013**, *476*, 228–243. [CrossRef]
16. Mei, Y.; Anagnostou, E.N.; Shen, X.; Nikolopoulos, E.I. Decomposing the satellite precipitation error propagation through the rainfall-runoff processes. *Adv. Water Resour.* **2017**, *109*, 253–266. [CrossRef]
17. Qi, W.; Liu, J.G.; Yang, H.; Sweetapple, C. An ensemble-based dynamic Bayesian averaging approach for discharge simulations using multiple global precipitation products and hydrological models. *J. Hydrol.* **2018**, *558*, 405–420. [CrossRef]
18. Yi, L.; Zhang, W.C.; Wang, K. Evaluation of heavy precipitations dynamically downscaled by WRF 4D-Var data assimilation system with TRMM 3B42 and GPM IMERG over the Huaihe River Basin, China. *Remote Sens.* **2018**, *10*. [CrossRef]
19. Chen, Y.J.; Ebert, E.; Walsh, K.E.; Davidson, N. Evaluation of TRMM 3B42 precipitation estimates of tropical cyclone rainfall using PACRAIN data. *J. Geophys. Res.-Atmos.* **2013**, *118*, 2184–2196. [CrossRef]
20. McCabe, M.F.; Rodell, M.; Alsdorf, D.E.; Miralles, D.G.; Uijlenhoet, R.; Wagner, W.; Lucieer, A.; Houborg, R.; Verhoest, N.E.C.; Franz, T.E.; et al. The future of earth observation in hydrology. *Hydrol. Earth Syst. Sci.* **2017**, *21*, 3879–3914. [CrossRef] [PubMed]
21. Steiner, M.; Smith, J.A.; Burges, S.J.; Alonso, C.V.; Darden, R.W. Effect of bias adjustment and rain gauge data quality control on radar rainfall estimation. *Water Resour. Res.* **1999**, *35*, 2487–2503. [CrossRef]
22. Lorenz, C.; Kunstmann, H. The hydrological cycle in three state-of-the-art reanalyses: Intercomparison and performance analysis. *J. Hydrometeorol.* **2012**, *13*, 1397–1420. [CrossRef]
23. Huffman, G.J.; Adler, R.F.; Arkin, P.; Chang, A.; Ferraro, R.; Gruber, A.; Janowiak, J.; McNab, A.; Rudolf, B.; Schneider, U. The Global Precipitation Climatology Project (GPCP) combined precipitation dataset. *Bull. Am. Meteorol. Soc.* **1997**, *78*, 5–20. [CrossRef]

24. Joyce, R.J.; Janowiak, J.E.; Arkin, P.A.; Xie, P.P. CMORPH: A method that produces global precipitation estimates from passive microwave and infrared data at high spatial and temporal resolution. *J. Hydrometeorol.* **2004**, *5*, 487–503. [CrossRef]
25. Garstang, M.; Kummerow, C.D. The joanne simpson special issue on the Tropical Rainfall Measuring Mission (TRMM). *J. Appl. Meteorol.* **2000**, *39*, 1961. [CrossRef]
26. Hou, A.Y.; Kakar, R.K.; Neeck, S.; Azarbarzin, A.A.; Kummerow, C.D.; Kojima, M.; Oki, R.; Nakamura, K.; Iguchi, T. The global precipitation measurement mission. *Bull. Am. Meteorol. Soc.* **2014**, *95*, 701–722. [CrossRef]
27. Zhou, X.; Luo, Y.L.; Guo, X.L. Application of a CMORPH-a WS merged hourly gridded precipitation product in analyzing charateristics of short-duration heavy rainfall over southern China. *J. Trop. Meteorol.* **2015**, *31*, 333–344.
28. Gaona, M.F.R.; Overeem, A.; Leijnse, H.; Uijlenhoet, R. First-year evaluation of GPM rainfall over the netherlands: IMERG Day 1 final run (VO3D). *J. Hydrometeorol.* **2016**, *17*, 2799–2814. [CrossRef]
29. Schmidli, J.; Goodess, C.M.; Frei, C.; Haylock, M.R.; Hundecha, Y.; Ribalaygua, J.; Schmith, T. Statistical and dynamical downscaling of precipitation: An evaluation and comparison of scenarios for the European Alps. *J. Geophys. Res.-Atmos.* **2007**, *112*. [CrossRef]
30. Zhang, X.X.; Anagnostou, E.; Frediani, M.; Solomos, S.; Kallos, G. Using NWP simulations in satellite rainfall estimation of heavy precipitation events over mountainous areas. *J. Hydrometeorol.* **2013**, *14*, 1844–1858. [CrossRef]
31. Dee, D.P.; Uppala, S.M.; Simmons, A.J.; Berrisford, P.; Poli, P.; Kobayashi, S.; Andrae, U.; Balmaseda, M.A.; Balsamo, G.; Bauer, P.; et al. The ERA-Interim reanalysis: Configuration and performance of the data assimilation system. *Q. J. R. Meteorol. Soc.* **2011**, *137*, 553–597. [CrossRef]
32. Koizumi, K.; Ishikawa, Y.; Tsuyuki, T. Assimilation of precipitation data to the JMA mesoscale model with a four-dimensional variational method and its impact on precipitation forecasts. *Sola* **2005**, *1*, 45–48. [CrossRef]
33. Mazrooei, A.; Sinha, T.; Sankarasubramanian, A.; Kumar, S.; Peters-Lidard, C.D. Decomposition of sources of errors in seasonal streamflow forecasting over the US Sunbelt. *J. Geophys. Res.-Atmos.* **2015**, *120*. [CrossRef]
34. Chen, J.; Brissette, F.P.; Leconte, R. Uncertainty of downscaling method in quantifying the impact of climate change on hydrology. *J. Hydrol.* **2011**, *401*, 190–202. [CrossRef]
35. Chen, J.; Brissette, F.P.; Chaumont, D.; Braun, M. Performance and uncertainty evaluation of empirical downscaling methods in quantifying the climate change impacts on hydrology over two North American river basins. *J. Hydrol.* **2013**, *479*, 200–214. [CrossRef]
36. Bardossy, A.; Pegram, G. Downscaling precipitation using regional climate models and circulation patterns toward hydrology. *Water Resour. Res.* **2011**, *47*. [CrossRef]
37. Huang, X.Y.; Xiao, Q.N.; Barker, D.M.; Zhang, X.; Michalakes, J.; Huang, W.; Henderson, T.; Bray, J.; Chen, Y.S.; Ma, Z.Z.; et al. Four-dimensional variational data assimilation for WRF: Formulation and preliminary results. *Mon. Weather Rev.* **2009**, *137*, 299–314. [CrossRef]
38. Lei, L.; Stauffer, D.R.; Deng, A. A hybrid nudging-ensemble Kalman filter approach to data assimilation in WRF/DART. *Q. J. R. Meteorol. Soc.* **2012**, *138*, 2066–2078. [CrossRef]
39. Lorenc, A.C.; Bowler, N.E.; Clayton, A.M.; Pring, S.R.; Fairbairn, D. Comparison of hybrid-4DEnVar and hybrid-4DVar data assimilation methods for global NWP. *Mon. Weather Rev.* **2015**, *143*, 212–229. [CrossRef]
40. Buehner, M.; Houtekamer, P.L.; Charette, C.; Mitchell, H.L.; He, B. Intercomparison of variational data assimilation and the ensemble Kalman filter for global deterministic NWP. *Part II: One-month experiments with real observations. Mon. Weather Rev.* **2010**, *138*, 1567–1586. [CrossRef]
41. Buehner, M.; Houtekamer, P.L.; Charette, C.; Mitchell, H.L.; He, B. Intercomparison of variational data assimilation and the ensemble Kalman filter for global deterministic NWP. *Part I: Description and single-observation experiments. Mon. Weather Rev.* **2010**, *138*, 1550–1566. [CrossRef]
42. Black, T.L. The new NMC mesoscale ETA model–description and forecast examples. *Weather Forecast.* **1994**, *9*, 265–278. [CrossRef]
43. Dudhia, J.; Klemp, J.; Skamarock, W.; Dempsey, D.; Janjic, Z.; Benjamin, S.; Brown, J. A collaborative effort towards a future community mesoscale model (WRF). In Proceedings of the 12th Conference on Numerical Weather Prediction, Phoenix, AZ, USA, 11–16 January 1998; pp. 242–243.

44. Saito, K.; Fujita, T.; Yamada, Y.; Ishida, J.I.; Kumagai, Y.; Aranami, K.; Ohmori, S.; Nagasawa, R.; Kumagai, S.; Muroi, C.; et al. The operational JMA nonhydrostatic mesoscale model. *Mon. Weather Rev.* **2006**, *134*, 1266–1298. [CrossRef]
45. Molteni, F.; Buizza, R.; Palmer, T.N.; Petroliagis, T. The ECMWF ensemble prediction system: Methodology and validation. *Q. J. R. Meteorol. Soc.* **1996**, *122*, 73–119. [CrossRef]
46. Zubieta, R.; Getirana, A.; Espinoza, J.C.; Lavado-Casimiro, W.; Aragon, L. Hydrological modeling of the Peruvian-Ecuadorian Amazon Basin using GPM-IMERG satellite-based precipitation dataset. *Hydrol. Earth Syst. Sci.* **2017**, *21*. [CrossRef]
47. Xu, H.L.; Xu, C.Y.; Saelthun, N.R.; Zhou, B.; Xu, Y.P. Evaluation of reanalysis and satellite-based precipitation datasets in driving hydrological models in a humid region of Southern China. *Stoch. Environ. Res. Risk Assess.* **2015**, *29*, 2003–2020. [CrossRef]
48. Wu, H.; Adler, R.F.; Tian, Y.D.; Gu, G.J.; Huffman, G.J. Evaluation of quantitative precipitation estimations through hydrological modeling in IFloodS River basins. *J. Hydrometeorol.* **2017**, *18*, 529–553. [CrossRef]
49. Rasmussen, S.H.; Christensen, J.H.; Drews, M.; Gochis, D.J.; Refsgaard, J.C. Spatial-scale characteristics of precipitation simulated by regional climate models and the implications for hydrological modeling. *J. Hydrometeorol.* **2012**, *13*, 1817–1835. [CrossRef]
50. Parkes, B.L.; Wetterhall, F.; Pappenberger, F.; He, Y.; Malamud, B.D.; Cloke, H.L. Assessment of a 1-h gridded precipitation dataset to drive a hydrological model: A case study of the summer 2007 floods in the Upper Severn, UK. *Hydrol. Res.* **2013**, *44*, 89–105. [CrossRef]
51. Lin, C.A.; Wen, L.; Lu, G.H.; Wu, Z.Y.; Zhang, J.Y.; Yang, Y.; Zhu, Y.F.; Tong, L.Y. Atmospheric-hydrological modeling of severe precipitation and floods in the Huaihe River Basin, China. *J. Hydrol.* **2006**, *330*, 249–259. [CrossRef]
52. Liechti, T.C.; Matos, J.P.; Boillat, J.L.; Schleiss, A.J. Comparison and evaluation of satellite derived precipitation products for hydrological modeling of the Zambezi River Basin. *Hydrol. Earth Syst. Sci.* **2012**, *16*, 489–500. [CrossRef]
53. Lauri, H.; Rasanen, T.A.; Kummu, M. Using reanalysis and remotely sensed temperature and precipitation data for hydrological modeling in monsoon climate: Mekong river aase study. *J. Hydrometeorol.* **2014**, *15*, 1532–1545. [CrossRef]
54. Georgakakos, K.P.; Kavvas, M.L. Precipitation analysis, modeling, and prediction in hydrology. *Rev. Geophys.* **1987**, *25*, 163–178. [CrossRef]
55. Nguyen, T.H.M.; Masih, I.; Mohamed, Y.A.; van der Zaag, P. Validating rainfall-runoff modelling using satellite-based and reanalysis precipitation products in the Sre Pok catchment, the Mekong river basin. *Geosciences* **2018**, *8*, 164. [CrossRef]
56. Ottle, C.; Vidalmadjar, D. Assimilation of soil-moisture inferred from infrared remote-sensing in a hydrological model over the HAPEX-MOBILHY region. *J. Hydrol.* **1994**, *158*, 241–264. [CrossRef]
57. Western, A.W.; Grayson, R.B.; Green, T.R. The Tarrawarra project: High resolution spatial measurement, modelling and analysis of soil moisture and hydrological response. *Hydrol. Process.* **1999**, *13*, 633–652. [CrossRef]
58. Wanders, N.; Bierkens, M.F.P.; de Jong, S.M.; de Roo, A.; Karssenberg, D. The benefits of using remotely sensed soil moisture in parameter identification of large-scale hydrological models. *Water Resour. Res.* **2014**, *50*, 6874–6891. [CrossRef]
59. Draper, C.; Mahfouf, J.F.; Calvet, J.C.; Martin, E.; Wagner, W. Assimilation of ASCAT near- surface soil moisture into the SIM hydrological model over France. *Hydrol. Earth Syst. Sci.* **2011**, *15*, 3829–3841. [CrossRef]
60. Lopez, P.L.; Wanders, N.; Schellekens, J.; Renzullo, L.J.; Sutanudjaja, E.H.; Bierkens, M.F.P. Improved large-scale hydrological modelling through the assimilation of streamflow and downscaled satellite soil moisture observations. *Hydrol. Earth Syst. Sci.* **2016**, *20*, 3059–3076. [CrossRef]
61. Baguis, P.; Roulin, E. Soil Moisture Data Assimilation in a Hydrological Model: A Case Study in Belgium Using Large-Scale Satellite Data. *Remote Sens.* **2017**, *9*, 820. [CrossRef]
62. Santi, E.; Paloscia, S.; Pettinato, S.; Notarnicola, C.; Pasolli, L.; Pistocchi, A. Comparison between SAR Soil Moisture Estimates and Hydrological Model Simulations over the Scrivia Test Site. *Remote Sens.* **2013**, *5*, 4961–4976. [CrossRef]

63. Liu, S.; Mo, X.; Zhao, W.; Naeimi, V.; Dai, D.; Shu, C.; Mao, L. Temporal variation of soil moisture over the Wuding River basin assessed with an eco-hydrological model, in-situ observations and remote sensing. *Hydrol. Earth Syst. Sci.* **2009**, *13*, 1375–1398. [CrossRef]
64. Bertoldi, G.; Della Chiesa, S.; Notarnicola, C.; Pasolli, L.; Niedrist, G.; Tappeiner, U. Estimation of soil moisture patterns in mountain grasslands by means of SAR RADARSAT2 images and hydrological modeling. *J. Hydrol.* **2014**, *516*, 245–257. [CrossRef]
65. Trudel, M.; Leconte, R.; Paniconi, C. Analysis of the hydrological response of a distributed physically-based model using post-assimilation (EnKF) diagnostics of streamflow and in situ soil moisture observations. *J. Hydrol.* **2014**, *514*, 192–201. [CrossRef]
66. Koch, J.; Cornelissen, T.; Fang, Z.F.; Bogena, H.; Diekkruger, B.; Kollet, S.; Stisen, S. Inter-comparison of three distributed hydrological models with respect to seasonal variability of soil moisture patterns at a small forested catchment. *J. Hydrol.* **2016**, *533*, 234–249. [CrossRef]
67. Iacobellis, V.; Gioia, A.; Milella, P.; Satalino, G.; Balenzano, A.; Mattia, F. Inter-comparison of hydrological model simulations with time series of SAR-derived soil moisture maps. *Eur. J. Remote Sens.* **2013**, *46*, 739–757. [CrossRef]
68. Xiong, L.H.; Yang, H.; Zeng, L.; Xu, C.Y. Evaluating Consistency between the Remotely Sensed Soil Moisture and the Hydrological Model-Simulated Soil Moisture in the Qujiang Catchment of China. *Water* **2018**, *10*, 291. [CrossRef]
69. Khan, U.; Ajami, H.; Tuteja, N.K.; Sharma, A.; Kim, S. Catchment scale simulations of soil moisture dynamics using an equivalent cross-section based hydrological modelling approach. *J. Hydrol.* **2018**, *564*, 944–966. [CrossRef]
70. Tang, G.Q.; Ma, Y.Z.; Long, D.; Zhong, L.Z.; Hong, Y. Evaluation of GPM Day-1 IMERG and TMPA version-7 legacy products over Mainland China at multiple spatiotemporal scales. *J. Hydrol.* **2016**, *533*, 152–167. [CrossRef]
71. Pombo, S.; de Oliveira, R.P.; Mendes, A. Validation of remote-sensing precipitation products for Angola. *Meteorol. Appl.* **2015**, *22*, 395–409. [CrossRef]
72. Pan, X.D.; Li, X.; Cheng, G.D.; Hong, Y. Effects of 4D-Var data assimilation using remote sensing precipitation products in a WRF model over the complex terrain of an arid region river basin. *Remote Sens.* **2017**, *9*, 963. [CrossRef]
73. Lin, L.F.; Ebtehaj, A.M.; Bras, R.L.; Flores, A.N.; Wang, J.F. Dynamical precipitation downscaling for hydrologic applications using WRF 4D-Var data assimilation: Implications for GPM era. *J. Hydrometeorol.* **2015**, *16*, 811–829. [CrossRef]
74. Rogelis, M.C.; Werner, M. Streamflow forecasts from WRF precipitation for flood early warning in mountain tropical areas. *Hydrol. Earth Syst. Sci.* **2018**, *22*, 853–870. [CrossRef]
75. Pennelly, C.; Reuter, G.; Flesch, T. Verification of the WRF model for simulating heavy precipitation in Alberta. *Atmos. Res.* **2014**, *135*, 172–192. [CrossRef]
76. Bukovsky, M.S.; Karoly, D.J. Precipitation simulations using WRF as a nested regional climate model. *J. Appl. Meteorol. Climatol.* **2009**, *48*, 2152–2159. [CrossRef]
77. Yuan, Z.; Yang, Z.Y.; Zheng, X.D.; Yuan, Y. Spatial and temporal variations of precipitation in Huaihe river basin in recent 50 years. *South-to-North Water Divers. Water Sci. Technol.* **2012**, *10*. [CrossRef]
78. Xia, J.; She, D.X.; Zhang, Y.Y.; Du, H. Spatio-temporal trend and statistical distribution of extreme precipitation events in Huaihe River Basin during 1960–2009. *J. Geogr. Sci.* **2012**, *22*, 195–208. [CrossRef]
79. Kleczek, M.A.; Steeneveld, G.J.; Holtslag, A.A.M. Evaluation of the Weather Research and Forecasting mesoscale model for GABLS3: Impact of boundary-layer schemes, boundary conditions and spin-up. *Bound.-Layer Meteor.* **2014**, *152*, 213–243. [CrossRef]
80. Srinivas, D.; Rao, D.V.B. Implications of vortex initialization and model spin-up in tropical cyclone prediction using Advanced Research Weather Research and Forecasting Model. *Nat. Hazards* **2014**, *73*, 1043–1062. [CrossRef]
81. Veerse, F.; Thepaut, J.N. Multiple-truncation incremental approach for four-dimensional variational data assimilation. *Q. J. R. Meteorol. Soc.* **1998**, *124*, 1889–1908. [CrossRef]
82. Sharifi, E.; Steinacker, R.; Saghafian, B. Assessment of GPM-IMERG and Other Precipitation Products against Gauge Data under Different Topographic and Climatic Conditions in Iran: Preliminary Results. *Remote Sens.* **2016**, *8*, 135. [CrossRef]

83. Michaelides, S.; Levizzani, V.; Anagnostou, E.; Bauer, P.; Kasparis, T.; Lane, J.E. Precipitation: Measurement, remote sensing, climatology and modeling. *Atmos. Res.* **2009**, *94*, 512–533. [CrossRef]
84. Jiang, S.; Ren, L.; Yong, B.; Hong, Y.; Yang, X.; Yuan, F. Evaluation of latest TMPA and CMORPH precipitation products with independent rain gauge observation networks over high-latitude and low-latitude basins in China. *Chin. Geogr. Sci.* **2016**, *26*, 439–455. [CrossRef]
85. Asong, Z.E.; Razavi, S.; Wheater, H.S.; Wong, J.S. Evaluation of Integrated Multisatellite Retrievals for GPM (IMERG) over Southern Canada against Ground Precipitation Observations: A Preliminary Assessment. *J. Hydrometeorol.* **2017**, *18*, 1033–1050. [CrossRef]
86. Liu, Y.B.; Wu, G.P.; Ke, C.Q. *Hydrological Remote Sensing*; Science Press: Beijing, China, 2016; ISBN 978-7-03-049302-6.
87. Wang, Z.L.; Zhong, R.D.; Lai, C.G.; Chen, J.C. Evaluation of the GPM IMERG satellite-based precipitation products and the hydrological utility. *Atmos. Res.* **2017**, *196*, 151–163. [CrossRef]
88. Liu, Z. Comparison of Integrated Multisatellite Retrievals for GPM (IMERG) and TRMM Multisatellite Precipitation Analysis (TMPA) Monthly Precipitation Products: Initial Results. *J. Hydrometeorol.* **2016**, *17*, 777–790. [CrossRef]
89. Skamarock, W.C.; Klemp, J.B. A time-split nonhydrostatic atmospheric model for weather research and forecasting applications. *J. Comput. Phys.* **2008**, *227*, 3465–3485. [CrossRef]
90. Zhang, X.Z.; Xiong, Z.; Zheng, J.Y.; Ge, Q.S. High-resolution precipitation data derived from dynamical downscaling using the WRF model for the Heihe River Basin, northwest China. *Theor. Appl. Climatol.* **2018**, *131*, 1249–1259. [CrossRef]
91. Pieri, A.B.; von Hardenberg, J.; Parodi, A.; Provenzale, A. Sensitivity of Precipitation Statistics to Resolution, Microphysics, and Convective Parameterization: A Case Study with the High-Resolution WRF Climate Model over Europe. *J. Hydrometeorol.* **2015**, *16*, 1857–1872. [CrossRef]
92. Cardoso, R.M.; Soares, P.M.M.; Miranda, P.M.A.; Belo-Pereira, M. WRF high resolution simulation of Iberian mean and extreme precipitation climate. *Int. J. Climatol.* **2013**, *33*, 2591–2608. [CrossRef]
93. Lim, K.S.S.; Hong, S.Y. Development of an effective double-moment cloud microphysics scheme with prognostic Cloud Condensation Nuclei (CCN) for weather and climate models. *Mon. Weather Rev.* **2010**, *138*, 1587–1612. [CrossRef]
94. Mlawer, E.J.; Taubman, S.J.; Brown, P.D.; Iacono, M.J.; Clough, S.A. Radiative transfer for inhomogeneous atmospheres: RRTM, a validated correlated-k model for the longwave. *J. Geophys. Res.-Atmos.* **1997**, *102*, 16663–16682. [CrossRef]
95. Dudhia, J. Numerical study of convection observed during the winter monsoon experiment using mesoscale two-dimensional model. *J. Atmos. Sci.* **1989**, *46*, 3077–3107. [CrossRef]
96. Chen, F.; Dudhia, J. Coupling an advanced land surface-hydrology model with the Penn State-NCAR MM5 modeling system. *Part I: Model implementation and sensitivity. Mon. Weather Rev.* **2001**, *129*, 569–585. [CrossRef]
97. Hong, S.-Y.; Noh, Y.; Dudhia, J. A new vertical diffusion package with an explicit treatment of entrainment processes. *Mon. Weather Rev.* **2006**, *134*, 2318–2341. [CrossRef]
98. Grell, G.A.; Devenyi, D. A generalized approach to parameterizing convection combining ensemble and data assimilation techniques. *Geophys. Res. Lett.* **2002**, *29*. [CrossRef]
99. Barker, D.; Huang, X.Y.; Liu, Z.Q.; Auligne, T.; Zhang, X.; Rugg, S.; Ajjaji, R.; Bourgeois, A.; Bray, J.; Chen, Y.S.; et al. The Weather Research and Forecasting model's community variational/ensemble data assimilation system WRFDA. *Bull. Am. Meteorol. Soc.* **2012**, *93*, 831–843. [CrossRef]
100. Barker, D.M.; Huang, W.; Guo, Y.R.; Bourgeois, A.J.; Xiao, Q.N. A three-dimensional variational data assimilation system for MM5: Implementation and initial results. *Mon. Weather Rev.* **2004**, *132*, 897–914. [CrossRef]
101. Courtier, P.; Thepaut, J.N.; Hollingsworth, A. A strategy for operational imlementation of 4D-Var, using an incremental approach. *Q. J. R. Meteorol. Soc.* **1994**, *120*, 1367–1387. [CrossRef]
102. Lorenc, A.C. Modelling of error covariances by 4D-Var data assimilation. *Q. J. R. Meteorol. Soc.* **2003**, *129*, 3167–3182. [CrossRef]
103. Parrish, D.F.; Derber, J.C. The national-meteorological-centers spectral statistical-interpolation analysis system. *Mon. Weather Rev.* **1992**, *120*, 1747–1763. [CrossRef]

104. Reichle, R.H.; De Lannoy, G.J.M.; Liu, Q.; Ardizzone, J.V.; Colliander, A.; Conaty, A.; Crow, W.; Jackson, T.J.; Jones, L.A.; Kimball, J.S.; et al. Assessment of the SMAP level-4 surface and root-zone soil moisture product using in situ measurements. *J. Hydrometeorol.* **2017**, *18*, 2621–2645. [CrossRef]
105. Niu, G.Y.; Yang, Z.L.; Dickinson, R.E.; Gulden, L.E. A simple TOPMODEL-based runoff parameterization (SIMTOP) for use in global climate models. *J. Geophys. Res.-Atmos.* **2005**, *110*. [CrossRef]
106. Yi, L.; Zhang, W.C.; Yan, C.A. A modified topographic index that incorporates the hydraulic and physical properties of soil. *Hydrol. Res.* **2017**, *48*, 370–383. [CrossRef]
107. Yong, B. *Development of a land-Surface Hydrological Model TOPX and Its Coupling Study with Regional Climate Model RIEMS*; Nanjing University: Nanjing, China, 2007.
108. Wu, Z.Y. Study on Quantitative Rainfall and Real Time Flood Forecasting. Ph.D. Thesis, Hohai University, Nanjing, China, 2007.
109. Zhao, R.J. *Watershed Hydrological Simulation-Xin'anjiang Model and Shanbei Model*; Water Resources and Electric Power Press: Beijing, China, 1984.
110. Lu, G.H.; Wu, Z.Y.; He, H. *Hydrologic Cycle Process and Quantitative Prediction*; Science Press: Beijing, China, 2010; ISBN 978-7-03-026608-8.
111. Nash, L.L.; Gleick, P.H. Sensitivity of streamflow in the Colorado basin to climatic changes. *J. Hydrol.* **1991**, *125*, 221–241. [CrossRef]
112. Taylan, E.D.; Damcayiri, D. The Prediction of Precipitations of Isparta Region By Using IDW and Kriging. *Teknik Dergi* **2016**, *27*, 7551–7559.
113. Accadia, C.; Mariani, S.; Casaioli, M.; Lavagnini, A.; Speranza, A. Sensitivity of precipitation forecast skill scores to bilinear interpolation and a simple nearest-neighbor average method on high-resolution verification grids. *Weather Forecast.* **2003**, *18*, 918–932. [CrossRef]
114. Wilks, D.S. *Statistical Methods in the Atmospheric Sciences*; Academic Press: Cambridge, MA, USA, 2006; Volume 91, p. 627. ISBN 978-0-12-751996-1.
115. Cui, C.Y.; Xu, J.; Zeng, J.Y.; Chen, K.S.; Bai, X.J.; Lu, H.; Chen, Q.; Zhao, T.J. Soil moisture mapping from satellites: An intercomparison of SMAP, SMOS, FY3B, AMSR2, and ESA CCI over two dense network regions at different spatial scales. *Remote Sens.* **2018**, *10*. [CrossRef]
116. Huang, T.N.; Zheng, Y.F.; Duan, C.C.; Yin, J.F.; Wu, R.J. Comparative analysis of soil moisture retrieval by satellites in China. *Remote Sens. Inf.* **2017**, *32*, 25–33. [CrossRef]
117. Albergel, C.; Zakharova, E.; Calvet, J.C.; Zribi, M.; Parde, M.; Wigneron, J.P.; Novello, N.; Kerr, Y.; Mialon, A.; Fritz, N.E.D. A first assessment of the SMOS data in southwestern France using in situ and airborne soil moisture estimates: The CAROLS airborne campaign. *Remote Sens. Environ.* **2011**, *115*, 2718–2728. [CrossRef]
118. Chen, C.; Chen, Q.; Duan, Z.; Zhang, J.; Mo, K.; Li, Z.; Tang, G. Multiscale Comparative Evaluation of the GPM IMERG v5 and TRMM 3B42 v7 Precipitation Products from 2015 to 2017 over a Climate Transition Area of China. *Remote Sens.* **2018**, *10*. [CrossRef]
119. Kumar, S.; Godrej, A.N.; Grizzard, T.J. Watershed size effects on applicability of regression-based methods for fluvial loads estimation. *Water Resour. Res.* **2013**, *49*, 7698–7710. [CrossRef]
120. Lee, K.T.; Huang, J.-K. Influence of storm magnitude and watershed size on runoff nonlinearity. *J. Earth Syst. Sci.* **2016**, *125*, 777–794. [CrossRef]
121. Black, P.E.; Cronn, J.W. Hydrograph responses to watershed model size and similitude relations. *J. Hydrol.* **1975**, *26*, 255–266. [CrossRef]
122. Zhou, S.M.; Lei, T.W.; Warrington, D.N.; Lei, Q.X.; Zhang, M.L. Does watershed size affect simple mathematical relationships between flow velocity and discharge rate at watershed outlets on the Loess Plateau of China. *J. Hydrol.* **2012**, *444*, 1–9. [CrossRef]

remote sensing

MDPI

Article

Exploiting Satellite-Based Surface Soil Moisture for Flood Forecasting in the Mediterranean Area: State Update Versus Rainfall Correction

Christian Massari *, Stefania Camici, Luca Ciabatta and Luca Brocca

Research Institute for Geo-Hydrological Protection, National Research Council, Via della Madonna Alta 126, 06128 Perugia, Italy; stefania.camici@irpi.cnr.it (S.C.); luca.ciabatta@irpi.cnr.it (L.C.); luca.brocca@irpi.cnr.it (L.B.)

* Correspondence: christian.massari@irpi.cnr.it; Tel.: +39-0755014417; Fax: +39-0755014420

Received: 22 December 2017; Accepted: 10 February 2018; Published: 13 February 2018

Abstract: Many satellite soil moisture products are today globally available in near real-time. These observations are of paramount importance for enhancing the understanding of the hydrological cycle and particularly useful for flood forecasting purposes. In recent decades, several studies assimilated satellite soil moisture observations into rainfall-runoff models to improve their flood forecasting skills. The rationale is that a better representation of the catchment states leads to a better stream flow estimation. By exploiting the strong physical connection between the soil moisture dynamic and rainfall, some recent studies demonstrated that satellite soil moisture observations can be also used for enhancing the quality of rainfall observations. Given that the quality of the rainfall is one of the main drivers of the hydrological model uncertainty, this begs the question—to what extent updating soil moisture states leads to better flood forecasting skills than correcting rainfall forcing? In this study, we try to answer this question by using rainfall-runoff observations from 10 catchments throughout the Mediterranean area and a continuous rainfall-runoff model—MISDc—forced with reanalysis- and satellite-based rainfall observations. Satellite soil moisture retrievals from the Advanced SCATterometer (ASCAT) are either assimilated into MISDc model via the Ensemble Kalman filter to update model states or, alternatively, used to correct rainfall observations derived from a reanalysis and a satellite-based product through the integration with soil moisture-based rainfall estimates. 4–9 years (depending on the catchment) of stream flow observations are organized into calibration and validation periods to test the two different schemes. Results show that the rainfall correction is favourable if the target is the predictions of high flows while for low flows there is a small advantage of the state correction scheme with respect to the rainfall correction. The improvements for high flows are particularly large when the quality of the rainfall is relatively poor with important implications for large-scale flood forecasting in the Mediterranean area.

Keywords: floods soil moisture; rainfall; data assimilation; rainfall correction; remote sensing; Mediterranean basin

1. Introduction

The value of soil moisture observations for hydrological modelling is unquestionable. Soil moisture influences the partitioning of rainfall into evapotranspiration, infiltration and runoff, hence it is an important factor for determining the magnitude of flood events (e.g., [1,2]). Continuous hydrological models simulate the spatio-temporal evolution of soil moisture for single or multiple soil layers and use this information for predicting the occurrence and the magnitude of floods. A rainfall event occurring in wet and dry conditions shows large differences in terms of hydrologic response, determining the triggering or not of a potentially catastrophic flood event. Therefore, soil moisture observations represent a crucial information for improving hydrological predictions both for small [3,4] and large [5–7] river basins.

Measurements of soil moisture can be obtained from both in situ and satellite sensors (see [8] and [9] for two recent reviews), however, in situ observations have been rarely ingested into hydrological models [4,10,11] because their relatively scarce availability in many parts of the world. On the contrary, in recent years, it has been observed a proliferation of studies using coarser satellite soil moisture products for improving model flood forecasting skills [1,3,12–18]. These studies have demonstrated a general positive impact of satellite soil moisture observations with improvements that have ranged from minor [3,11] to significant [14,16,18] or no improvements. Thanks to the recent availability of high resolution soil moisture observations from Sentinel-1—and its synergy with the soil moisture observations derived from the Soil Moisture Active and Passive (SMAP, [19]) mission [20]—there is high chance that the number of hydrological assimilation studies (involving also smaller catchments) will further increase in the near future (e.g., [21]).

Thanks to the close connection between rainfall and soil moisture, coarse resolution satellite observations were also recently employed by some authors for correcting rainfall [22–27]. Corrected rainfall observations (through soil moisture) were then used for forcing hydrological models to explore whether this correction is more beneficial than the classical state update for improving flood forecasting skills [28–30].

Satellite soil moisture observations can therefore be ingested in hydrological models through two possible approaches: via state update through classical data assimilation (i.e., variational and sequential techniques) and via rainfall correction (or a combination of both as in [4,28]). In this respect, Crow, W.T. et al. [28] demonstrated the potential of a dual data assimilation (i.e., state correction plus rainfall correction) through a synthetic experiment by using the Ensemble Kalman Filter (EnKF, [31]). The authors highlighted however, that such dual (and simultaneous) use of soil moisture retrievals can conceivably lead to correlation between forecasting and observations errors within the EnKF, and, consequently, to sub-optimal filter performance.

Chen, F. et al. [29] applied the same approach of [28] to 13 basins in the central United States by using the real-time version of the Tropical Rainfall Measuring Mission Multi-satellite Precipitation Analysis (TMPA 3B42RT, [32]) as rainfall forcing and ASCAT [33] and SMOS [34] soil moisture observations for rainfall/state correction. In particular, for rainfall correction, they used the Soil Moisture Analysis Rainfall Tool (SMART, [22]) whereas the state correction was performed through the EnKF. The authors showed that the state correction is overall better than rainfall correction but the latter is able to improve stream flow simulations during high-flow periods better than the state correction. [30] tested the same concept to four large basins in Australia by only changing the hydrological model (i.e., PDM, Probability Distributed Model) and by including the AMSR-E (Advanced Microwave Scanning Radiometer—Earth Observing System) soil moisture product together with SMOS and ASCAT. They confirmed the previous results obtained by [33], also highlighting some of the limitations.

In this study, we further investigate the state/rainfall correction approaches and try to address the following research questions:

1. To what extent updating soil moisture states leads to better flood predictions than the correction of the rainfall?
2. How much these improvements are affected by the underlying accuracy of the original rainfall product used for forcing the hydrological model?
3. What is the impact of the basin size and the climate conditions on the results?

Here, MISDc [35] hydrological model is used for stream flow simulations and the EnKF for correcting the model states within the first of the two approaches (approach SM-corr from here onward). The rainfall correction of a satellite-(TMPA 3B42RT) and a reanalysis-based (ERA-Interim, [36]) product is performed by their integration with rainfall estimates obtained thought the inversion of ASCAT soil moisture observations via SM2RAIN [24]. The corrected rainfall products are then used as alternative to the states correction to force MISDc within the second of the two approaches (approach P-corr from here onward). Daily stream flow observations from 10 basins (4–9 years of data) throughout the Mediterranean area are used as benchmark to address the above research questions. The application of the two schemes to a reanalysis and a satellite-based rainfall product aims to simulate a scenario of ground data scarcity as happens in many regions of the world [37].

With respect to previous studies, three important novelties are introduced: (1) a different rainfall correction algorithm (i.e., the integration of the rainfall observations is carried out with SM2RAIN), (2) a different study area—the Mediterranean area, and) the use of both satellite- and a reanalysis rainfall products which—given the different accuracy of the products—allows to test the two different schemes (i.e., the SM-corr and P-corr) as a function of the underlying rainfall quality.

The paper is organized as follows. Section 2 contains the description of the study area and the datasets used to carry out the analysis. Section 3 provides a description of the hydrological model, the data assimilation and the rainfall correction schemes as well as the performance metrics used for evaluating the results. Results and discussion are contained in Section 4 and final remarks are presented in the Conclusion section.

2. Material

2.1. Study Area

The study is carried out over the Mediterranean where flood and flash flood events cause significant economic and social losses. Mediterranean is characterized by varied and contrasting topography including the Alpine Mountains in the Italian and Balkan peninsulas, Northern Spain, and Southern France. Due to the topographic complexity, the climate includes hot-dry summers and humid-cold winters. A clear contrast exists between the more-rainy northern part of the study region (Southern Europe) and the drier southern area (North Africa, Iberian Peninsula) and between the western sides (rainsides) of the Iberian, Italian and Balkan peninsulas and their eastern sides (rainshadows). The mean annual precipitation averaged over the study area is equal to 593 ± 203 mm/year, characterized by a strong spatial variability which ranges from 20–40 mm/year in North Africa to 1500–2000 mm/year over the Alps. A significant seasonal variability exists, with the late autumn and early winter months (September to November) being the wettest where floods usually occur.

Table 1 summarizes the main characteristics of the selected study catchments, with area ranging from nearly 450 km^2 for the Kolpa river basin in Slovenia to about 5000 km^2 for the Tevere River basin in central Italy, mean basin elevation ranges from 197 m a.s.l. (lowland basin) to 1362 m a.s.l. (mountainous basin). Given the different climatic and physiographic conditions that characterize the selected catchments, they can be considered a representative sample of the catchments located in the Mediterranean (Figure 1).

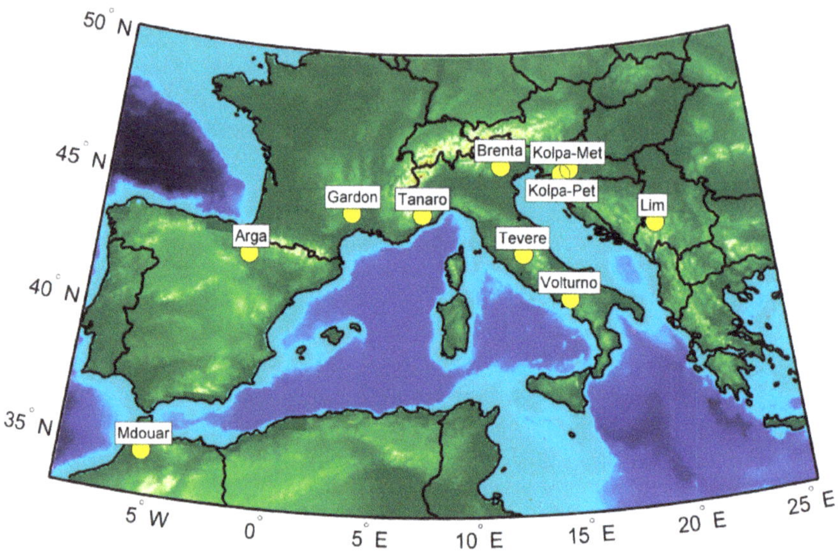

Figure 1. Location of the investigated catchments in the Mediterranean area.

Table 1. Main characteristics of the investigated catchments. Cfb: temperate warm summer, Dfb: Cold Warm summer; Csa: Temperate dry and hot summer; Csb: Temperate dry and warm summer according to the Köppen classication.

ID#	Basin	Station	Country	Area (km^2)	Mean Elev. (m)	Annual Rainfall (mm)	Daily Temp (°C)	Climate Type	Calibration Period	Validation Period
1	Kolpa	Petrina	Slovenia	460	629	1304	8	Cfb	2007–2009	2010–2012
2	Arga	Arazuri	Spain	810	559	609	13	Cfb	2007–2011	2012–2014
3	Brenta	Berzizza	Italy	1506	1362	701	10	Dfb	2010–2011	2012–2013
4	Gardon	Russan	France	1530	514	679	13	Csb	2008–2011	2012–2013
5	Mdouar	Elmakhazine	Morocco	1800	304	561	18	Csa	2007–2009	2010–2011
6	Kolpa	Metlika	Slovenia	2002	197	920	11	Cfb	2007–2010	2011–2012
7	Volturno	Solopaca	Italy	2580	611	455	15	Csa	2010–2011	2012–2013
8	Lim	Prijepolje	Serbia	3160	612	668	9	Cfb	2007–2008	2009–2010
9	Tanaro	Asti	Italy	3230	927	630	11	Cfb	2010–2011	2012–2013
10	Tevere	M. Molino	Italy	4820	435	710	14	Csa	2007–2011	2012–2015

2.2. Datasets

2.2.1. Soil Moisture Observations

In this study, we used the soil moisture observations derived from the Advanced SCATterometer (ASCAT) onboard the Metop-A and -B satellites [33]. The version of the product is the "H111" which is obtained from the combination of Metop-A and -B satellites and distributed within the "EUMETSAT Satellite Application Facility on Support to Operational Hydrology and Water Management (H-SAF, http://hsaf.meteoam.it/)". H111 is globally available since 2007 with a spatial sampling of 12.5 km and a nearly daily temporal resolution. ASCAT observations were selected because they (1) cover the periods where discharge observations are available, (2) are available in near-real time and (3) are characterized by a relatively good performance in the Mediterranean (e.g., [38,39]). Only for the purpose of characterizing the ASCAT error (see Section 3.2.1) we used the satellite soil moisture observations derived from passive product of the European Space Agency (ESA) Climate Change Initiative (CCI) (http://www.esa-soilmoisture-cci.org/) which is available from 1978 until 2016 with a spatial sampling of 0.25° and a daily temporal sampling [40]. The choice of this dataset—CCI_{pas}—is driven by its independence with respect to ASCAT observations and its full availability during the period of analysis. Both the ASCAT and CCI_{pas} products are spatially resampled over the catchment boundaries to provide watershed-scale average of soil moisture.

2.2.2. Rainfall and Temperature Data

As ground-based rainfall and temperature products, we used the European daily high-resolution gridded data sets of precipitation and air temperature E-OBS [41], developed as part of the EU-FP6 ENSEMBLES project. The rainfall and temperature provided by E-OBS are available for the period 1950 up to now. Given the relatively high density of the rain gauges and thermometers used by this product, E-OBS dataset can be considered as a high-quality meteorological dataset [42]. Rainfall observations derived from E-OBS will be referred for simplicity as EOBS from here onward and will be used for MISDc parameter calibration and for benchmarking non-gauge rainfall observations and stream flow simulations (via MISDc).

As satellite-based rainfall datasets, we used two products. The first product is the real-time version of the Tropical Rainfall Measuring Mission Multi-Satellite Precipitation Analysis (TMPA 3B42RT) which is available from 1997 onward with 3-hourly temporal resolution and a spatial resolution of 0.25° for the 50° north-south latitude band. The retrieval algorithm takes advantages of multiple sensors, by blending polar microwave satellite sensors with geostationary infrared data (e.g., [32]) and does not contain ground-based rain gauge observations. For simplicity, the satellite-based rainfall product is referred to as 3B42RT hereinafter. The second product is the European SM2RAIN-ASCAT dataset currently available from 2007 to 2015, with a spatial/temporal sampling of 0.25 degree/1-day (data of this product are accessible at [43]). Rainfall estimates derived from SM2RAIN-ASCAT are obtained by inverting the soil moisture observations derived from the H111 dataset (see Section 2.2.1) through SM2RAIN [24]. SM2RAIN is an algorithm for estimating rainfall accumulations from soil moisture observations. The method has demonstrated to provide accumulated rainfall estimates with an accuracy comparable (and higher depending on the regions) to state-of-the-art satellite rainfall products [24,26,44]. For further details on the method and its application, the reader is referred to [24].

In addition to satellite-based rainfall products, we used a reanalysis product derived from the ERA-Interim atmospheric ocean and land reanalysis, ERA-Interim ([36], http://www.ecmwf.int) of the European Centre for Medium-Range Weather Forecasts (ECMWF). ERA-Interim precipitation (ERA from here onward) is available from 1 January 1979 to now. In this study, daily precipitation values are obtained from the temporal aggregation of ERA-Interim 12-hourly-precipitation accumulation estimates. ERA-Interim pixels falling inside the catchment boundaries are selected from the bi-linearly interpolated 0.25° grid obtained directly from the ECMWF API.

As for soil moisture observations, rainfall and temperature data obtained from the foregoing products are averaged at the watershed scale for all the study catchments.

2.2.3. Stream Flow Data

Daily stream flow data are available for all the study catchments as described in Table 1. The discharge dataset ranges from 4 to 9 years depending on the catchment with a mean length of nearly 6 years and were collected from local authorities and from the Global Runoff Data Center (GRDC, http://www.bafg.de/GRDC/EN/01_GRDC/grdc_node.html) for the stream flow data of two Kolpa and Lim catchments.

3. Methods

In this study, the hydrological simulations are obtained through a modified version of the MISDc hydrological model [35,45]. Two data assimilation strategies are used to ingest satellite soil moisture observations into MISDc. The first one is the classical EnKF where the model states are updated by weighing the relative accuracy of ASCAT soil moisture observations with the model predictions by using a Montecarlo based approach. The second one, already used in [4,25], is based on a simple static integration scheme.

3.1. MISDc

MISDc—"Modello Idrologico Semi-Distribuito in continuo"—is a continuous rainfall-runoff model developed by [35] for the operational forecasting of flood events in the Tevere River Basin (central Italy). In this paper, a two-layers version of the model is used. With respect to the previous version, it includes a snow module and a different infiltration equation. The model uses as input daily rainfall and air temperature data and simulates the temporal evolution of two independent soil water states W_1 and W_2. Water is extracted from the first layer by evapotranspiration which is calculated by a linear function between the potential evaporation (estimated via the Blaney and Criddle relation modified by [45,46] and the soil saturation. A non-linear relation proposed by [47] calculates percolation from the surface to the root zone layer. The rainfall excess is calculated by a power law relationship as a function of the first layer soil saturation while base flow is a non-linear function of the soil moisture content of the third layer [48].

Three different components contribute to generate runoff: the surface runoff, the saturation excess from the surface and the deep layer and the sub-surface runoff component. The first two are summed and routed to the outlet by the Geomorphological Instantaneous Unit Hydro-graph (GIUH) while the subsurface runoff is transferred to the outlet section by a linear reservoir approach. For both routing schemes, the lag time is evaluated by the relationship proposed by [49]. Full details on model equations are already given in [35] and, hence, are not repeated here.

The 10 model parameters of the model are shown in Table 2 along with the assumed range of variability.

Table 2. List of the calibrated parameter of MISDc model.

Parameter	Description	Range of Variability	Unit
$Wmax_1$	Maximum water capacity of the first layer	150	mm
$Wmax_2$	Maximum water capacity of the second layer	300–4000	mm
m_1	Exponent of drainage for 1st layer	2–10	-
m_2	Exponent of drainage for 2nd layer	5–20	-
Ks_1	Hydraulic conductivity of the 1st layer	0.1–20	mm/day
Ks_2	Hydraulic conductivity of the 2nd layer	0.01–45	mm/day
γ	Coefficient lag-time relationship	0.5–3.5	-
Kc	Parameter of potential evapotranspiration	0.4–2	-
α	Exponent of the infiltration relationship	1–15	-
C_m	Snow module parameter degree-day	0.004–3	°C/day

3.2. Soil Moisture Data Assimilation

3.2.1. Pre-Processing of Soil Moisture Observations and Error Estimation

Satellite soil moisture observations are representative of a shallow soil layer of 2–3 cm while the model first layer water capacity is 150 mm (which roughly corresponds to 300 mm by assuming a reasonable range of porosity between 0.45 and 0.5). Therefore, prior to use them for the hydrological data assimilation they require a pre-processing step to address the depth mismatch with the model state [13,14,16]. For this purpose, the recursive formulation of the Exponential filter [50,51] was used to obtain the so-called Soil Water Index (SWI) based on a single parameter, T, named characteristic time length. T was optimized by maximizing the correlation between SWI obtained from ASCAT (SWI_{ASCAT}) and the model state of the first model layer (W_1). Before the assimilation, the satellite soil moisture observations were bias corrected to the model climatology by the quantile mapping approach [52]. A 2nd order polynomial function was used for mapping SWI data to the model, thus obtaining SWI^*_{ASCAT}. The same processing steps applied to ASCAT were applied to the CCI_{pas} to obtain SWI^*_{CCIpas}. This was done for obtaining the same climatology and dynamic range of W_1 and reduce the impact caused by the different vertical representativeness when used within the Triple Collocation (TC, [53]) analysis (see below).

Given the relatively poor presence of ground monitoring stations of soil moisture, the estimation of the satellite soil moisture error variance needed for data assimilation is not an easy task. In this respect, TC has demonstrated to be a reliable technique [54] to estimate the error variance of three independent soil moisture datasets provided that they are characterized by zero cross correlation errors. Here, we follow the approach of [29] and applied TC to soil moisture observations derived from the triplets built among SWI^*_{ASCAT}, SWI^*_{CCIpas} and W_1 to calculate the SWI^*_{ASCAT} error variance ($\sigma*_{ASCAT}$). In practice, the three datasets are decomposed into the corresponding climatology anomalies time series by subtracting the long-term 31-day moving average from the raw time series. This guarantees the estimation of only random error sources and a more accurate observation error variance estimate [55]. The application of TC is only performed for the calculation of the scaled error variance of SWI^*_{ASCAT} which is then used within the SM-corr approach.

The calibration of the parameters involved in the pre-processing steps (T and the coefficients of the quantile mapping) along with the estimation of the ASCAT error variance were carried out during the calibration period in order to maintain a rigorous separation with the validation period and guarantees a more objective evaluation of the different methodologies (see Section 3.5).

3.2.2. The Ensemble Kalman Filter

In hydrological data assimilation, EnKF (and its variations) have been largely used because of their computational efficiency and flexibility. The EnKF is based upon Monte Carlo method and the Kalman filter formulation to approximate the true probability distribution of the model state, conditioned on a series of observations of the model states. The EnKF was introduced as an alternative to the traditional Extended Kalman filter which has been shown to be problematic because of the strongly nonlinear dynamics of hydrological models.

Being $\boldsymbol{Y}(t_k)$ the vector of system states at time step t_k, $\boldsymbol{Y}(t) = [W_1(t_k), W_2(t_k)]^T$ obtained via a generic model and $\boldsymbol{Z}_k$ the observation vector at time t_k, then, the optimal updating of $\boldsymbol{Y}_k$, can be expressed as:

$$\boldsymbol{Y}_k^{i+} = \boldsymbol{Y}_k^{i-} + \boldsymbol{G}_k\left(\boldsymbol{Z}_k + v_k^i - \boldsymbol{H}_k\boldsymbol{Y}_k^{i-}\right) \quad (1)$$

where $\boldsymbol{Y}_k^{i-}$ and $\boldsymbol{Y}_k^{i+}$ refer to the forecast and analysis states for the ith ensemble member, respectively, $\boldsymbol{H}_k$ is the observation operator that maps the model states to the observations, v_k is a synthetically generated error added to the observation $\boldsymbol{Z}_k$ and represents the uncertainties of the observation process that is assumed to be a mean-zero Gaussian random variable with variance $\boldsymbol{R}_k$. $\boldsymbol{G}_k$ is the Kalman gain:

$$\boldsymbol{G}_k = \boldsymbol{P}_k^{i-}\boldsymbol{H}_k^T\left(\boldsymbol{H}_k\boldsymbol{P}_k^{i-}\boldsymbol{H}_k^T + \boldsymbol{R}_k^-\right)^{-1} \quad (2)$$

where $\boldsymbol{P}_k^{i-}$ is 2×2 covariance matrix of forecast error obtained from the N-member ensemble of background predictions:

$$\boldsymbol{P}_k^{i-}\frac{\left[\boldsymbol{Y}_k^{i-} - \left\langle\boldsymbol{Y}_k^{i-}\right\rangle\right]\left[\boldsymbol{Y}_k^{i-} - \left\langle\boldsymbol{Y}_k^{i-}\right\rangle\right]^T}{N-1} \quad (3)$$

Given the preprocessing applied to the satellite SM products (see Section 3.2.1), the observation operator in Equation (2) reduces to $\boldsymbol{H} = [1\ 0]^T$ while the observation error covariance matrix, $\boldsymbol{R}_k$, reduces to σ^{*2}_{ASCAT} since only pre-processed soil moisture derived from ASCAT is assimilated. The single deterministic EnKF prediction (i.e., the "analysis") is calculated by averaging model state predictions, $\boldsymbol{Y}_k^{i-}$, and the consequent stream flow at each time step across the N members of the ensemble. In Equation (3), $\left\langle\boldsymbol{Y}_k^{i-}\right\rangle$ denotes the mean of $\boldsymbol{Y}_k^{i-}$. The covariance matrix of the forecast error was obtained by perturbing rainfall and temperature data along with the model soil moisture predictions (see next section).

3.2.3. Filter Calibration

In this study, we adopted a multiplicative model error [56] for perturbing rainfall observations (assuming no autocorrelation in the rainfall error) and an additive perturbation for temperature [13] and soil moisture predictions. These perturbations aimed to represent the main sources of model error, coming from the forcing data, the model parameters and the model structure. Rainfall was perturbed by a log-normally distributed, unit mean, spatially homogeneous and temporally uncorrelated multiplicative random noise with standard deviation equal to σ_P, whereas for temperature and soil moisture a zero-mean normally distributed additive error was chosen with standard deviations equal to σ_P and σ_M, respectively. σ_T was chosen equal to 1 °C while $\sigma_M = 10^{-3}$. σ_P was made variable between 10^{-5} and 2 by assuming that the main error of the model is associated to the uncertainty in the precipitation forcing. The optimal value σ_P was selected by picking the one that, by running the filter during the calibration period, minimizes the root mean square error (RMSE) between simulated (ensemble-averaged) and observed discharge time series and ensures that innovations (the second term in square brackets in Equation (1)) have zero mean and are serially uncorrelated [57]. Note that, given the model non-linearity, the satisfaction of the latter criterion was very difficult to obtain and thus was not always guaranteed. In addition, the model error calibration based on the minimization of the RMSE as done here, assumes that the error in stream flow observations is negligible. However, if significant errors are present in observed stream flow, this procedure may be sub-optimal and the filter inflated by these errors. Alternative procedures that guarantee a more optimal filter performance are also possible [56] but are beyond to scope of the paper and are not treated here. The ensemble size was set to 50 members, more numerous ensembles were also tested but did not provide significant changes therefore $N = 50$ was finally set to speed up the calculations.

3.3. Rainfall Correction

3.3.1. Pre-Processing of Rainfall Observations

Prior to run the model with satellite-based rainfall estimates we bias-corrected SM2RAIN-ASCAT, ERA and 3B42RT with E-OBS rainfall observations. The main reason of the bias correction of satellite-based rainfall estimates is that—due to the indirect nature of the measurement—they are potentially affected by significant biases that can seriously impact the quality of the hydrological simulations [58]. For that, we used the same quantile mapping approach used for soil moisture bias correction [52] with a 2nd order polynomial function (different orders were preliminary tested and the 2nd order was found to perform the best). The bias corrected products will be referred to $P_{SM2RAIN\text{-}ASC}$ for SM2RAIN-ASCAT, P_{ERA} for ERA-Interim and P_{3B42RT} for 3B42RT from here onward. To maintain a consistent notation, the original EOBS rainfall product will be also denoted as P_{EOBS} in the following. As for the pre-processing of soil moisture the bias correction calibration parameters were determined during the calibration period (see Section 3.5) and then used in validation.

3.3.2. Rainfall Integration

The merging between $P_{SM2RAIN\text{-}ASC}$ and the specific rainfall product (i.e., P_{3B42RT} or P_{ERA}) was carried out by a simple Newtonian nudging scheme [4]:

$$P_{COR}(t) = P_{SM2RAIN-ASC}(t) + K[P^*(t) - P_{SM2RAIN-ASC}(t)] \quad (4)$$

where t is time, P_{COR} is the corrected rainfall product ($P_{ERA+\ SM2RAIN\text{-}ASC}$ or $P_{3B42RT+SM2RAIN\text{-}ASC}$), P^* is P_{3B42RT} or P_{ERA}. K is a static weighting parameter estimated during the calibration period by minimizing the RMSE between simulated stream flow time series and observations. K gives the relative weight of $P_{SM2RAIN\text{-}ASC}$ with respect to the satellite (reanalysis) rainfall product. K equal to 1 means that the error in satellite (reanalysis) rainfall is much lower than $P_{SM2RAIN\text{-}ASC}$ and no correction is performed while K equal to 0 means that $P_{SM2RAIN\text{-}ASC}$ error is much lower than satellite (reanalysis) rainfall, thus only $P_{SM2RAIN\text{-}ASC}$ is used. To maintain a similar methodology approach with the one used in Section 3.2.3, K was calibrated by minimizing the RMSE between simulated and observed discharge time series during the calibration period. The calibrated K for each basin were then used in validation. Being calibrated based on the RMSE between simulated and observed stream flow time series, the determination of K is subjected to the same limitations described in Section 3.2.3 (i.e., it can be inflated by errors contained in observed stream flow).

3.4. Performance Metrics

The hydrological performance of the different simulations was assessed through three different metrics specifically targeted for floods. The first metric—to evaluate the prediction of high flows—is the Nash–Sutcliffe efficiency (NSE, [59]) adapted to high-flow conditions (ANSE, [60]):

$$ANSE = 1 - \frac{\sum_{t=1}^{Nt}(Q_{obs} + \langle Q_{obs}\rangle)(Q_{sim} + Q_{obs})^2}{\sum_{t=1}^{Nt}(Q_{obs} + \langle Q_{obs}\rangle)(\langle Q_{obs}\rangle + Q_{obs})^2} \tag{5}$$

The second metric—specifically designed for low flows—is the NSE calculated on logarithmic discharges:

$$NS_{lnQ} = 1 - \frac{\sum_{t=1}^{Nt}[log(Q_{sim} + \varepsilon) - log(Q_{obs} + \varepsilon)]^2}{\sum_{t=1}^{Nt}[log(Q_{obs} + \varepsilon) - log(\langle Q_{obs}\rangle + \varepsilon)]^2} \tag{6}$$

In Equations (5) and (6) the sum is carried over the length N_t of the observed, Q_{obs}, and simulated Q_{sim}, discharges vectors. The term ε in Equation (6), was arbitrarily chosen as a small fraction of the inter-annual mean discharge (e.g., $\langle Q_{obs}\rangle/40$) and was introduced to avoid problems with nil observed or simulated discharges.

The third metric, the Kling-Gupta efficiency [61] was adopted for assessing the simulation performance for a wider range of flow conditions (but with a tendency to privilege high flow given that it was introduced for flood assessment). The KGE is a performance indicator based on the equal weighting of three sub-components: linear correlation (r), bias ratio ($\beta = \mu_{sim}/\mu_{obs}$), and variability ($\delta = CV_{sim}/CV_{obs}$), between Q_{sim} and Q_{obs}. KGE is defined as follows:

$$KGE = 1 - \sqrt{(r-1)^2 + (\beta - 1)^2 + (\delta - 1)^2} \tag{7}$$

being μ and CV the mean and the coefficient of variation. $ANSE$, NS_{logQ} and KGE vary between $-\infty$ and 1 with values equal to one denoting perfect agreement between stream flow observations and simulations.

3.5. Method Implementation

The implementation of the two schemes (i.e., SM-corr and P-corr) was carried out by organizing the datasets in two sub-periods; the first sub-period was used for model parameters calibration and for the calibration of the parameters associated to the data assimilation/integration schemes. In particular, about 50–60% of discharge observations—depending on the basin—were assigned to the calibration period by guaranteeing a minimum time period of two years (see Table 1). The remainder of the period was used for validation.

The calibration of the model parameters was carried out by maximizing the KGE index between observed and simulated discharge time series obtained by forcing MISDc model with P_{EOBS} rainfall and temperature (see Figure 2a) through a standard gradient-based automatic optimization algorithm [62]. The calibrated parameters were then used within all the simulations involving P_{3B42RT} and P_{ERA}. As denoted in Sections 3.2.1, 3.2.3 and 3.3.1, during calibration, we determined the (1) optimal rainfall bias correction parameters, (2) the parameter K related to the rainfall integration, (3) the data assimilation parameters (i.e., the characteristic time length T and the parameters associated to the bias correction of soil moisture), (4) the satellite soil moisture observation error (σ^*_{SWI}) and (5) the forecast model error. The calibrated integration/assimilation parameters were then used during the validation periods for obtaining discharge simulations for SM-corr and P-corr for a total of six different runs (Figure 2b). That is, the two off-line simulations obtained by forcing the model with P_{ERA} (OLM) and P_{3B42RT} (OLS), the two data assimilation experiments where ASCAT soil moisture observations were assimilated into MISDc forced with P_{ERA} (DAM) and P_{3B42RT} (DAS) and the two integration experiments where corrected rainfall $P_{SM2RAIN\text{-}ASC + ERA}$ (RCM) and $P_{SM2RAIN\text{-}ASC + 3B42RT}$ (RCS) were used to force MISDc. As a baseline for comparing the performance of satellite and reanalysis rainfall products, also the runs in which MISDc was forced with P_{EOBS} were considered (OLG in Figure 2b).

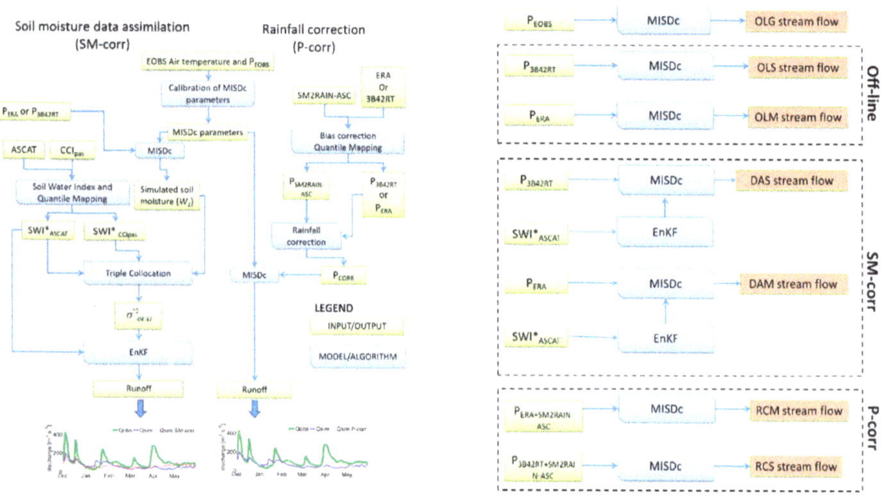

(**a**) (**b**)

Figure 2. (**a**) Flowchart illustrating the main implementation steps of the soil moisture data assimilation (SM-corr, left) and the rainfall correction (P-corr, right) methods. (**b**) Simulation runs used in the study.

4. Results and Discussion

4.1. MISDc Model Calibration and Validation Forced with Ground-Based Data

The calibrated parameters of MISDc (obtained by forcing the model with P_{EOBS}) provide a median KGE efficiency index equal to 0.692 (i.e., always above 0.6 except for Volturno, see Table 1). In this respect, MISDc performance can be considered relatively good and between the intermediate (0.75 > KGE > 0.5) and good (KGE > 0.75) level as identified in [63]. This ensures the reliability of the model for stream flow simulations. Based on Table 1 (and Table 3) it can be observed that cold and more humid catchments generally perform better than warm and drier ones.

Table 3. Kling-Gupta performance index obtained during calibration by forcing the model with P_{EOBS} (CALIBRATION in dark grey) and during the validation period (VALIDATION in light grey and blue) for (1) MISDc model forced with P_{EOBS} (OLG in white), (2) MISDc model forced with P_{ERA} and P_{3B42RT} (OLM, OLS), (3) the state correction scheme (DAM, DAS) and (4) the rainfall correction scheme (RCM, RCS). Numbers in bold refer to the best score obtained in validation among OLG, OLM, DAM, RCM, OLS, DAS and RCS.

BASIN	CALIBRATION	VALIDATION						
		OLG	OLM	DAM	RCM	OLS	DAS	RCS
Kolpa@Petrina	0.817	**0.637**	0.510	0.510	0.508	0.426	0.425	0.389
Arga	0.770	**0.536**	0.419	0.373	0.438	0.135	0.143	0.530
Brenta	0.701	**0.414**	0.379	0.398	0.366	0.328	0.313	0.321
Gardon	0.665	**0.736**	0.716	0.716	0.689	0.537	0.536	0.480
Mdouar	0.683	**0.562**	−1.085	−1.320	−0.376	0.234	0.379	0.136
Kolpa@Metilka	0.709	**0.796**	0.656	0.655	0.624	0.588	0.363	0.510
Volturno	0.416	0.426	0.193	0.187	0.228	0.090	0.093	**0.508**
Lim	0.680	0.420	0.526	0.524	**0.714**	0.279	0.165	0.617
Tanaro	0.713	**0.262**	0.152	0.197	0.152	0.121	0.097	0.121
Tevere	0.603	0.417	0.327	0.434	0.474	0.299	0.320	**0.701**
Median	0.692	0.481	0.399	0.416	0.456	0.289	0.317	0.494

4.2. Satellite Soil Moisture Pre-Processing and Filter Calibration

Figure 3a shows the parameter T that maximises the correlation coefficient between the modelled soil moisture of the first layer W_1 and the SWI^*_{ASCAT} for all the analysed catchments for P_{ERA} and P_{3B42RT}. T is lower than 20 days for most of the catchments except Arga and Volturno where it reaches a value of about 60 days. These results are consistent with the range of values found in previous studies (e.g., [13,16,30,64]). There is not a specific pattern that is possible to identify for the study catchments because T variations are not only related to the specific catchment hydrology but also to the model and the satellite observation quality

Figure 3b shows the observation error variances of SWI^*_{ASCAT} obtained by considering the triplets among SWI^*_{ASCAT}, SWI_{CCIpas} and the soil moisture simulated by MISDc model forced with P_{ERA} (P_{3B42RT}). The error variances found with the two triplets maintain a similar comparative relationship among basins showing smaller values for drier and warm catchments (Tevere, Arga, Mdouar) and larger values for more cold and humid (mountainous) catchments (Kolpa@Petrina, Gardon, Lim). The relatively better performance of ASCAT in semi-arid environments is consistent with the results of [64,65].

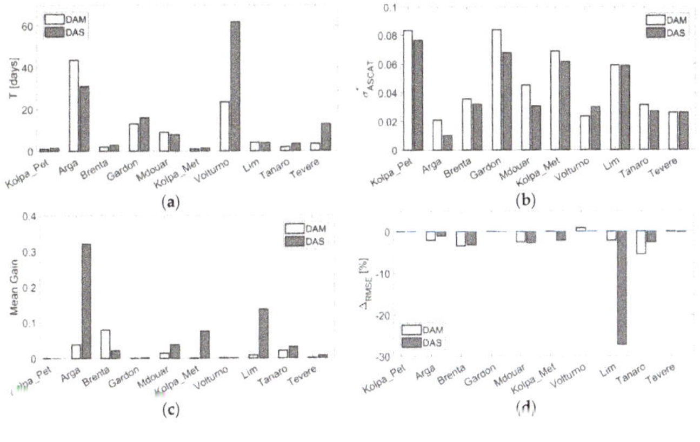

Figure 3. Characteristic time length T (**a**), SWI^*_{ASCAT} error standard deviation (**b**), mean Kalman Gain G (**c**), and % improvement in RMSE between simulated and observed stream flow (**d**) obtained during the calibration period within the SM-corr approach by using P_{3B42RT} (DAS) and P_{ERA} (DAM) for forcing MISDc model.

Figure 3c,d plot the mean Kalman gain, G, for the first model layer and the RMSE changes obtained after the assimilation of SWI^*_{ASCAT} during the calibration period. The reduction in RMSE is relatively low and less than the 10% found by [29] with some catchments characterized by no-improvements.

4.3. Rainfall Correction Calibration

Figure 4a summarises the values of the parameter K obtained during the calibration period for all the investigated catchments while Figure 4b shows the reduction in RMSE between observed and simulated stream flow after integrating P_{ERA} and P_{3B42RT} with $P_{SM2RAIN\text{-}ASC}$ through Equation (4). It can be seen that K is significantly higher for RCM (mean $K = 0.77$) with respect to RCS (mean $K = 0.42$) suggesting a higher quality of P_{ERA} with respect to P_{3B42RT}. It can be also seen that lower K values (i.e., which means that $P_{SM2RAIN\text{-}ASC}$ is weighed more with respect to the counterpart product in Equation (4) provide a larger decrease in RMSE and this reduction is generally larger for RCS with respect to RCM.

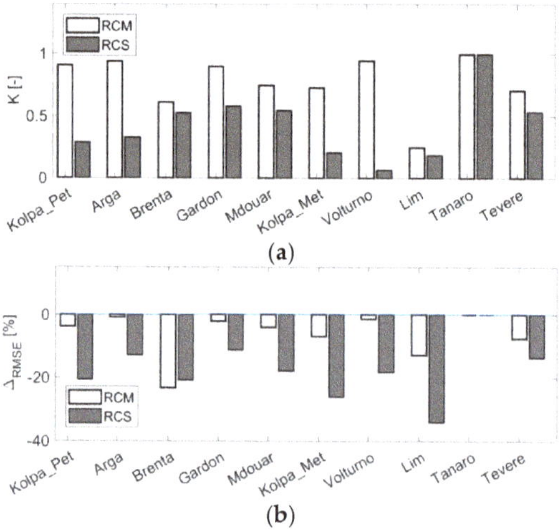

Figure 4. Values of the calibrated gain parameter K for P_{ERA} and the P_{3B42RT} obtained during the calibration period (**a**) and % reduction in RMSE between observed and simulated stream flow during the calibration period (**b**). K close to zero indicates that more weight is assigned to $P_{SM2RAIN\text{-}ASCAT}$ dataset according to Equation (4).

4.4. Rainfall Evaluation

Given the relatively high density of rain gauges used in EOBS dataset, P_{EOBS} can be considered a good reference for evaluating the performance of P_{3B42RT} and P_{ERA} (and their associated integrated products) over the study catchments. Figure 5 shows the scatter plots of the correlations between P_{3B42RT}, P_{ERA}, $P_{ERA+SM2RAIN\text{-}ASC}$, $P_{3B42RT+SM2RAIN\text{-}ASC}$ and P_{EOBS} for each catchment in Table 1 during the validation period.

Panel a compares the correlation between P_{ERA} and P_{3B42RT}. It can be seen that P_{ERA} performs relatively better than P_{3B42RT} for almost all catchments thus confirming the previous results in terms of K obtained in calibration. Similar results were also found by [44] who observed a higher quality of ERA-Interim in Europe with respect to 3B42RT. The only exception is the Mdouar catchment. Here the reanalysis product performs relatively worse than the satellite-based one. A possible reason of the lower performance is related to the type of precipitation that characterizes this area (stratiform vs. convective precipitation) as also found in [66–68].

The integration between $P_{SM2RAIN\text{-}ASC}$ and P_{ERA} (P_{3B42RT})—which is based on minimization of the RMSE between observed and simulated stream flow during the calibration period—indirectly leads to increased rainfall quality both during the calibration (not shown) and the validation periods (Figure 5b,c). In Figure 5b,c, the Lim catchment (#8), is the only catchment where the integration provides a significant deterioration of the correlation. A slight deterioration is also observed for Kolpa@Petrina (#1) and Kolpa@Metilka (#6) when integrating P_{3B42RT} and $P_{SM2RAIN\text{-}ASC}$. However, these deteriorations (including the one of Lim catchments) are significantly lower when the rainfall products are compared in terms of RMSE (not shown). Possible reasons of the deteriorations are related to ASCAT error that, in this catchment is relatively higher (see Figure 3) and to the SM2RAIN limitations when the soil is close to saturation (see below).

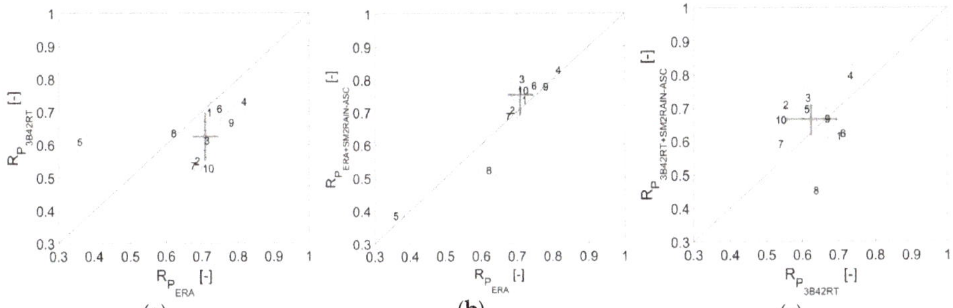

Figure 5. (**a**) comparison of correlations R between P_{ERA} and P_{3B42RT} obtained with P_{EOBS} for all study catchments during the validation period; (**b**) same as panel (**a**) but between P_{ERA} and $P_{ERA+SM2RAIN\text{-}ASC}$; (**c**) same as panel (**a**) but between P_{3B42RT} and $P_{3B42RT+SM2RAIN\text{-}ASC}$. The points where the lines cross refer to the medians while the line edges represent the 25th and the 75th percentiles.

4.5. Stream Flow Evaluation

Table 3 shows the performance obtained during the validation period for the off-line (OLM, OLS), SM-corr (DAM, DAS) and P-corr (RCM, RCS) runs along with the stream flow simulations obtained by forcing the model with P_{EOBS} during the validation period (i.e., OLG run).

OLG performs relatively well for 5 out of 10 catchments (Kolpa@Petrina, Kolpa@Metilka, Gardon, Mdouar, Lim) with median KGE equal to 0.481. For the others catchments stream flow simulations are poor with the Tanaro basin providing the worst result. The performance for OLM and OLS are in general lower than OLG with median KGE equal to 0.399 and 0.289, respectively (i.e., below the intermediate level of 0.5). For OLM, the simulation for Mdouar provides KGE lower than zero due to the poor quality of P_{ERA} precipitation in this basin (as seen in Section 4.4). The topographic complexity of the study area along with the strong non-stationary performance of satellite-based rainfall products over time caused by the season and by the variable number of satellite microwave passes used for the retrieval of precipitation [42,69,70] are the main causes of the low scores obtained in the stream flow simulations with P_{3B42RT} [71]. In practice, the satellite precipitation error has both (1) a direct effect on stream flow estimates by determining under(over) estimations due to the erroneous instantaneous precipitation and (2) an indirect effect on the state estimation that propagates in time for several days/months causing additional stream flow errors. The low scores of the stream flow estimates derived from satellite-based rainfall observations are in line with those found in many other studies in literature [30,37,71,72]. In the latter, it was who found that reanalysis-based rainfall products generally outperform satellite-based ones in hydrological modelling.

SM-corr (DAM and DAS) has generally a small positive impact on stream flow simulations in terms of KGE, with an increase in median KGE from 0.399 to 0.416 (about 4%) for DAM and from 0.289 to 0.317 for DAS (10%). The catchments that benefit more from the SM-corr scheme are

Brenta, Tanaro and Tevere for DAM and Arga, Mdouar and Tevere for DAS (although for some of them KGE remains very low). One main reason for the small or no improvements observed for some catchments (e.g., Kolpa@Metilka, Kolpa@Petrina and Gardon) can be due to the specific runoff generation mechanism/model structure/data assimilation configuration. In this respect, we found that the correction of the model state in the root-zone layer for these catchments is always very small, meaning that the model is characterized by a weak coupling between the first and the second layer. Therefore, if the runoff generation mechanism is mainly associated to the root zone, then the assimilation of surface observations has a negligible impact on stream flow simulations. Similar results were also found in [11,13,15]. Other possible reasons are directly related to the ASCAT quality itself and to the different pre-processing steps that characterize the state correction scheme [16,30]. In this respect, the application of TC, the rescaling procedure and the filter calibration were applied to the calibration period and it is not guaranteed that they are optimal also in a different period (e.g., Lim catchment). This is particularly true for the satellite soil moisture error variance and for the bias which are characterized by a non-stationary behaviour over time [73]. Moreover, the introduction of auto correlated errors in the observation error derived from the application of the Exponential Filter [30] is another reason of a potential sub-optimal performance.

For P-corr, the increase in median KGE is from 0.399 to 0.456 (14%) for RCM and from 0.317 to 0.494 (71%) for RCS indicating that the precipitation correction has a remarkable impact on the stream flow simulations. In particular, RCS provides median KGE larger than the performance obtained in OLG (i.e., the model forced with P_{EOBS}) and those obtained with RCM. Here, KGE is equal 0.494 for OLS and 0.481 for OLG with Volturno and Tevere being the best among OLG, OLS, DAS, DAM. In a recent study by [72] it was found that gauge-based and reanalysis products generally outperform satellite based products for flood simulations [72]. We found that the correction of 3B42RT with SM2RAIN rainfall estimates performs better than simulations using gauge-based observations (i.e., EOBS). This encouraging result demonstrates the potentiality to improve operational stream flow forecasting by using remotely sensed surface soil moisture.

With respect to the results obtained with OLM, both RCM and RCS are able to increase KGE scores in drier and warm catchments (e.g., Arga, Volturno, Lim and Tevere) while in more humid and cold basins we observed deteriorations or no improvement (Kolpa@Petrina, Kolpa@Metilka Tanaro and Brenta). One reason for the deterioration is the problem of saturation associated to the SM2RAIN rainfall estimates. In practice, when surface soil moisture reaches the saturation (which occurs mode frequently in more humid climates) SM2RAIN can no longer reliably inverts soil moisture to precipitation [24] and can provides significant underestimation of the rainfall events. Similar issues were found in [11] by correcting rainfall from 3B42RT via the SMART algorithm. Another possible reason for the deteriorations is related to the static nature of the integration scheme combined with the variable performance in time of satellite-based and reanalysis products. For the latter, the variable performance in time of precipitation depends on its underlying nature (stratiform or convective, [74] with potential underestimation of total precipitation for convection dominated conditions (which occur during summer/earlier autumn in the Mediterranean area, [75]). In practice, both for satellite- and reanalysis-based products it is not guarantee that K values found during the calibration period are optimal also during the validation period. In this respect, more optimal and dynamic (as a function of the current retrieval error) integration strategies will likely lead to better results.

To analyse the impact of the SM-corr and P-corr schemes on low and high flows we plotted the box plots of ANSE and NS_{lnQ} indexes in Figure 6. For ANSE, both the SM-corr and P-corr schemes improve the model performance obtained in the off-line simulations with a clear advantage of the P-corr scheme. In median, the enhancements obtained for RCM and RCS are larger with respect to DAM and DAS and allow to obtain ANSE values close to the ones obtained in OLG (note that as for the KGE results presented in Table 3 some catchments—not shown—have ANSE values larger than the ones obtained in OLG). For low flow conditions, the performance of the model is generally lower and cannot be observed a clear advantage of one technique with respect to the other. However, in median, DAM and DAS provide a slightly better performance (note that for DAM, these scores are above those obtained in OLG). These results are consistent with those found by [11] where the rainfall correction scheme improved better high flows with respect to state correction. They are less consistent with the results of [30] who showed a higher positive impact on the stream flow prediction of the state correction with respect to the forcing correction scheme both for high and low flows (for the latter the improvements were higher though). The smaller increments obtained in the DAM and DAS cases are in line with those found in [28] and lower with respect to the ones in [11] where the state correction scheme implemented via EnKF benefited from the correction of the ensemble perturbation bias.

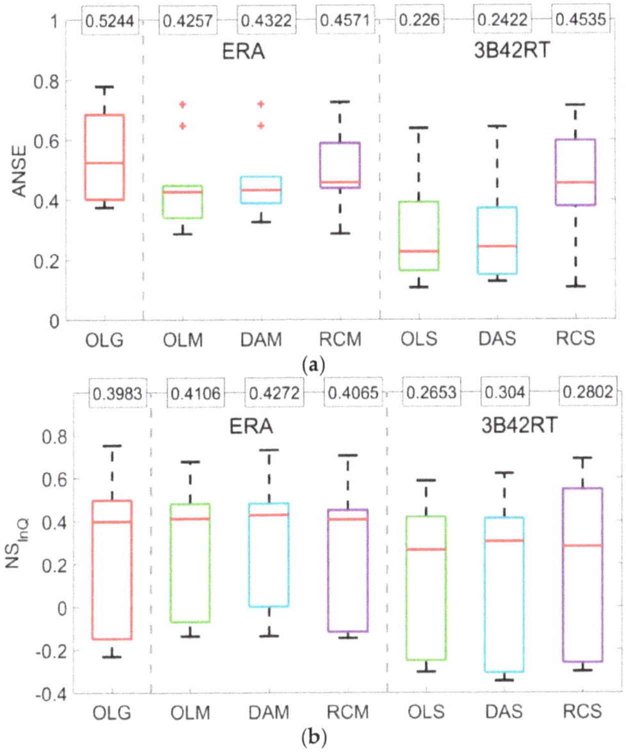

Figure 6. Summary of the results in terms of ANSE (Nash–Sutcliffe efficiency for high-flow conditions) and NS_{lnQ} (NS adapted for low flow conditions) for all the investigated basins during the validation period. Red box plots refer to the results obtained by forcing MISDc with P_{EOBS} datasets; the number in the square boxes represent the median values. Results are shown for the off-line simulations (OLM, OLS), for the SM-corr scheme (DAM, DAS) and for the P-corr scheme (RCM, RCS).

Figure 7 shows the stream flow simulations for the OLM, DAM and RCM (first column) and OLS, DAS and RCS (second column) during the calibration (white background) and validation periods (grey background) for two representative catchments (Kolpa@Petrina and Tevere). In each panel, the upper plot shows the comparison in terms of stream flow whereas the bottom one displays the RMSE variation in time between simulate and observed stream flow smoothed time series (a moving mean of 60 days was chosen for sake of visualization). It can be seen that for Kolpa@Petrina catchment both the SM-corr and P-corr schemes fail to improve stream flow simulations with negligible effect on the time series. Some improvements can be seen between OLS and RCS until January 2011 but strong deteriorations occur from January 2012 onward with severe underestimation of the flow peaks during winter 2012 likely due to the issues of saturation discussed before (K for RCS is very small in this catchment and the integrated rainfall largely consists of $P_{SM2RAIN\text{-}ASC}$ rainfall estimates). This assumption is supported by the detrimental effect of the integration on the precipitation for this catchment as plotted in Figure 5c.

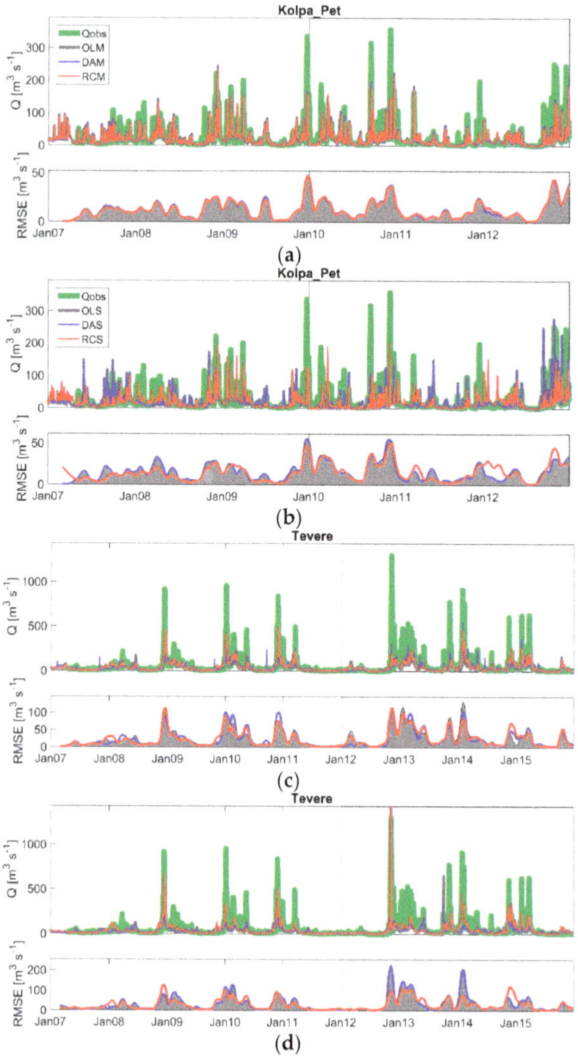

Figure 7. Stream flow simulations in the calibration and validation period for Kolpa-Pet (panels **a**,**b**) and Tevere (panel **c**,**d**) catchments. For each catchment, the results for OLM, OLS, DAM, DAS, RCM and RCS are shown. In each panel, the upper plot shows the comparison in terms of stream flow while the bottom one shows the smoothed time series of the RMSE (by using a moving windows of 60 days for sake of visualization) between observed (Q_{obs}) and the three simulated stream flow. The white background refers to the calibration period and the grey background to the validation.

For Tevere, we can see a more marked effect of the SM-corr and P-corr schemes with better skills for DAM for low flow conditions with respect to RCM and vice versa for high flows. The advantage of the P-corr scheme is more remarkable when the satellite precipitation product is considered. Here, RCS outperforms DAS both during the validation and the calibration period for high flow conditions, while it shows some deteriorations in terms of low flows. RCS presents also deteriorations which are particularly relevant during January 2015 (also present for RCM simulation in panel c). The climate (more humid for Kolpa@Petrina) and the hydrologic differences between the two catchments (Kolpa@Petrina is characterized by a higher baseflow component with respect to Tevere) might explain the different benefit of the state and rainfall correction schemes on the simulations.

5. Conclusions

The current availability of different satellite soil moisture products, together with the well-known importance of soil moisture observations for flood prediction, asks for the development of optimal approaches to be implemented for the full exploitation of such products. In this study, we compared two approaches for exploiting satellite soil moisture retrievals derived from the scatterometer ASCAT, namely, the state-correction scheme implemented through classical data assimilation via the EnKF and the rainfall correction scheme through the integration of satellite (and reanalysis) rainfall observations with SM2RAIN rainfall estimates. The experiments were conducted in the Mediterranean on 10 catchments of different size and different climatic conditions. A rigorous separation between calibration and validation was adopted in order to remove possible interferences of the calibration steps necessary for the selection of the parameters associated with the two schemes. Ground-based observations of discharge and rainfall were used for model calibration and for benchmarking the different simulation runs. A satellite- and reanalysis based rainfall product (i.e., 3B42RT and ERA-interim) were used to simulate a scenario of ground data scarcity as happens in many developing countries. For these two products, we performed the state and the rainfall correction schemes.

Based on the obtained results, the following main conclusions can be drawn:

1. The gauge-based rainfall dataset (EOBS) performs satisfactorily well over the Mediterranean area with median ANSE and KGE values close to 0.5 (in validation) for the investigated catchments while P_{ERA} and P_{3B42RT} provide poorer stream flow predictions.
2. The soil moisture correction produces an overall slight improvement in terms of median KGE and ANSE scores (4.25% and 1.5% for ERA-Interim and 9.6% and 7.6% for 3B42RT, respectively) whereas the rainfall correction provides a much larger impact with an increase in KGE and ANSE values equal to 14.81 and 7.3% for ERA and 71.8 and 100% for 3B42RT, respectively. In summary, the impact of the rainfall correction for flood simulation is much larger than the soil moisture correction and is consistently higher when the quality of the non-corrected rainfall forcing is poor. Conversely, for low flows, the soil moisture correction schemes provide slight better results but these improvements are limited.
3. After the rainfall correction, the simulation run using the satellite-based product (i.e., 3B42RT) shows KGE scores larger than those obtained by using ground-based observations (EOBS). This is an encouraging result that demonstrates the potentiality to improve operational stream flow forecasting by using remotely sensed surface soil moisture.
4. The climate, the specific catchment hydrology/model configuration/data assimilation set up and the pre-processing steps associated with the two schemes exert a remarkable effect on the results that complicates the answer to weather is preferable correcting rainfall or updating the model states.

These results, which go in the same direction of few previous studies found in literature, can be very useful for advancing the understanding of the optimal use of satellite products for hydrology since they involve an area (i.e., the Mediterranean) which is very sensitive to climate change and that was not tested yet. Being subjected to many assumptions and limited in what concern the choice of the catchments, the analysis periods and the type of satellite soil moisture product (i.e., ASCAT), there is a moderate risk that these results can be case specific therefore the comparison of the two approaches over other regions, and by using other soil moisture and precipitation products is recommended and will be the object of future investigations.

Acknowledgments: The study was primarily supported by the ESA WACMOS-MED project (contract ESA/AO/1-8173/15/I-SBo) and partly by EUMETSAT Satellite Application Facility in Support of Operational Hydrology and Water Management (H-SAF) project and the Italian Civil Protection Department. The authors wish to thank Umbria Region, the Italian Civil Protection Department and Simone Gabellani (CIMA foundation) for providing stream flow data in Italy, Yves Tramblay (IRD) for the stream flow data in Morocco, Javier Loizu (Public University of Navarre) for the stream flow data of Arga catchment, Isabelle Braud (IRSTEA) for the stream flow data of Gardon catchment, and the Global Runoff Data Centre (GRDC) for the stream flow data of Kolpa and Lim catchments.

Author Contributions: Christian Massari led the manuscript, made the elaborations and wrote part of the paper. Stefania Camici, Luca Ciabatta and Luca Brocca prepared the data and helped the writing of the manuscript.

Conflicts of Interest: The authors declare no conflict of interest.

References

1. Beck, H.E.; de Jeu, R.A.; Schellekens, J.; van Dijk, A.I.; Bruijnzeel, L.A. Improving curve number based storm runoff estimates using soil moisture proxies. *IEEE J. Sel. Top. Appl. Earth Obs. Remote Sens.* **2009**, *2*, 250–259. [CrossRef]
2. Koster, R.D.; Mahanama, S.P.; Livneh, B.; Lettenmaier, D.P.; Reichle, R.H. Skill in stream flow forecasts derived from large-scale estimates of soil moisture and snow. *Nat. Geosci.* **2010**, *3*, 613–616. [CrossRef]
3. Matgen, P.; Fenicia, F.; Heitz, S.; Plaza, D.; de Keyser, R.; Pauwels, V.R.; Wagner, W.; Savenije, H. Can ASCAT-derived soil wetness indices reduce predictive uncertainty in well-gauged areas? A comparison with in situ observed soil moisture in an assimilation application. *Adv. Water Resour.* **2012**, *44*, 49–65. [CrossRef]
4. Massari, C.; Brocca, L.; Moramarco, T.; Tramblay, Y.; Lescot, J.F.D. Potential of soil moisture observations in flood modelling: Estimating initial conditions and correcting rainfall. *Adv. Water Resour.* **2014**, *74*, 44–53. [CrossRef]
5. Crow, W.T.; Zhan, X. Continental-scale evaluation of remotely sensed soil moisture products. *IEEE Geosci. Remote Sens. Lett.* **2007**, *4*, 451–455. [CrossRef]
6. Renzullo, L.J.; Van Dijk, A.I.; Perraud, J.M.; Collins, D.; Henderson, B.; Jin, H.; Smith, A.B.; McJannet, D.L. Continental satellite soil moisture data assimilation improves root-zone moisture analysis for water resources assessment. *J. Hydrol.* **2014**, *519*, 2747–2762. [CrossRef]
7. López López, P.; Wanders, N.; Schellekens, J.; Renzullo, L.J.; Sutanudjaja, E.H.; Bierkens, M.F.P. Improved large-scale hydrological modelling through the assimilation of stream flow and downscaled satellite soil moisture observations. *Hydrol. Earth Syst. Sci.* **2016**, *20*, 3059–3076. [CrossRef]
8. Romano, N. Soil moisture at local scale: Measurements and simulations. *J. Hydrol.* **2014**, *516*, 6–20. [CrossRef]
9. Mohanty, B.P.; Cosh, M.H.; Lakshmi, V.; Montzka, C. Soil moisture remote sensing: State-of-the-science. *Vadose Zone J.* **2017**, *16*, 1–9. [CrossRef]
10. Brocca, L.; Melone, F.; Moramarco, T.; Singh, V.P. Assimilation of observed soil moisture data in storm rainfall-runoff modeling. *J. Hydrol. Eng.* **2009**, *14*, 153–165. [CrossRef]
11. Chen, F.; Crow, W.T.; Starks, P.J.; Moriasi, D.N. Improving hydrologic predictions of a catchment model via assimilation of surface soil moisture. *Adv. Water Resour.* **2011**, *34*, 526–536. [CrossRef]
12. Crow, W.T.; Bindlish, R.; Jackson, T.J. The added value of spaceborne passive microwave soil moisture retrievals for forecasting rainfall-runoff partitioning. *Geophys. Res. Lett.* **2005**, *32*. [CrossRef]
13. Brocca, L.; Moramarco, T.; Melone, F.; Wagner, W.; Hasenauer, S.; Hahn, S. Assimilation of surface-and root-zone ASCAT soil moisture products into rainfall–runoff modeling. *IEEE Trans. Geosci. Remote Sens.* **2011**, *50*, 2542–2555. [CrossRef]
14. Alvarez-Garreton, C.; Ryu, D.; Western, A.W.; Su, C.H.; Crow, W.T.; Robertson, D.E.; Leahy, C. Improving operational flood ensemble prediction by the assimilation of satellite soil moisture: Comparison between lumped and semi-distributed schemes. *Hydrol. Earth Syst. Sci.* **2015**, *19*, 1659–1676. [CrossRef]
15. Lievens, H.; Tomer, S.K.; Al Bitar, A.; De Lannoy, G.J.; Drusch, M.; Dumedah, G.; Franssen, H.J.; Kerr, Y.H.; Martens, B.; Pan, M.; et al. SMOS soil moisture assimilation for improved hydrologic simulation in the Murray Darling Basin, Australia. *Remote Sens. Environ.* **2015**, *168*, 146–162. [CrossRef]

16. Massari, C.; Brocca, L.; Tarpanelli, A.; Moramarco, T. Data assimilation of satellite soil moisture into rainfall-runoff modelling: A complex recipe? *Remote Sens.* **2015**, *7*, 11403–11433. [CrossRef]
17. Cenci, L.; Laiolo, P.; Gabellani, S.; Campo, L.; Silvestro, F.; Delogu, F.; Boni, G.; Rudari, R. Assimilation of H-SAF soil moisture products for flash flood early warning systems. Case study: Mediterranean catchments. *IEEE J. Sel. Top. Appl. Earth Obs. Remote Sens.* **2016**, *9*, 5634–5646. [CrossRef]
18. Laiolo, P.; Gabellani, S.; Campo, L.; Silvestro, F.; Delogu, F.; Rudari, R.; Pulvirenti, L.; Boni, G.; Fascetti, F.; Pierdicca, N.; et al. Impact of different satellite soil moisture products on the predictions of a continuous distributed hydrological model. *Int. J. Appl. Earth Obs. Geoinf.* **2016**, *48*, 131–145. [CrossRef]
19. Brown, M.E.; Escobar, V.; Moran, S.; Entekhabi, D.; O'Neill, P.E.; Njoku, E.G.; Doorn, B.; Entin, J.K. NASA's soil moisture active passive (SMAP) mission and opportunities for applications users. *Bull. Am. Meteorol. Soc.* **2013**, *94*, 1125–1128. [CrossRef]
20. Lievens, H.; Reichle, R.H.; Liu, Q.; De Lannoy, G.J.M.; Dunbar, R.S.; Kim, S.B.; Das, N.N.; Cosh, M.; Walker, J.P.; Wagner, W. Joint Sentinel-1 and SMAP data assimilation to improve soil moisture estimates. *Geophys. Res. Lett.* **2017**, *44*, 6145–6153. [CrossRef]
21. Cenci, L.; Pulvirenti, L.; Boni, G.; Chini, M.; Matgen, P.; Gabellani, S.; Campo, L.; Silvestro, F.; Versace, C.; Campanella, P.; et al. Satellite soil moisture assimilation: Preliminary assessment of the Sentinel 1 potentialities. In Proceedings of the 2016 IEEE International Geoscience and Remote Sensing Symposium (IGARSS), Beijing, China, 10–15 July 2016; pp. 3098–3101.
22. Crow, W.T.; van Den Berg, M.J.; Huffman, G.J.; Pellarin, T. Correcting rainfall using satellite-based surface soil moisture retrievals: The Soil Moisture Analysis Rainfall Tool (SMART). *Water Resour. Res.* **2011**, *47*. [CrossRef]
23. Pellarin, T.; Louvet, S.; Gruhier, C.; Quantin, G.; Legout, C. A simple and effective method for correcting soil moisture and precipitation estimates using AMSR-E measurements. *Remote Sens. Environ.* **2013**, *136*, 28–36. [CrossRef]
24. Brocca, L.; Ciabatta, L.; Massari, C.; Moramarco, T.; Hahn, S.; Hasenauer, S.; Kidd, R.; Dorigo, W.; Wagner, W.; Levizzani, V. Soil as a natural rain gauge: Estimating global rainfall from satellite soil moisture data. *J. Geophys. Res. Atmos.* **2014**, *119*, 5128–5141. [CrossRef]
25. Ciabatta, L.; Brocca, L.; Massari, C.; Moramarco, T.; Puca, S.; Rinollo, A.; Gabellani, S.; Wagner, W. Integration of satellite soil moisture and rainfall observations over the Italian territory. *J. Hydrometeorol.* **2015**, *16*, 1341–1355. [CrossRef]
26. Koster, R.D.; Brocca, L.; Crow, W.T.; Burgin, M.S.; De Lannoy, G.J. Precipitation estimation using L-band and C-band soil moisture retrievals. *Water Resour. Res.* **2016**, *52*, 7213–7225. [CrossRef]
27. Román-Cascón, C.; Pellarin, T.; Gibon, F.; Brocca, L.; Cosme, E.; Crow, W.; Fernández-Prieto, D.; Kerr, Y.H.; Massari, C. Correcting satellite-based precipitation products through SMOS soil moisture data assimilation in two land-surface models of different complexity: API and SURFEX. *Remote Sens. Environ.* **2017**, *200*, 295–310. [CrossRef]
28. Crow, W.T.; Ryu, D. A new data assimilation approach for improving runoff prediction using remotely-sensed soil moisture retrievals. *Hydrol. Earth Syst. Sci.* **2009**, *13*, 1–16. [CrossRef]
29. Chen, F.; Crow, W.T.; Ryu, D. Dual forcing and state correction via soil moisture assimilation for improved rainfall–runoff modeling. *J. Hydrometeorol.* **2014**, *15*, 1832–1848. [CrossRef]
30. Alvarez-Garreton, C.; Ryu, D.; Western, A.W.; Crow, W.T.; Su, C.H.; Robertson, D.R. Dual assimilation of satellite soil moisture to improve stream flow prediction in data-scarce catchments. *Water Resour. Res.* **2016**, *52*, 5357–5375. [CrossRef]
31. Evensen, G. *Data Assimilation: The Ensemble Kalman Filter*; Springer: Berlin, Germany, 2009.
32. Huffman, G.J.; Bolvin, D.T.; Nelkin, E.J.; Wolff, D.B.; Adler, R.F.; Gu, G.; Hong, Y.; Bowman, K.P.; Stocker, E.F. The TRMM multisatellite precipitation analysis (TMPA): Quasi-global, multiyear, combined-sensor precipitation estimates at fine scales. *J. Hydrometeorol.* **2007**, *8*, 38–55. [CrossRef]
33. Wagner, W.; Hahn, S.; Kidd, R.; Melzer, T.; Bartalis, Z.; Hasenauer, S.; Figa-Saldaña, J.; de Rosnay, P.; Jann, A.; Schneider, S.; et al. The ASCAT soil moisture product: A review of its specifications, validation results, and emerging applications. *Meteorol. Z.* **2013**, *22*, 5–33. [CrossRef]
34. Kerr, Y.H.; Waldteufel, P.; Wigneron, J.P.; Delwart, S.; Cabot, F.; Boutin, J.; Escorihuela, M.J.; Font, J.; Reul, N.; Gruhier, C.; et al. The SMOS mission: New tool for monitoring key elements of the global water cycle. *Proc. IEEE* **2010**, *98*, 666–687. [CrossRef]

35. Brocca, L.; Melone, F.; Moramarco, T. Distributed rainfall-runoff modelling for flood frequency estimation and flood forecasting. *Hydrol. Process.* **2011**, *25*, 2801–2813. [CrossRef]
36. Dee, D.P.; Uppala, S.M.; Simmons, A.J.; Berrisford, P.; Poli, P.; Kobayashi, S.; Andrae, U.; Balmaseda, M.A.; Balsamo, G.; Bauer, P.; et al. The ERA-Interim reanalysis: Configuration and performance of the data assimilation system. *Q. J. R. Meteorol. Soc.* **2011**, *137*, 553–597. [CrossRef]
37. Beck, H.E.; Vergopolan, N.; Pan, M.; Levizzani, V.; van Dijk, A.I.; Weedon, G.P.; Brocca, L.; Pappenberger, F.; Huffman, G.J.; Wood, E.F. Global-scale evaluation of 22 precipitation datasets using gauge observations and hydrological modeling. *Hydrol. Earth Syst. Sci.* **2017**, *21*, 6201–6217. [CrossRef]
38. Brocca, L.; Hasenauer, S.; Lacava, T.; Melone, F.; Moramarco, T.; Wagner, W.; Dorigo, W.; Matgen, P.; Martínez-Fernández, J.; Llorens, P.; et al. Soil moisture estimation through ASCAT and AMSR-E sensors: An intercomparison and validation study across Europe. *Remote Sens. Environ.* **2011**, *115*, 3390–3408. [CrossRef]
39. Paulik, C.; Dorigo, W.; Wagner, W.; Kidd, R. Validation of the ASCAT Soil Water Index using in situ data from the International Soil Moisture Network. *Int. J. Appl. Earth Obs. Geoinf.* **2014**, *30*, 1–8. [CrossRef]
40. Dorigo, W.; Wagner, W.; Albergel, C.; Albrecht, F.; Balsamo, G.; Brocca, L.; Chung, D.; Ertl, M.; Forkel, M.; Gruber, A.; et al. ESA CCI soil moisture for improved Earth system understanding: State-of-the art and future directions. *Remote Sens. Environ.* **2017**, *203*, 185–215. [CrossRef]
41. Haylock, M.R.; Hofstra, N.; Klein Tank, A.M.G.; Klok, E.J.; Jones, P.D.; New, M. A European daily high-resolution gridded data set of surface temperature and precipitation for 1950–2006. *J. Geophys. Res. Atmos.* **2008**, *113*, 1–12. [CrossRef]
42. Stampoulis, D.; Anagnostou, E.N. Evaluation of Global Satellite Rainfall Products over Continental Europe. *J. Hydrometeorol.* **2012**, *13*, 588–603. [CrossRef]
43. SM2RAIN-ASCAT (1 January 2007–31 December 2015) European Rainfall Dataset (0.25 Degree/Daily). Available online: http://dx.doi.org/10.13140/RG.2.1.4068.6481/1 (accessed on 13 February 2018).
44. Massari, C.; Crow, W.; Brocca, L. An assessment of the performance of global rainfall estimates without ground-based observations. *Hydrol. Earth Syst. Sci.* **2017**, *21*, 4347–4361. [CrossRef]
45. Masseroni, D.; Cislaghi, A.; Camici, S.; Massari, C.; Brocca, L. A reliable rainfall–runoff model for flood forecasting: Review and application to a semi-urbanized watershed at high flood risk in Italy. *Hydrol. Res.* **2017**, *48*, 726–740. [CrossRef]
46. Doorenbos, J.; Pruitt, W.O. Background and Development of Methods to Predict Reference Crop Evapotranspiration (ETo). In *Crop Water Requirements. FAO Irrigation and Drainage Paper No. 24*; FAO: Rome, Italy, 1977; Appendix II; pp. 108–119.
47. Famiglietti, J.S.; Wood, E.F. Multiscale modeling of spatially variable water and energy balance processes. *Water Resour. Res.* **1994**, *30*, 3061–3078. [CrossRef]
48. Andreadis, K.M.; Clark, E.A.; Wood, A.W.; Hamlet, A.F.; Lettenmaier, D.P. Twentieth-century drought in the conterminous United States. *J. Hydrometeorol.* **2005**, *6*, 985–1001. [CrossRef]
49. Melone, F.; Corradini, C.; Singh, V.P. Lag prediction in ungauged basins: An investigation through actual data of the upper Tevere River valley. *Hydrol. Processes* **2002**, *16*, 1085–1094. [CrossRef]
50. Wagner, W.; Lemoine, G.; Rott, H. A method for estimating soil moisture from ERS scatterometer and soil data. *Remote Sens. Environ.* **1999**, *70*, 191–207. [CrossRef]
51. Albergel, C.; Rüdiger, C.; Pellarin, T.; Calvet, J.C.; Fritz, N.; Froissard, F.; Suquia, D.; Petitpa, A.; Piguet, B.; Martin, E. From near-surface to root-zone soil moisture using an exponential filter: An assessment of the method based on in-situ observations and model simulations. *Hydrol. Earth Syst. Sci.* **2008**, *12*, 1323–1337. [CrossRef]
52. Reichle, R.H.; Koster, R.D. Bias reduction in short records of satellite soil moisture. *Geophys. Res. Lett.* **2004**, *31*, L19501. [CrossRef]
53. Stoffelen, A. Toward the true near-surface wind speed: Error modeling and calibration using triple collocation. *J. Geophys. Res. Oceans* **1998**, *103*, 7755–7766. [CrossRef]
54. Gruber, A.; Su, C.H.; Zwieback, S.; Crow, W.; Dorigo, W.; Wagner, W. Recent advances in (soil moisture) triple collocation analysis. *Int. J. Appl. Earth Obs. Geoinf.* **2016**, *45*, 200–211. [CrossRef]
55. Draper, C.; Mahfouf, J.F.; Calvet, J.C.; Martin, E.; Wagner, W. Assimilation of ASCAT near-surface soil moisture into the SIM hydrological model over France. *Hydrol. Earth Syst. Sci.* **2011**, *15*, 3829. [CrossRef]
56. Tian, Y.; Huffman, G.J.; Adler, R.F.; Tang, L.; Sapiano, M.; Maggioni, V.; Wu, H. Modeling errors in daily precipitation measurements: Additive or multiplicative? *Geophys. Res. Lett.* **2013**, *40*, 2060–2065. [CrossRef]

57. Crow, W.T.; Van Loon, E. Impact of Incorrect Model Error Assumptions on the Sequential Assimilation of Remotely Sensed Surface Soil Moisture. *J. Hydrometeorol.* **2006**, *7*, 421–432. [CrossRef]
58. Maggioni, V.; Massari, C. On the performance of satellite precipitation products in riverine flood modeling: A review. *J. Hydrol.* **2018**, *558*, 214–224. [CrossRef]
59. Nash, J.; Sutcliffe, J. River flow forecasting through conceptual models part I—A discussion of principles. *J. Hydrol.* **1970**, *10*, 282–290. [CrossRef]
60. Hoffmann, L.; El Idrissi, A.; Pfister, L.; Hingray, B.; Guex, F.; Musy, A.; Humbert, J.; Drogue, G.; Leviandier, T. Development of regionalized hydrological models in an area with short hydrological observation series. *River Res. Appl.* **2004**, *20*, 243–254. [CrossRef]
61. Gupta, H.V.; Kling, H.; Yilmaz, K.K.; Martinez, G.F. Decomposition of the mean squared error and NSE performance criteria: Implications for improving hydrological modelling. *J. Hydrol.* **2009**, *377*, 80–91. [CrossRef]
62. Bober, W. *Introduction to Numerical and Analytical Methods with MATLAB® for Engineers and Scientists*; CRC Press: Boca Raton, FL, USA, 2013.
63. Thiemig, V. The Development of Pan-African Flood Forecasting and the Exploration of Satellite Based Precipitation Estimates. Ph.D. Thesis, Utrecht University, Utrecht, The Netherlands, 2014.
64. Loizu, J.; Massari, C.; Álvarez-Mozos, J.; Tarpanelli, A.; Brocca, L.; Casalí, J. On the assimilation set-up of ASCAT soil moisture data for improving stream flow catchment simulation. *Adv. Water Resour.* **2018**, *111*, 86–104. [CrossRef]
65. Fascetti, F.; Pierdicca, N.; Pulvirenti, L.; Crapolicchio, R.; Muñoz-Sabater, J. A comparison of ASCAT and SMOS soil moisture retrievals over Europe and Northern Africa from 2010 to 2013. *Int. J. Appl. Earth Obs. Geoinf.* **2016**, *45*, 135–142. [CrossRef]
66. Scholze, M.; Buchwitz, M.; Dorigo, W.; Guanter, L.; Quegan, S. Reviews and syntheses: Systematic Earth observations for use in terrestrial carbon cycle data assimilation systems. *Biogeosci. Discuss.* **2017**, *14*, 3401–3429. [CrossRef]
67. Peña Arancibia, J.L.; van Dijk, A.I.J.M.; Renzullo, L.J.; Mulligan, M. Evaluation of precipitation estimation accuracy in reanalyses, satellite products, and an ensemble method for regions in Australia and Sout and East Asia. *J. Hydrometeorol.* **2013**, *14*, 1323–1333. [CrossRef]
68. Xie, P.; Joyce, R.J. Integrating information from satellite observations and numerical models for improved global precipitation analyses. In *Remote Sensing of the Terrestrial Water Cycle, Geophysical Monograph Series*; Lakshmi, V., Alsdorf, D., Anderson, M., Biancamaria, S., Cosh, M., Entin, J., Huffman, G., Kustas, W., van Oevelen, P., Painter, T., Eds.; John Wiley & Sons, Inc.: Hoboken, NJ, USA, 2014; Chapter 3.
69. Ebert, E.E.; Janowiak, J.E.; Kidd, C. Comparison of near-real-time precipitation estimates from satellite observations and numerical models. *Bull. Am. Meteorol. Soc.* **2007**, *88*, 47–64. [CrossRef]
70. Ciabatta, L.; Marra, A.C.; Panegrossi, G.; Casella, D.; Sanò, P.; Dietrich, S.; Massari, C.; Brocca, L. Daily precipitation estimation through different microwave sensors: Verification study over Italy. *J. Hydrol.* **2017**, *545*, 436–450. [CrossRef]
71. Ciabatta, L.; Brocca, L.; Massari, C.; Moramarco, T.; Gabellani, S.; Puca, S.; Wagner, W. Rainfall-runoff modelling by using SM2RAIN-derived and state-of-the-art satellite rainfall products over Italy. *Int. J. Appl. Earth Obs. Geoinf.* **2016**, *48*, 163–173. [CrossRef]
72. Beck, H.E.; van Dijk, A.I.; Levizzani, V.; Schellekens, J.; Miralles, D.G.; Martens, B.; de Roo, A. MSWEP: 3-hourly 0.25 global gridded precipitation (1979–2015) by merging gauge, satellite, and reanalysis data. *Hydrol. Earth Syst. Sci.* **2017**, *21*, 589. [CrossRef]
73. Su, C.-H.; Ryu, D. Multi-scale analysis of bias correction of soil moisture. *Hydrol. Earth Syst. Sci.* **2015**, *19*, 17–31. [CrossRef]
74. De Leeuw, J.; Methven, J.; Blackburn, M. Evaluation of ERA-Interim reanalysis precipitation products using England and Wales observations. *Q. J. R. Meteorol. Soc.* **2015**, *141*, 798–806. [CrossRef]
75. Rebora, N.; Molini, L.; Casella, E.; Comellas, A.; Fiori, E.; Pignone, F.; Siccardi, F.; Silvestro, F.; Tanelli, S.; Parodi, A. Extreme rainfall in the mediterranean: What can we learn from observations? *J. Hydrometeorol.* **2013**, *14*, 906–922. [CrossRef]

remote sensing

MDPI

Article

Quantifying Drought Propagation from Soil Moisture to Vegetation Dynamics Using a Newly Developed Ecohydrological Land Reanalysis

Yohei Sawada

Meteorological Research Institute, Japan Meteorological Agency, Tsukuba 305-0052, Japan; ysawada@mri-jma.go.jp; Tel.: +81-029-853-8552

Received: 23 June 2018; Accepted: 27 July 2018; Published: 30 July 2018

Abstract: Despite the importance of the interaction between soil moisture and vegetation dynamics to understand the complex nature of drought, few land reanalyses explicitly simulate vegetation growth and senescence. In this study, I provide a new land reanalysis which explicitly simulates the interaction between sub-surface soil moisture and vegetation dynamics by the sequential assimilation of satellite microwave brightness temperature observations into a land surface model (LSM). Assimilating satellite microwave brightness temperature observations improves the skill of a LSM to simultaneously simulate soil moisture and the seasonal cycle of leaf area index (LAI). By analyzing soil moisture and LAI simulated by this new land reanalysis, I identify the drought events which significantly damage LAI on the climatological day-of-year of the LAI's seasonal peak and quantify drought propagation from soil moisture to LAI in the global snow-free region. On average, soil moisture in the shallow soil layers (0–0.45 m) quickly recovers from the drought condition before the climatological day-of-year of the LAI's seasonal peak while soil moisture in the deeper soil layer (1.05–2.05 m) and LAI recover from the drought condition approximately 100 days after the climatological day-of-year of the LAI's seasonal peak.

Keywords: drought; soil moisture; vegetation; land data assimilation; microwave remote sensing

1. Introduction

Drought is one of the costliest natural disasters in the world [1], so mitigating negative impacts of catastrophic drought events on society is a grand challenge for hydrological researchers and practitioners. Drought monitoring and prediction capabilities contribute to building resilience against severe drought events. Many previous studies have provided the important contributions to drought monitoring [2–4], and drought prediction [5–7]. In those literatures, precipitation, soil moisture, and vegetation condition were recognized as the key variables to be observed and/or simulated for drought monitoring and prediction.

Understanding the complex nature of drought from historical drought events is strongly needed to improve drought monitoring and early warning and help decision makers consider the adaptation plan against severe drought events. Drought is a complicated multi-spatiotemporal phenomenon and an integrated process which has meteorological, hydrological, and ecological aspects. According to the conceptual model proposed by Wilhite and Glantz [8], extreme climate conditions such as precipitation deficiency and high temperature induce soil water deficiency which causes plant water stress and agricultural drought. Then, streamflow and groundwater level respond to soil water deficiency, which causes hydrological drought. Soil moisture has an important role in this drought propagation since it controls both agricultural and hydrological aspects of drought. Quantifying the drought propagation is important to deliver the effective information of the drought progress to decision makers in a wide variety of sectors and has been intensively investigated.

Sawada et al. [9] found that hydrological drought quantified by river discharge and groundwater level lasted longer than agricultural drought quantified by vegetation dynamics in Medjerda river basin, Tunisia. Sepulcre-Canto et al. [10] developed a drought indicator in Europe considering the drought propagation from soil moisture deficits to crop failures. Akarsh and Mishra [11] analyzed the drought propagation from soil moisture in shallow soil layers to vegetation condition in India. These previous studies indicated that the interaction between soil moisture and vegetation dynamics is important to understand the drought propagation. However, published knowledge is limited to regional applications and the holistic view of the drought propagation from soil moisture to vegetation dynamics has yet to be obtained.

To identify and quantify the drought propagation, land reanalysis is useful since it provides spatially and temporally homogeneous terrestrial water cycle data. Land reanalysis is generated by driving a land surface model (LSM) with meteorological forcings from atmospheric reanalysis with adjustments using in situ observation networks. For example, the Global Land Data Assimilation System (GLDAS) [12] provides land surface variables such as soil moisture globally from 1948 to present by driving multiple LSMs. MERRA-Land [13] provides global estimates of soil moisture, heat fluxes, snow, and runoff from 1979 to present as a part of the Modern-Era Retrospective Analysis for Research and Applications (MERRA) reanalysis. ERA-Interim/Land [14] is generated by a single 32-year (1979–2010) simulation of the European Centre for the Medium Range Weather Forecasts (ECMWF) land surface model driven by meteorological forcings from the ERA-Interim atmospheric reanalysis. These land reanalysis datasets were widely used for drought identification and quantification [11,15–20].

However, there are limitations in the existing land reanalysis datasets for the quantification of the drought propagation. First, few land reanalyses explicitly simulate vegetation growth and senescence. The interaction between soil moisture and vegetation dynamics is not currently considered by the LSMs which are used to generate global land reanalysis datasets. The decline of vegetation activity due to drought cannot be identified and quantified by the existing global land reanalysis datasets although the ecohydrological process is important in the drought propagation.

Second, few satellite observations are assimilated into LSMs in the current global land reanalyses. There are important satellite observations which can improve the simulation of a LSM. For example, passive and active microwave remote sensing provides the all-sky observation of land surface soil moisture, temperature, and vegetation water content [21–27]. Although many previous studies have developed Land Data Assimilation Systems (LDASs) to assimilate these satellite microwave observations into a LSM [28–35] and some previous studies have successfully monitored regional drought events using LDASs [7,36], they have yet to be fully applied to generate land reanalysis for the quantification of the drought propagation.

There are several LDASs which can accurately simulate both soil moisture and vegetation dynamics. Barbu et al. [37] assimilated satellite-observed surface soil moisture and leaf area index (LAI) jointly into a LSM which can simultaneously calculate soil moisture and vegetation dynamics. Liu et al. [38] assimilated both active and passive microwave observations into a LSM and demonstrated that data assimilation can improve the simulation of soil moisture and biomass. Sawada and Koike [39] and Sawada et al. [40] developed the Coupled Land and Vegetation Data Assimilation System (CLVDAS) which can improve the skill of a LSM to simulate both soil moisture and vegetation dynamics by assimilating satellite microwave brightness temperature observations. These ecohydrological LDASs are the promising tools to monitor the drought propagation considering the interaction between soil moisture and vegetation dynamics. However, the skill of the ecohydrological LDASs to simultaneously estimate surface soil moisture, sub-surface soil moisture, and vegetation dynamics has yet to be fully validated on continental and global scales.

This study aims to quantify the drought propagation from soil moisture to vegetation dynamics in the global snow-free region. First, I develop a verified land reanalysis, which includes both soil moisture and vegetation condition, by driving CLVDAS. Then, I obtain the holistic view of the integrated drought propagation from soil moisture deficits to vegetation degradation using soil moisture and LAI simulated by the new land reanalysis.

2. Materials and Methods

2.1. Data and Study Area

As observed meteorological forcings, the GLDAS v2.1 data [12,41] were used. GLDAS provides the complete set of meteorological forcings which are needed to run LSMs. The meteorological forcings used in this study were surface pressure, precipitation, surface air temperature, relative humidity, incoming solar radiation, incoming longwave radiation, and wind speed. The horizontal resolution is 0.25 degree. The temporal resolution is 3-hourly and the data were linearly interpolated into hourly data since the time step of the LSM used in this study is set to 1 h.

As microwave brightness temperature observation, the Advanced Microwave Scanning Radiometer for Earth Observing System (AMSR-E) L3 product [42] was used. Brightness temperature observations at 6.925 and 10.25 GHz were used for data assimilation. Both horizontally and vertically polarized observations were used. I used only night scene data since the effect of surface temperature errors on the estimation of microwave brightness temperature is small at night [43]. The native horizontal resolution of the L3 product is 0.1 degree and the data were resampled to 0.25 degree by box averaging to match them to the resolution of the GLDAS meteorological forcings and the LSM. The temporal resolution is approximately 2-daily.

The International Satellite Land Surface Climatology Project 2 soil data [44] were used to derive soil texture. The Food and Agricultural Organization global dataset [45] was used to obtain the parameters for the water retention curve.

The Global Land Surface Satellite LAI (GLASS LAI) product [46] was used to evaluate the skill of the land reanalysis to simulate phenology. The GLASS LAI was generated from MODerate resolution Imaging Spectroradiometer (MODIS) visible and infrared observations [47]. Xiao et al. [46] reported that GLASS LAI could reproduce field-measured LAI with the accuracy of R^2 = 0.87 and RMSE = 0.64 [m^2/m^2]. The native horizontal resolution of the GLASS LAI product is 1 km and I resampled it into 0.25 degree by box averaging. The temporal resolution is 8-daily and I generated the monthly data.

The European Space Agency Climate Change Initiative Soil Moisture (ESA CCI SM) v3.2 product [48–50] was used to evaluate the skill of the land reanalysis to simulate surface soil moisture. This product was generated by blending many existing satellite-observed soil moisture products. The combined passive and active microwave products were used. Although Dorigo et al. [48] reported that the correlation coefficient between this product and in situ surface soil moisture observations is approximately 0.6 on average, significantly low correlations are also reported in some in situ observation sites. The horizontal resolution of the ESA CCI SM is 0.25 degree and the temporal resolution is approximately 2-daily.

To evaluate the skill of the land reanalysis to simulate soil moisture from surface to 2m depth, I used the in situ soil moisture observations archived at International Soil Moisture Network (ISMN) [51]. I used the network of the African Monsoon Multidisciplinary Analysis—Coupling the Tropical Atmosphere and the Hydrological Cycle (AMMA-CATCH) [52], CARBOAFRICA [53], and Dahra [54]. The land cover of the AMMA-CATCH Belefoungou site is crop and the land cover of the other sites is semi-arid savanna. The annual maximum LAIs in Dahra, CARBOAFRICA, AMMA-CATCH Belefoungou and Tondikiboro are approximately 1.5, 0.6, 2.5, and 0.6, respectively. Figure 1 shows the locations of the in situ soil moisture observation sites. I chose these sites because they provided both surface and subsurface (more than 1 m depth) in situ soil moisture observations with the sufficient observation periods. In addition, as mentioned below, I did not provide the land reanalysis in the pixels in which there is snowfall from 2003 to 2010. These criteria excluded many in situ soil moisture observations in ISMN. The native temporal resolutions of the point-scale observations were resampled to daily. Although the in situ soil moisture observations provide hourly data, here I focused on the daily-scale soil moisture dynamics since soil moisture dynamics in an hourly scale is not important for the application to drought analysis. In addition, the skill to simulate the hourly-scale soil moisture dynamics may be strongly degraded by the bias of the timing of precipitation in the GLDAS data, which cannot be corrected by the LDAS used in this study.

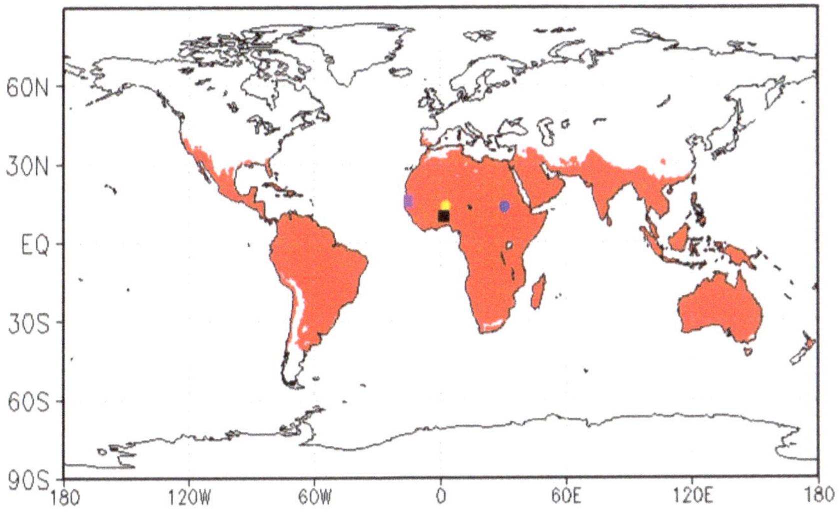

Figure 1. Study area (red) and the locations of the in situ observation sites of the DAHRA (purple rectangle), CARBOAFRICA (blue circle), AMMA-CATCH Belefoungou in Benin (black rectangle), and AMMA-CATCH Tondikiboro in Niger (yellow circle) networks.

The Emergency Events Database (EM-DAT: http://www.emdat.de) was used as reference data of historical drought events. The EM-DAT provides information on historic natural and technological disasters reported by different institutions across the world. Please note that the database may not include all disasters and the reported damages may sometimes be biased.

The study area was the global snow-free region (Figure 1). I identified the snow-free region using the GLDAS snowfall data. There are two reasons why I focused only on the snow-free region. First, snowmelt may have an important role in the drought propagation in some regions. The drought propagation may be more complicated in snowfall regions than in snow-free regions. In snow-free regions, I can focus only on the interaction between soil moisture and vegetation dynamics to quantify the drought propagation. Second, the performance of my LDAS (see Section 2.2) may be degraded by snow because snow accumulation makes it difficult to simulate microwave radiative transfer. The tau-omega model of the radiative transfer model (RTM) (Equation (9) in Section S2 of the Supplementary Materials) does not currently consider the effect of snow accumulation on microwave radiative transfer. Considering the effect of snow accumulation in the RTM may bring additional errors in the land reanalysis, which was avoided in this study. Please note that the study area includes the arid and semi-arid area in which the effects of vegetation on microwave signals are small so that I can evaluate the impact of the data assimilation on the simulation of surface soil moisture.

2.2. Coupled Land and Vegetation Data Assimilation System (CLVDAS)

CLVDAS [39,40] has been developed to improve the skill of a LSM to simultaneously simulate soil moisture and vegetation dynamics. The modules of CLVDAS include a LSM, a microwave Radiative Transfer Model (RTM), and a particle filtering (PF) data assimilation system. The LSM of CLVDAS is EcoHydro-SiB, which can simultaneously simulate surface soil moisture, sub-surface soil moisture, and vegetation dynamics (see Section S1 of the Supplementary Materials). EcoHydro-SiB can explicitly simulate vegetation growth and senescence. The RTM of CLVDAS can estimate microwave emissivity on land (see Section S2 of the Supplementary Materials). Microwave brightness temperature at C- and X-bands is sensitive to both surface soil moisture and vegetation water content [23] and insensitive to atmospheric condition. Therefore, the RTM can estimate satellite level microwave brightness temperature from surface soil moisture, surface soil temperature, and vegetation water content simulated by EcoHydro-SiB. The PF can assimilate satellite microwave brightness temperature observations into the LSM and improve the skill of the LSM to simulate their state variables (see Section S3 of the Supplementary Materials). The detailed description of the CLVDAS is not included in this main manuscript and can be found in the Supplementary Materials since the description of all CLVDAS's modules has already been published. Please refer to previous papers [9,39,40] for the complete description of CLVDAS (i.e., the LSM, the RTM, and the PF data assimilation system) and its parameter variables. References of the Supplementary Materials are also useful to understand the LSM, the RTM, and the PF data assimilation system.

2.3. ECoHydrological Land reAnalysis (ECHLA)

I generated a new land reanalysis dataset, called the ECoHydrological Land reAnalysis (ECHLA), by driving CLVDAS. First, I run EcoHydro-SiB without data assimilation (the open loop (OL) experiment). Initial conditions are 80-year spun-up results of the OL experiment. Second, comparing simulated brightness temperature at 6.9 GHz and 10.65 GHz by the RTM with observed brightness temperature by AMSR-E from 2004 to 2006, I optimized the surface soil roughness parameter of the RTM (rms height: see a previous paper [55]) in each pixel. Although there is another soil roughness parameter, correlation length, in the RTM, I did not optimize it in this study. This is because the combination of large rms height and large correlation length has a similar effect to the combination of small rms height and small correlation length so that it is practically difficult to optimize both of soil roughness parameters simultaneously [39]. The optimization was implemented using the Shuffled Complex Evolution method (SCE-UA) [56]. The parameter optimization method in CLVDAS was described previously [39]. Third, I run EcoHydro-SiB with the assimilation of the AMSR-E observed brightness temperature observations by the PF (the DA experiment). The output of the DA experiment is the dataset of ECHLA.

The horizontal resolution of ECHLA was set to 0.25 degree. The depth of the first soil layer was set to 0.05 m and the depths of the other soil layers were set to 0.1 m. The total soil depth was set to 2.05 m since the previous studies revealed that most of root biomass was distributed from 0 to 2 m depth soil [57]. The temporal resolution of EcoHydro-SiB was set to hourly and I analyzed the daily data.

The observation errors for $T_b^{V,6.9GHz}$, $T_b^{H,6.9GHz}$, $T_b^{V,10.65GHz}$, and $T_b^{H,10.65GHz}$ are set to 9.9 K, 14 K, 8.3 K, and 13 K, respectively ($T_b^{V,6.9GHz}$ is the vertically polarized brightness temperature at 6.9 GHz, for instance). It is generally difficult to objectively determine the observation errors especially for satellite brightness temperature observations [58]. In this study, these variables are set to the mean root-mean-square differences between observed brightness temperature and simulated brightness temperature by the OL experiment averaged in all pixels of the study area. I assumed that the observation errors are uncorrelated. The duration of the data assimilation window was set to 5 days so that past 5 days' and present observations were assimilated to adjust the model state at the present time. The ensemble size was set to 100 and the ensemble mean was used for the drought identification and quantification.

Figure 2 summarizes production steps of ECHLA. The LSM, EcoHydro-SiB, estimated land state variables for 5 days (from t_0 to t_e in Figure 2) using GLDAS meteorological forcing data. Using the simulated land state variables, such as surface soil moisture, LAI, and temperature, the RTM estimated brightness temperatures at 6.9 GHz and 10.65 GHz from t_0 to t_e. Then, the PF adjusted the land state variables at the end of the data assimilation window (t_e) by comparing simulated and AMSR-E observed brightness temperatures from t_0 to t_e. The PF assimilated all available observations in the assimilation window. All state variables (i.e., soil moisture in all soil columns, biomass pools, and temperature) were adjusted. The adjusted biomass pools included the leaf, stem, and root biomass and LAI was also updated by calculating it from the adjusted leaf biomass (see also the Supplementary Materials). Since all state variables raised above are included in particles, they are updated by the PF process although they are not directly observed. The adjusted land state variables were used for the initial condition of the next data assimilation cycle.

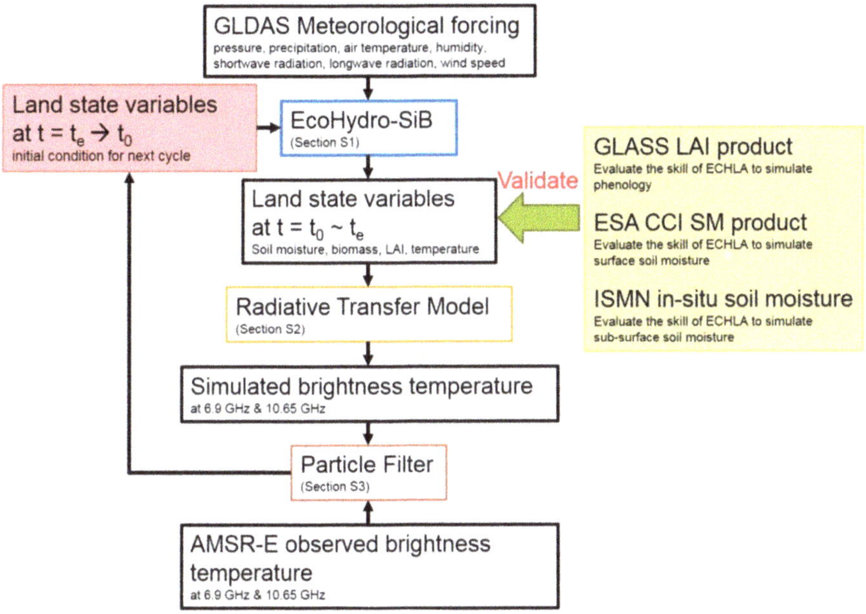

Figure 2. Schematic of production steps of ECHLA. See Sections 2.2 and 2.3, and the Supplementary Materials for details.

The study period was set to 2003–2010, which is nearly identical to the period of the AMSR-E operational observation. Although the study period was set to the short duration in this first trial of generating the new land reanalysis, CLVDAS and the existing satellite observation missions have a potential to extend the period (see Section 4.2).

2.4. Drought Identification and Quantification

I identified and quantified the drought propagation using soil moisture and LAI of ECHLA. The drought index used in this study was the Standardized Anomaly (SA) [9,59]:

$$SA(t) = \frac{x(t) - \overline{x}}{\sigma} \quad (1)$$

where $SA(t)$ is the SA at time t, $x(t)$ is soil moisture or LAI, $\overline{x}$ is the temporal mean of x, and σ is the standard deviation of x. To calculate (1), I used daily data. For instance, SA of LAI on 1 January 2004 is calculated by the mean and standard deviation of LAI on January 1 in each year (i.e., from 2003 to 2010).

In this paper, I identified drought events using the time series of the SA of LAI. I focused on the vegetation degradation on the day of the seasonal peak as an indicator of drought because it is strongly related to the degradation of the agricultural activities. First, I analyzed the time series of LAI in each pixel using the TIMESAT software [60] and calculated the timing of the LAI's seasonal peak. TIMESAT can fit the time series of LAI to a polynomial and analytically obtain the timing of the peak of a fitted polynomial. I obtained the climatological day-of-year of the LAI's seasonal peak averaged from 2003 to 2010 (Figure 3a).

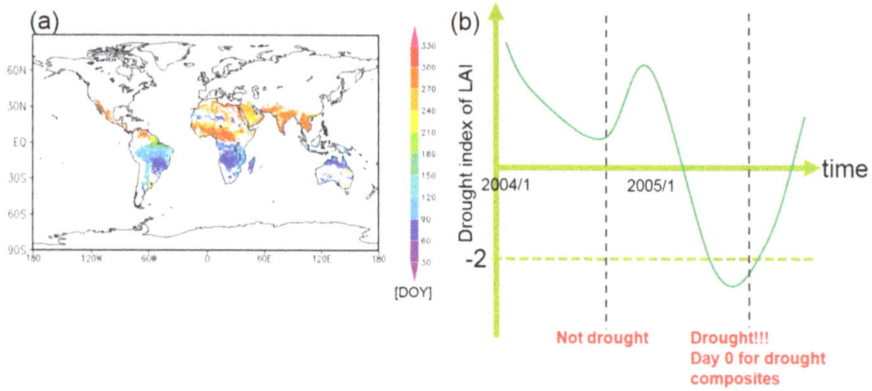

Figure 3. (**a**) The climatological day-of-year (DOY) of the ECoHydrological Land reAnalysis (ECHLA)-simulated leaf area index's (LAI's) seasonal peak averaged from 2003 to 2010. (**b**) Schematic of the drought detection method. Black dashed line shows the timing of the climatological LAI's seasonal peak. See Section 2.4 for details.

Then, I evaluated the SA of LAI on the climatological day-of-year of the LAI's seasonal peak every year in each pixel to identify drought events. The threshold of the LAI SA to identify drought events was set to −2.0. This threshold of SA is often used as "extremely dry" in the literature (e.g., [59]). If the LAI SA is less than −2.0 on the climatological day-of-year of the LAI's seasonal peak, I considered that the drought event occurred (Figure 3b). The 2-year time series of the SAs of LAI, surface soil moisture at 0–0.05 [m] depth (called surface), soil moisture averaged from 0.05 to 0.45 [m] (root 1), soil moisture averaged from 0.45 to 1.05 [m] (root2), and soil moisture averaged from 1.05 to 2.05 [m] (deep) in the identified drought events were extracted. For instance, in the pixel whose climatological day-of-year of the LAI's seasonal peak is 1 September, I evaluated the SA of LAI on 1 September every year. I found that the SA of LAI was less than −2 on 1 September 2005 (Figure 3b) so that I extract the time series from 1 September 2004 to 1 September 2006 in order to calculate the statistics of the drought event (see below). Please note that I did not assume that the drought event lasted whole two years. I calculated intensity as a minimum value of the SA in each drought event, duration as a length of each drought event when the SA is less than −1.0 and start and end dates of each drought event as timing when the SA goes below and above −1.0, respectively. This threshold of SA (−1.0) is often used as "moderately dry" in the literature (e.g., [59]). If there are multiple drought events when the SA is less than −1.0 during the extracted 2-year period, I used the event which has the longest duration.

3. Results

3.1. Validation of ECHLA

In this subsection, I evaluated the performance of ECHLA to accurately reproduce soil moisture and LAI in the global snow-free region. The correlation coefficient, root-mean-square-error (RMSE), and unbiased RMSE (ubRMSE) were used as metrics.

Figure 4 shows the performance of ECHLA to reproduce the optically observed LAI product (GLASS LAI). Both GLASS LAI and ECHLA LAI are temporally resampled to the monthly data. The correlation coefficients between ECHLA LAI and GLASS LAI in most pixels are higher than 0.6 (Figure 4a) so that ECHLA can accurately reproduce the seasonal cycle of LAI. The small correlations can be found in the rainforest regions. Data assimilation greatly contributes to the high performance of ECHLA to reproduce phenology. Figure 4c shows the difference of correlation coefficients against GLASS LAI between the DA experiment and the OL experiment (see Section 3.2).

Assimilating microwave brightness temperature observations significantly increases correlation coefficients. However, ECHLA has large RMSEs against GLASS LAI in many regions (Figure 4b) and data assimilation increases them (Figure 4d). Although the absolute value of the optically observed LAI cannot be reliable and directly compared with the model estimation [61] (see also Section 4.1), there may be limitations in the parameterizations to relate LAI to microwave signals.

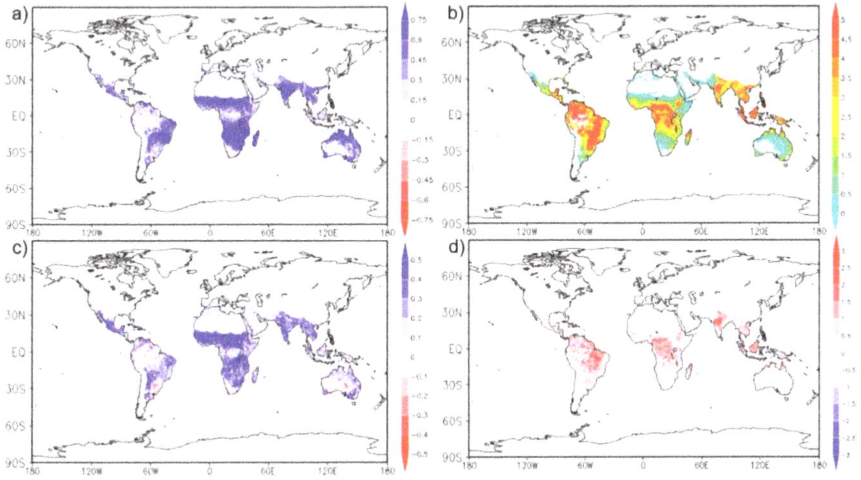

Figure 4. Performance of ECHLA to reproduce satellite-observed LAI (GLASS LAI). (**a**) Correlation coefficient and (**b**) RMSE [m^2/m^2] of simulated LAI by the DA experiment. The differences of (**c**) correlation coefficient and (**d**) RMSE between the DA experiment and the OL experiment.

Figure 5 shows the performance of ECHLA to reproduce the satellite observed surface soil moisture product (ESA CCI). ECHLA surface soil moisture is positively correlated with ESA CCI surface soil moisture in most pixels (Figure 5a). Data assimilation has no significant impacts on correlation coefficients (Figure 5c). Although ECHLA has large RMSEs ($\geq$0.1 m^3/m^3) against ESA CCI in many regions (Figure 5b), data assimilation generally decreases RMSEs in many pixels (Figure 5d). The improvement of simulating surface soil moisture by data assimilation can be found in Sahel, South Africa, and North Australia. The performance is degraded by data assimilation in the moderately and densely vegetated area such as Ethiopia, South East Asia, and south Amazon (see also Section 4.1).

Tables 1–4 show the performance of ECHLA to reproduce the in-situ observed soil moisture at the depths of 0–2 m. In all four in situ observation sites, data assimilation improves RMSE and ubRMSE of the surface soil moisture simulation. In the DAHRA site of West Africa, data assimilation decreases RMSEs and increases correlation coefficients against in situ sub-surface soil moisture observations (Table 1). In this site, data assimilation successfully improves the simulation of sub-surface soil moisture, which is not directly observed by the satellite, by inversely estimating sub surface soil moisture from surface soil moisture and vegetation dynamics (see Section S3 and [40]). The simulation of sub-surface soil moisture is improved only by removing biases so that no significant improvement of ubRMSE can be found. The improvement of the sub-surface soil moisture simulation by data assimilation can also be found in the CARBO AFRICA site of East Africa (Table 2) although the skill of simulating deep-zone ($\geq$1.5 m) soil moisture is degraded when the correlation coefficient is evaluated. The improvement of the skill to simulate sub-surface soil moisture cannot be found in the AMMA-CATCH sites of the Sahel region although data assimilation reduces RMSE and ubRMSE for surface soil moisture (Tables 3 and 4; see also Section 4.1)

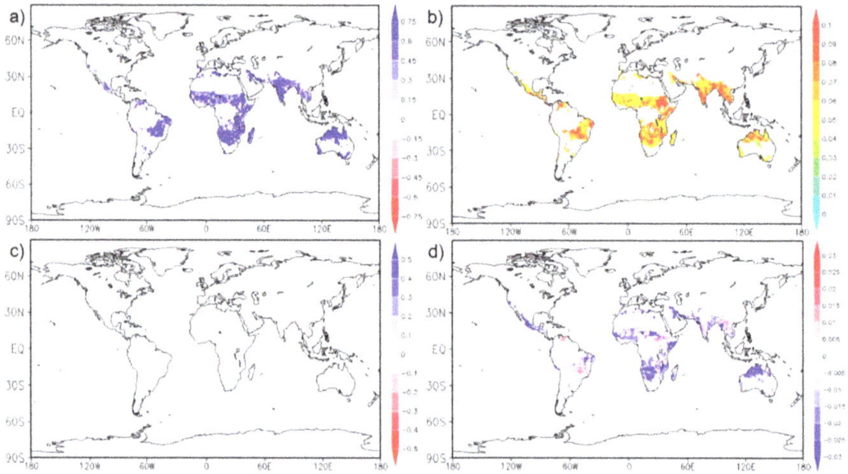

Figure 5. Performance of ECHLA to reproduce satellite-observed surface soil moisture (ESA CCI SM). (**a**) Correlation coefficient and (**b**) RMSE [m^3/m^3] of simulated surface soil moisture by the DA experiment. The differences of (**c**) correlation coefficient and (**d**) RMSE between the DA experiment and the OL experiment. ESA CCI SM does not provide surface soil moisture retrievals in the dense vegetated area and the performance of ECHLA was not evaluated there.

Table 1. RMSE, ubRMSE, and correlation coefficient of soil moisture simulated by the DA experiment and the OL experiment compared to the in situ observation data in the DAHRA site.

Depth [m]		RMSE [m^3/m^3]	ubRMSE [m^3/m^3]	R
0.05	DA	**0.079**	**0.057**	**0.42**
	OL	0.091	0.075	0.45
0.10	DA	**0.13**	**0.022**	**0.61**
	OL	0.16	0.023	0.42
0.30	DA	**0.13**	**0.021**	**0.61**
	OL	0.16	0.019	0.53
0.50	DA	**0.13**	**0.019**	**0.66**
	OL	0.16	0.018	0.54
1.00	DA	**0.13**	**0.021**	**0.54**
	OL	0.16	0.023	0.46

Table 2. Same as Table 1 but for the in situ observation data in the CARBO AFRICA site.

Depth [m]		RMSE [m^3/m^3]	ubRMSE [m^3/m^3]	R
0.05	DA	**0.086**	**0.035**	**0.67**
	OL	0.092	0.039	0.72
0.15	DA	**0.12**	**0.026**	**0.68**
	OL	0.15	0.022	0.56
0.30	DA	**0.092**	**0.023**	**0.66**
	OL	0.12	0.017	0.65

Table 2. *Cont.*

Depth [m]		RMSE [m^3/m^3]	ubRMSE [m^3/m^3]	R
0.60	DA	**0.082**	**0.022**	**0.56**
	OL	0.12	0.018	0.56
1.00	DA	**0.12**	**0.020**	**0.41**
	OL	0.16	0.020	0.45
1.50	DA	**0.12**	**0.024**	**0.20**
	OL	0.17	0.021	0.39
2.00	DA	**0.12**	**0.027**	**−0.03**
	OL	0.17	0.024	0.30

Table 3. Same as Table 1 but for the in situ observation data in the AMMA-CATCH Belefoungou site.

Depth [m]		RMSE [m^3/m^3]	ubRMSE [m^3/m^3]	R
0.05	DA	**0.064**	**0.051**	**0.86**
	OL	0.068	0.064	0.86
0.10	DA	**0.082**	**0.049**	**0.86**
	OL	0.086	0.051	0.85
0.20	DA	**0.072**	**0.049**	**0.84**
	OL	0.075	0.049	0.86
0.40	DA	**0.063**	**0.051**	**0.83**
	OL	0.059	0.049	0.88
0.60	DA	**0.053**	**0.046**	**0.85**
	OL	0.050	0.044	0.89
1.00	DA	**0.058**	**0.031**	**0.81**
	OL	0.049	0.024	0.91

Table 4. Same as Table 1 but for the in situ observation data in the AMMA-CATCH Tondikiboro site.

Depth [m]		RMSE [m^3/m^3]	ubRMSE [m^3/m^3]	R
0.05	DA	**0.078**	**0.044**	**0.65**
	OL	0.067	0.062	0.75
0.10	DA	**0.15**	**0.025**	**0.54**
	OL	0.15	0.020	0.66
0.4–0.7	DA	**0.16**	**0.021**	**0.53**
	OL	0.16	0.016	0.68
0.7–1.0	DA	**0.16**	**0.021**	**0.44**
	OL	0.17	0.016	0.61
1.05–1.35	DA	**0.19**	**0.018**	**0.34**
	OL	0.19	0.014	0.55

3.2. Identification and Quantification of the Drought Propagation

In this subsection, I identified and quantified the drought propagation using simulated soil moisture and LAI by ECHLA. By the method described in Section 2.4, I identified the drought events. By analyzing the time series of the soil moisture SAs and the LAI SAs in all identified drought events, I present the holistic view of the integrated drought propagation from soil moisture deficits to vegetation degradation.

In Figure 6, each dot shows the location of the identified drought event. The color of dots shows the year when the identified drought event occurred. In each dot, the LAI SA is less than −2.0 on the climatological day-of-year of the LAI's seasonal peak (Figure 3a) in the year shown by the color of the dot. The 3499 drought events (pixels) from 2004 to 2009 are identified.

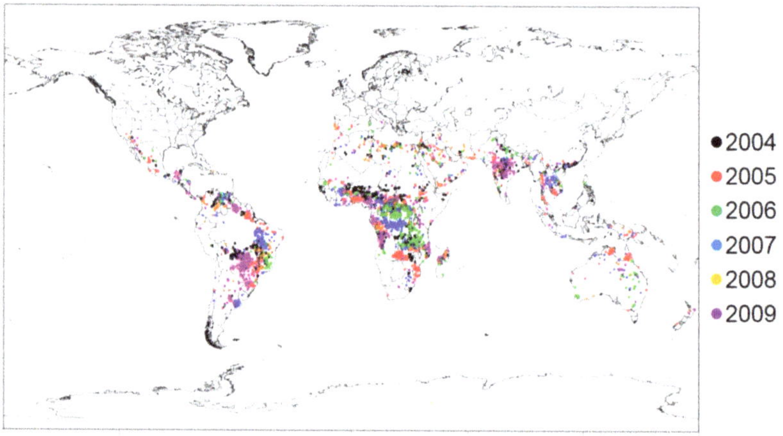

Figure 6. The locations of the identified drought events. Each dot shows the location of the identified drought event. The color of dots shows the year when the identified drought event occurred. The identified droughts shown by black, red, green, blue, yellow, and purple occurred in 2004, 2005, 2006, 2007, 2008, and 2009, respectively.

The identified drought events include the several major droughts listed in EM-DAT such as Brazil in 2005 and 2007, Paraguay in 2009, Zambia in 2005, Ethiopia in 2005, India in 2009, and Thailand in 2005 (Figure 6). ECHLA has the potential to provide the catalog of the historical drought events. However, Figure 6 shows many drought events which are not listed in EM-DAT. For example, there are many dots in the Sahel region although few droughts are reported in EM-DAT there from 2004 to 2009. This is mainly because the study period is too short to obtain enough data to identify the extreme drought events and I incorrectly recognize the relatively small declines of LAI as extremely dry events. In addition, there are many droughts which are listed in EM-DAT but not included in my drought catalog based on ECHLA because the criteria of the drought identification (i.e., the LAI SA is less than −2.0 on the climatological day-of-year of the LAI's seasonal peak) are strict.

Figure 7 shows the time series of the SAs of LAI, surface soil moisture at 0–0.05 m depth (called surface), soil moisture averaged from 0.05 to 0.45 m depth (root 1), soil moisture averaged from 0.45 to 1.05 m depth (root 2), and soil moisture averaged from 1.05 to 2.05 m depth (deep) in the selected drought events. In each drought event, I set the climatological day-of-year of the LAI's seasonal peak to day 0 and show the time series of the SAs over 2 years (from day −365 to day +365; see also Figure 3b). In all four drought events shown in Figure 7, the LAI SA has its minimum value on around day 0, which is less than −2.0, due to the criteria to identify the drought events. However, the processes of the decline and recovery of the LAI SA are different in the different drought events. In addition, there is the diversity of the progress of the soil moisture SAs among drought events.

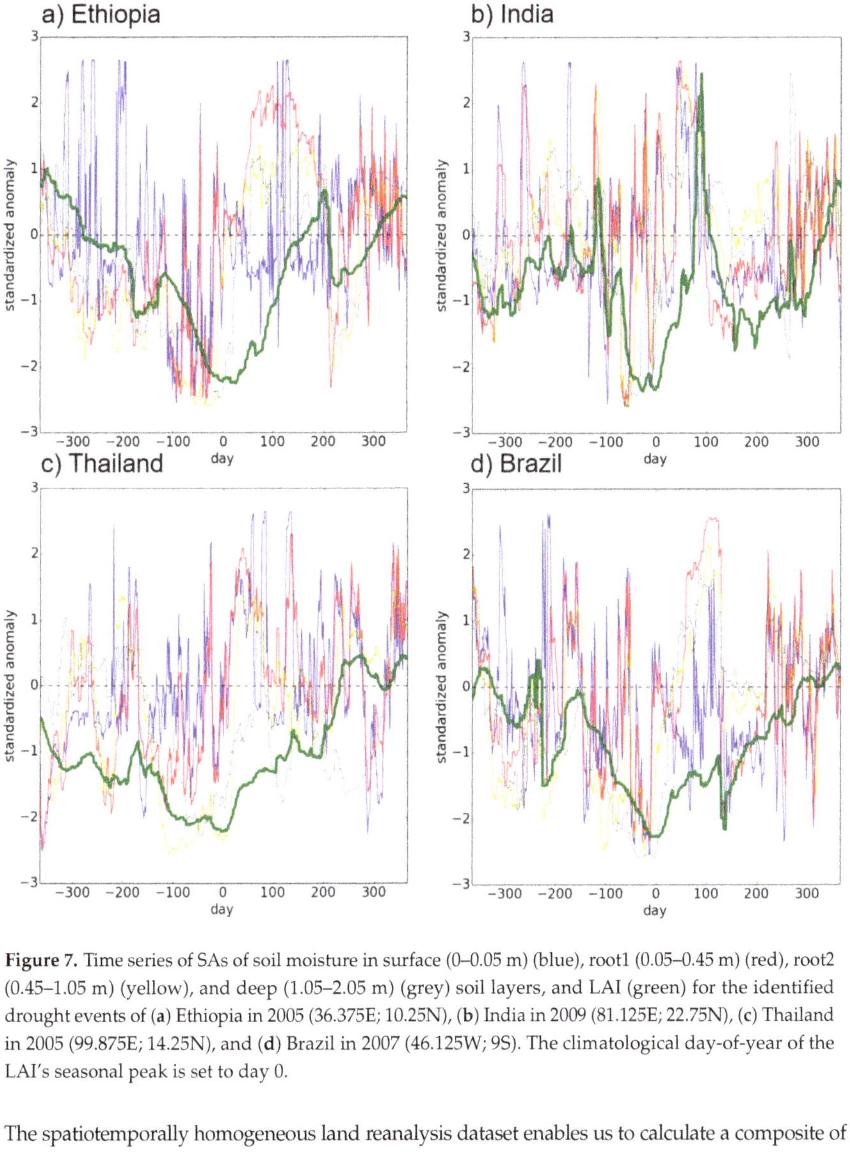

Figure 7. Time series of SAs of soil moisture in surface (0–0.05 m) (blue), root1 (0.05–0.45 m) (red), root2 (0.45–1.05 m) (yellow), and deep (1.05–2.05 m) (grey) soil layers, and LAI (green) for the identified drought events of (**a**) Ethiopia in 2005 (36.375E; 10.25N), (**b**) India in 2009 (81.125E; 22.75N), (**c**) Thailand in 2005 (99.875E; 14.25N), and (**d**) Brazil in 2007 (46.125W; 9S). The climatological day-of-year of the LAI's seasonal peak is set to day 0.

The spatiotemporally homogeneous land reanalysis dataset enables us to calculate a composite of the dynamics of soil moisture and vegetation in the extreme drought events. Figure 8 shows the composite of the LAI SA and the soil moisture SAs by averaging the SA time series of the 3499 identified drought events. Soil moisture in the shallow soil layers (i.e., surface, root 1, and root 2) has the quicker responses to rainfall deficits than soil moisture in the deep soil layer and LAI. Soil moisture in the depth of 0.05−1.05 m recovers from the drought condition quickly after day 0 and then becomes the wetter condition due to the reduction of transpiration caused by vegetation degradation.

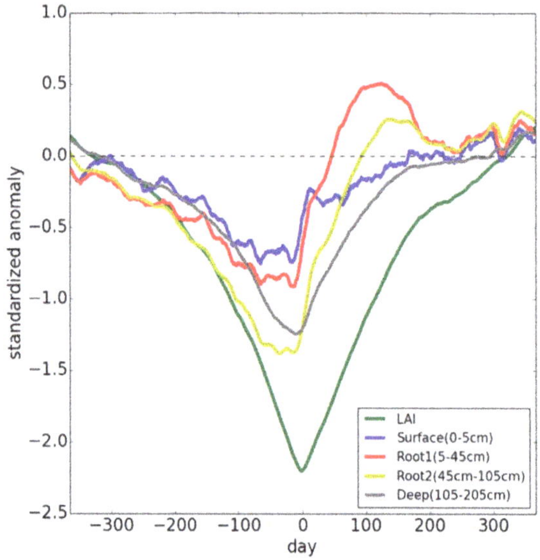

Figure 8. Composite of the identified drought events' SAs of LAI (green), surface (0–0.05 m) (blue), root1 (0.05–0.45 m) (red), root2 (0.45–1.05 m) (yellow), and deep (1.05–2.05 m) (grey) soil layers in the global snow-free region. The climatological day-of-year of the LAI's seasonal peak is set to day 0.

Figure 9 shows the boxplots of intensity, duration, and start and end date in the 3499 identified drought events (see Section 2.4). In most of the identified drought events with the LAI SA less than −2.0, the soil moisture SAs for all soil layers are also less than −2.0, which indicates that soil moisture in all soil layers is significantly degraded in the identified drought events (Figure 9a). The duration of the degradation of LAI and soil moisture in the deeper soil layers is longer than that of the soil moisture's degradation in the shallower soil layers (Figure 9b). Figure 9c quantifies the drought propagation in the 3499 identified drought events. Soil moisture in the shallow layers (surface and root1) recovers from the drought condition before the climatological day-of-year of the LAI's seasonal peak (i.e., day 0) in many drought events. On the other hand, soil moisture in the deeper soil layer (deep) recovers from the drought condition after the climatological day-of-year of the LAI's seasonal peak on average. LAI has a much slower response to the rainfall deficit than soil moisture. In the median values of the days when the SAs have their minimum values, soil moisture in surface, root1, root2, and deep soil layers proceeds LAI by 63 days, 64 days, 47 days, and 19 days, respectively.

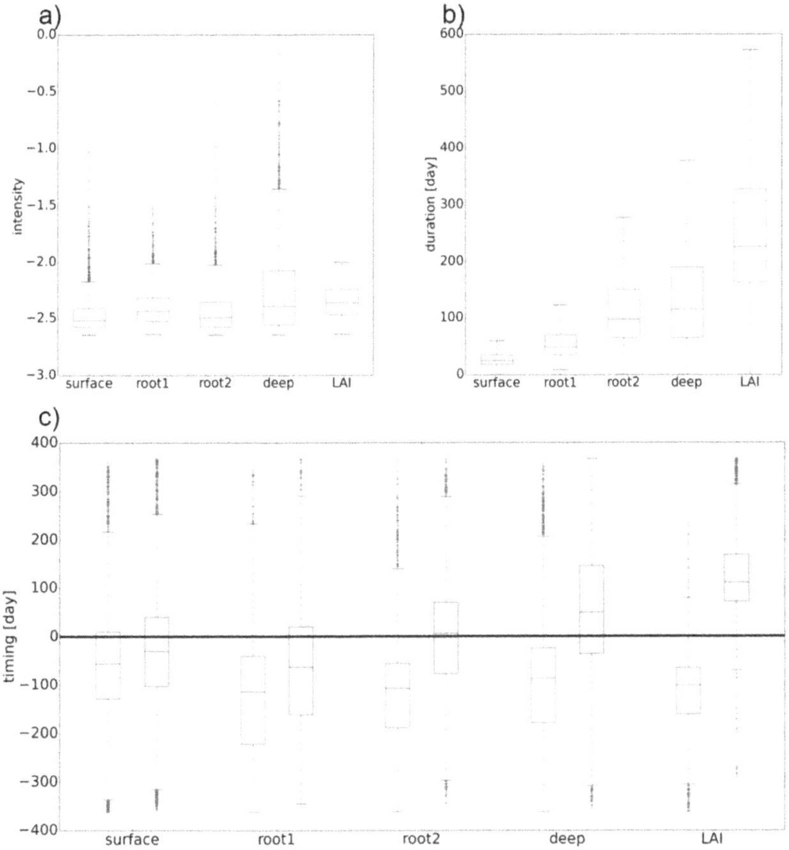

Figure 9. (**a**) Boxplot of intensity of SAs in the 3499 identified drought events. (**b**) Same as (a) but for duration. (**c**) Same as (a) but for the start date (blue) and the end date (red). Outliers are larger (smaller) than Q3 + interquartile range (Q1—interquartile range). See Section 2.4 for the definitions of intensity, duration, and the start and end date of droughts used in this study.

4. Discussions

4.1. *Performance of ECHLA*

I provided the new land reanalysis which explicitly simulates vegetation dynamics and evaluated its performance. It should be noted that there are three sources of errors, which need to be considered to interpret my verification study.

- Errors in meteorological forcings: The input of the LSM may be biased. The GLDAS meteorological forcings used in this study may have large biases especially in the poorly gauged regions.
- Errors in observations used for verification: The observations used for verification may be biased. The quality of the satellite products strongly depends on the skill of the algorithms to retrieve soil moisture and LAI from brightness temperature. Although in situ soil moisture observations may have relatively small instrument errors, they may not represent soil moisture in the coarse model grids.
- Errors in the data assimilation system: It includes the errors in the LSM, the RTM, and the data assimilation method.

Although I aim to quantify the third type of errors, it is generally difficult to decompose the bulk errors quantified by the metrics into the different types of errors.

Although there are high correlations between simulated LAI by ECHLA and satellite optically observed LAI, the absolute value of simulated LAI has the large deviation from that of satellite-observed LAI. Jarlan et al. [61] showed that simulated LAI by the ECMWF's CTESSEL LSM deviates from MODIS LAI by a factor of 2. It is difficult to directly compare the absolute value of LSM-simulated LAI with that of satellite-observed LAI. It may be caused partly by the uncertainties in the satellite-observed LAI products [62].

There are two possible reasons why the data assimilation cannot improve the simulation of the absolute value of LAI. First, to solve the radiative transfer in the canopy, the information of vegetation water content and other parameters related to optical properties of plants is needed although the LSM directly solves carbon cycle. Therefore, I need to rely on some empirical equations to convert the LSM's output to the RTM's input (see Equations (8)–(11) in the Supplementary Materials), which induces uncertainties in the LDAS. Second, as Sawada et al. [55] indicated, the sensitivity of C- and X-band brightness temperature to vegetation becomes small when LAI becomes larger than 4 so that it is difficult to quantitatively constrain LAI in the forested area by assimilating the microwave signals.

It should be noted that the improvement of the representation of phenology by data assimilation positively impacts the identification and quantification of the drought propagation since I focused on the seasonal cycle of vegetation activity to identify the drought events (see Section 2.4). The increase of correlation coefficients between simulated and satellite-observed LAI improves the skill to estimate the timing of the beginning, end, and peak of seasons. It brings the accurate retrievals of the climatological day-of-year of the LAI's seasonal peak, which positively impacts the identification and quantification of drought events based on the LAI's time series.

The significant improvement of the surface soil moisture simulation by data assimilation can be found in the sparsely vegetated area when I compared the ECHLA simulation with both satellite and in situ observations. However, the impact of data assimilation on the simulation of sub-surface soil moisture was marginal in this study. The performance of ECHLA to simulate surface soil moisture is compatible to the other land reanalysis studies [14,36]. Assimilating satellite microwave brightness temperature observations reduces RMSEs against satellite surface soil moisture observations and in-situ surface soil moisture observations. The degradation of the skill to reproduce the satellite-observed surface soil moisture data can be found in the moderately and densely vegetated area. In the moderately and densely vegetated area, the uncertainty in the parameterization of the vegetation effect on microwave radiative transfer may degrade the skill to simulate the contribution of surface soil moisture to microwave brightness temperature. It should be noted that the accuracy of the satellite surface soil moisture observation product used for verification may also be low in the moderately and densely vegetated area [55,63]. Data assimilation has marginal impacts on correlation coefficients. This may be because the seasonal cycle of surface soil moisture is largely driven by the meteorological forcings and I did not consider the errors in the meteorological forcings in the data assimilation system. It may be beneficial to use multiple meteorological forcing datasets to consider their errors.

Few previous studies have evaluated the skill of LDASs to simulate sub-surface soil moisture. Sawada et al. [40] have found that CLVDAS can improve sub-surface soil moisture by solving the inversion problem to estimate root-zone soil moisture from vegetation growth on land surface. While Sawada et al. [40] evaluated the performance of CLVDAS in a single meteorological observation site using accurate in situ meteorological forcings' observation, I found an improvement of the sub-surface soil moisture simulation in two of the four in situ observation sites in the semi-global application of this study. The improvement of the sub-surface soil moisture simulation contributes to the drought analysis since sub-surface soil moisture has an important role in the drought propagation. It should be noted that the errors in in situ sub-surface soil moisture observations may be larger than those in in situ surface soil moisture observations. When installing observation probes in deep soil, soil structure may be somewhat disturbed so that it is unclear if the in situ observation represents sub-surface soil moisture in the coarse model grid. In addition, there are uncertainties in the depths of soil moisture measurements. The degradation of correlation coefficients in the AMMA-CATCH Tondikiboro site by data assimilation may be caused partly by these representation errors in the in-situ observations.

To improve the skill of ECHLA to simulate soil moisture and LAI, the parameter optimization of the LSM should be implemented although I optimized only the parameter of the RTM in this paper. Sawada and Koike [39] optimized the unknown hydrological and ecological parameters of the LSM (e.g., hydraulic conductivity), by assimilating microwave brightness temperature observations and improved the skill of the LSM to simulate both surface soil moisture and LAI. Optimizing unknown parameters of LSMs by data assimilation has been intensively investigated [28–30]. Although the parameter optimization module is included in the current version of CLVDAS used in this study, I did not apply it to optimize the unknown parameters of the LSM in the semi-global application of this study because it is computationally demanding. I should explore the computationally efficient parameter optimization method, which can be applied globally, in the future

4.2. Conceptual Model of the Ecohydrological Drought Propagation

On average, the drought events identified in this study significantly damage both vegetation activity and soil water storage from the surface to deep soil layers. The degradation of vegetation dynamics lasts 150–350 days while the duration of soil moisture deficits is shorter than that of vegetation degradation. The identified drought events start approximately 100 days before the climatological day-of-year of the LAI's seasonal peak. Soil moisture in the shallow soil layers quickly recovers from the drought condition before the climatological day-of-year of the LAI's seasonal peak while soil moisture in the deep soil layers and vegetation dynamics recover from the drought condition approximately 100 days after the climatological day-of-year of the LAI's seasonal peak.

This drought propagation from soil moisture to vegetation has been found by previous studies. Sepulcre-Canto et al. [10] found that a soil moisture index calculated by a precipitation-runoff model is significantly correlated with the anomaly of the satellite-observed fraction of absorbed photosynthetically active radiation (fAPAR) when soil moisture proceeds fAPAR by 10–20 days in Europe. Akarsh and Mishra [11] showed that the root-zone (top 60 cm) soil moisture anomaly calculated by GLDAS is strongly correlated with the anomaly of satellite-observed normalized difference vegetation index (NDVI) when soil moisture proceeds NDVI by 1 month in India. I support these regional findings using a composite of soil moisture and LAI in the global snow-free region, which are simulated by the new land reanalysis.

It should be noted that the farmer's strategy of planting and harvesting was not considered in the LSM. Ignoring the farmer's adaptation strategy against severe drought events may overestimate the negative impact of water deficits in this study. In addition, I focused only on the effect of water deficits on agricultural activities and many other factors which harm agricultural activities were not considered. For instance, crops are greatly affected by floods and associated inundation. Any political and economic effects on agriculture are not considered in the drought identification and quantification method used in this study.

The major limitation of this study is that the study period is short to identify the extreme drought events. To obtain an inclusive drought catalog using ECHLA, I need to extend the study period which is currently identical to the AMSR-E operational period. Since I directly assimilate satellite-observed brightness temperature observations instead of derived products, it is straightforward to assimilate observations from the other satellites such as AMSR2, Soil Moisture, Ocean Salinity (SMOS), and Soil Moisture Active Passive (SMAP). However, it should be noted that changes in observing systems used for data assimilation may cause temporal inhomogeneity in the reanalysis dataset [64,65]. The period of ECHLA will be extended to obtain an inclusive drought catalog.

5. Conclusions

I provided a semi-global land reanalysis which explicitly calculates vegetation growth and senescence and is constrained by satellite microwave observations. I revealed that the new land reanalysis had the reasonable skill to simulate surface soil moisture, sub-surface soil moisture, and LAI and was useful to identify and quantify the drought propagation. By analyzing the SAs of soil moisture and LAI in the land reanalysis, I identified the severe drought events and quantified the drought propagation from surface soil moisture to sub-surface soil moisture and the corresponding vegetation dynamics. On average, the drought events identified in this study significantly damage both vegetation activity and soil water storage from the surface to deep soil layers. Soil moisture in the shallow soil layers quickly recovers from the drought condition while the deficits of soil moisture in the deep soil layers and vegetation dynamics lasted until approximately 100 days after the climatological day-of-year of the LAI's seasonal peak.

Supplementary Materials: The following are available online at http://www.mdpi.com/2072-4292/10/8/1197/s1, Text S1: Formulations of Coupled Land and Vegetation Data Assimilation System (CLVDAS).

Funding: This research was funded by the JSPS KAKENHI Grant Number JP17K18352, Data Integration and Analysis System (DIAS), JAXA, and the FLAGSHIP2020 project of the Ministry of Education, Culture, Sports, Science, and Technology.

Acknowledgments: The GLDAS data products were provided by the National Aeronautics and Space Administration (NASA) and can be downloaded at http://disc.sci.gsfc.nasa.gov/hydrology/data-holdings. The AMSR-E brightness temperature observation data products were provided by the Japan Aerospace Exploration Agency (JAXA) and can be downloaded at https://gcom-w1.jaxa.jp/auth.html. The GLASS LAI data products were provided by Global Land Cover Facility (GLCF) and can be downloaded at http://www.glcf.umd.edu/data/lai/. The ESA CCI surface soil moisture product was provided by ESA and can be downloaded at http://www.esa-soilmoisture-cci.org/. The in situ soil moisture observations were archived at International Soil Moisture Network (ISMN) https://ismn.geo.tuwien.ac.at/. The numerical experiments were implemented using the Oakleaf-FX supercomputer furnished by Supercomputing Division, Information Technology Center, the University of Tokyo. I thank three anonymous reviewers for their helpful comments.

Conflicts of Interest: The author declares no conflicts of interest.

References

1. Wilhite, D.A. *Drought: A Global Assessment*; Routledge: London, UK, 2000.
2. Svoboda, M.; LeComte, D.; Hayes, M.; Heim, R.; Gleason, K.; Angel, J.; Rippey, B.; Tinker, R.; Palecki, M.; Stooksbury, D.; et al. The drought monitor. *Bull. Am. Meteorol. Soc.* **2002**, *83*, 1181–1190. [CrossRef]
3. Anderson, W.; Zaitchik, B.F.; Hain, C.R.; Anderson, M.C.; Yilmaz, M.T.; Mecikalski, J.; Schultz, L. Towards an integrated soil moisture drought monitor for East Africa. *Hydrol. Earth Syst. Sci.* **2012**, *16*, 2893–2913. [CrossRef]
4. McNally, A.; Arsenault, K.; Kumar, S.; Shukla, S.; Peterson, P.; Wang, S.; Funk, C.; Peters-Lidard, C.D.; Verdin, J.P. A land data assimilation system for sub-Saharan Africa food and water security applications. *Sci. Data* **2017**, *4*. [CrossRef] [PubMed]
5. Yuan, X.; Wood, E.F. Multimodel seasonal forecasting of global drought onset. *Geophys. Res. Lett.* **2013**, *40*, 4900–4905. [CrossRef]
6. Pan, M.; Yuan, X.; Wood, E.F. A probabilistic framework for assessing drought recovery. *Geophys. Res. Lett.* **2013**, *14*, 3637–3642. [CrossRef]
7. Sawada, Y.; Koike, T. Towards ecohydrological drought monitoring and prediction using a land data assimilation system: A case study on the Horn of Africa drought (2010–2011). *J. Geophys. Res. Atmos.* **2016**, *121*, 8229–8242. [CrossRef]
8. Wilhite, D.A.; Glantz, M.H. Understanding: The Drought Phenomenon: The Role of Definitions. *Water Int.* **1985**, *10*. [CrossRef]
9. Sawada, Y.; Koike, T.; Jaranilla-Sanchez, P.A. Modeling hydrologic and ecologic responses using a new eco-hydrological model for identification of droughts. *Water Resour. Res.* **2014**, *50*. [CrossRef]

10. Sepulcre-Canto, G.; Horion, S.; Singleton, A.; Carrao, H.; Vogt, J. Development of a Combined Drought Indicator to detect agricultural drought in Europe. *Nat. Harzards Earth Syst. Sci.* **2012**, *12*, 3519–3531. [CrossRef]
11. Akarsh, A.; Mishra, V. Prediction of vegetation anomalies to improve food security and water management in India. *Geophys. Res. Lett.* **2015**, *42*, 5290–5298. [CrossRef]
12. Rodell, M.; Houser, P.R.; Jambor, U.; Gottschalck, J.; Mitchell, K.; Meng, C.-J.; Arsenault, K.; Cosgrove, B.; Radakovich, J.; Bosilovich, M.; et al. The global land data assimilation system. *Bull. Am. Meteorol. Soc.* **2004**, *85*, 381–394. [CrossRef]
13. Reichle, R.H.; Koster, R.D.; De Lannoy, G.J.M.; Forman, B.M.; Liu, Q.; Mahanama, S.P.P.; Touré, A. Assessment and Enhancement of MERRA Land Surface Hydrology Estimates. *J. Clim.* **2011**, *24*, 6322–6338. [CrossRef]
14. Balsamo, G.; Albergel, C.; Beljaars, A.; Boussetta, S.; Brun, E.; Cloke, H.; Dee, D.; Dutra, E.; Muñoz-Sabater, J.; Pappenberger, F.; et al. ERA-Interim/Land: A global land surface reanalysis data set. *Hydrol. Earth Syst. Sci.* **2015**, *19*, 389–407. [CrossRef]
15. Tang, J.; Cheng, H.; Liu, L. Assessing the recent droughts in Southwestern China using satellite gravimetry. *Water Resour. Res.* **2014**, *50*, 3030–3038. [CrossRef]
16. Leblanc, M.; Tregoning, J.P.; Ramillien, G.; Tweed, S.O.; Fakes, A. Basin-scale, integrated observations of the early 21st century multiyear drought in Southeast Australia. *Water Resour. Res.* **2009**, *45*, W04408. [CrossRef]
17. Asoka, A.; Glesson, T.; Wada, Y.; Mishra, V. Relative contribution of monsoon precipitation and pumping to changes in groundwater storage in India. *Nat. Geosci.* **2017**, *10*, 109–117. [CrossRef]
18. Yuan, X.; Ma, Z.; Pan, M.; Shi, C. Microwave remote sensing of short-term droughts during crop growing seasons. *Geophys. Res. Lett.* **2015**, *42*, 4394–4401. [CrossRef]
19. Hao, Z.; Aghakouchak, A. A Nonparametric Multivariate Multi-Index Drought Monitoring Framework. *J. Hydrometeorol.* **2014**, *15*, 89–101. [CrossRef]
20. Madadgar, S.; AghaKouchak, A.; Farahmand, A.; Davis, S.J. Probabilistic estimates of drought impacts on agricultural production. *Geophys. Res. Lett.* **2017**, *44*. [CrossRef]
21. Ulaby, F.; Moore, R.K.; Fung, A. (Eds.) *Microwave Remote Sensing: Active and Passive—Volume Scattering and Emission Theory*; Artech House: Dedham, MA, USA, 1986; Volume 3.
22. Paloscia, S.; Pampaloni, P. Microwave polarization index for monitoring vegetation growth. *IEEE Trans. Geosci. Remote* **1988**, *26*, 617–621. [CrossRef]
23. Owe, M.; de Jeu, R.; Walker, J. A methodology for surface soil moisture and vegetation optical depth retrieval using the microwave polarization difference index. *IEEE Trans. Geosci. Remote* **2001**, *39*, 1643–1654. [CrossRef]
24. Fujii, H.; Koike, T.; Imaoka, K. Improvement of the AMSR-E Algorithm for Soil Moisture Estimation by Introducing a Fractional Vegetation Coverage Dataset Derived from MODIS Data. *J. Remote Sens. Soc. Jpn.* **2009**, *29*, 282–292.
25. Holmes, T.R.H.; De Jeu, R.A.M.; Owe, M.; Dolman, A.J. Land surface temperature from Ka band (37 GHz) passive microwave observations. *J. Geophys. Res.* **2009**, *114*, D04113. [CrossRef]
26. Kerr, Y.H.; Font, J.; Martin-Neira, M.; Mecklenburg, S. Introduction to the Special Issue on the ESA's Soil Moisture and Ocean Salinity Mission (SMOS)—Instrument performance and first results. *IEEE Trans. Geosci. Remote Sens.* **2012**, *50*, 1351–1353. [CrossRef]
27. Chan, S.K.; Bindlish, R.; O'Neill, P.; Jackson, T.; Njoku, E.; Dunbar, S.; Chaubell, J.; Piepmeier, J.; Yueh, S.; Entekhabi, D.; et al. Development and assessment of the SMAP enhanced passive soil moisture product. *Remote Sens. Environ.* **2016**, *204*, 931–941. [CrossRef]
28. Yang, K.; Watanabe, T.; Koike, T.; Li, X.; Fujii, H.; Tamagawa, K.; Ma, Y.; Ishikawa, H. Auto-calibration System Developed to Assimilate AMSR-E Data into a Land Surface Model for Estimating Soil Moisture and the Surface Energy Budget. *J. Meteorol. Soc. Jpn.* **2007**, *85A*, 229–242. [CrossRef]
29. Tian, X.; Xie, Z.; Dai, A.; Shi, C.; Jia, B.; Chen, F.; Yang, K. A dual-pass variational data assimilation framework for estimating soil moisture profiles from AMSR-E microwave brightness temperature. *J. Geophys. Res.* **2009**, *114*, D16102. [CrossRef]
30. Yang, K.; Koike, T.; Kaihotsu, I.; Qin, J. Validation of a Dual-Pass Microwave Land Data Assimilation System for Estimating Surface Soil Moisture in Semiarid Regions. *J. Hydrometeorol.* **2009**, *10*, 780–793. [CrossRef]
31. Qin, J.; Liang, S.; Yang, K.; Kaihotsu, I.; Liu, R.; Koike, T. Simultaneous estimation of both soil moisture and model parameters using particle filtering method through the assimilation of microwave signal. *J. Geophys. Res.* **2009**, *114*, D15103. [CrossRef]

32. Kumar, S.; Reichle, V.R.H.; Koster, R.D.; Crow, W.T.; Peters-Lidard, C.D. Role of Subsurface Physics in the Assimilation of Surface Soil Moisture Observations. *J. Hydrometeorol.* **2009**, *10*, 1534–1547. [CrossRef]
33. Li, B.; Toll, D.; Zhan, X.; Cosgrove, B. Improving estimated soil moisture fields through assimilation of AMSR-E soil moisture retrievals with an ensemble Kalman filter and a mass conservation constraint. *Hydrol. Earth Syst. Sci.* **2012**, *16*, 105–119. [CrossRef]
34. Su, Z.; de Rosnay, P.; Wen, J.; Wang, L.; Zeng, Y. Evaluation of ECMWF's soil moisture analyses using observations on the Tibetan Plateau. *J. Geophys. Res. Atmos.* **2013**, *118*, 5304–5318. [CrossRef]
35. He, L.; Chen, J.M.; Liu, J.; Belair, S.; Luo, X. Assessment of SMAP soil moisture for global simulation of gross primary production. *J. Geophys. Res. Biogeosci.* **2017**, *122*, 1549–1563. [CrossRef]
36. Kumar, S.V.; Peters-Lidard, C.D.; Mocko, D.; Reichle, R.H.; Liu, Y.; Arsenault, K.R.; Xia, Y.; Ek, M.; Riggs, G.; Livneh, B.; et al. Assimilation of Remotely Sensed Soil Moisture and Snow Depth Retrievals for Drought Estimation. *J. Hydrometeorol.* **2014**, *15*, 2446–2469. [CrossRef]
37. Bardu, A.L.; Calvet, J.-C.; Mahfouf, J.-F.; Albergel, C.; Lafont, S. Assimilation of Soil Wetness Index and Leaf Area Index into the ISBA-A-gs land surface model: Grassland case study. *Biogeosciences* **2011**, *8*, 1971–1986. [CrossRef]
38. Liu, P.-W.; Bongiovanni, T.; Monsivais-Huertero, A.; Judge, J.; Steele-Dunne, S.; Bindlish, R.; Jackson, T.J. Assimilation of Active and Passive Microwave Observations for Improved Estimates of Soil Moisture and Crop Growth. *IEEE J. Sel. Top. Appl. Earth Observ. Remote Sens.* **2016**, *9*, 1357–1369. [CrossRef]
39. Sawada, Y.; Koike, T. Simultaneous estimation of both hydrological and ecological parameters in an ecohydrological model by assimilating microwave signal. *J. Geophys. Res. Atmos.* **2014**, *119*. [CrossRef]
40. Sawada, Y.; Koike, T.; Walker, J.P. A land data assimilation system for simultaneous simulation of soil moisture and vegetation dynamics. *J. Geophys. Res. Atmos.* **2015**, *120*. [CrossRef]
41. Sheffield, J.; Goteti, G.; Wood, E.F. Development of a 50-Year High-Resolution Global Dataset of Meteorological Forcings for Land Surface Modeling. *J. Clim.* **2006**, *19*, 3088–3111. [CrossRef]
42. Kachi, M.; Naoki, K.; Hori, M.; Imaoka, K. AMSR2 validation results. In Proceedings of the 2013 IEEE International Geoscience and Remote Sensing Symposium (IGARSS), Melbourne, VIC, Australia, 21–26 July 2013; pp. 831–834. [CrossRef]
43. Liu, Y.Y.; de Jeu, R.A.M.; McCabe, M.F.; Evans, J.P.; van Dijk, A.I.J.M. Global long-term passive microwave satellite-based retrievals of vegetation optical depth. *Geophys. Res. Lett.* **2011**, *38*, L18402. [CrossRef]
44. Global Soil Data Task Group, Global Gridded Surfaces of Selected Soil Characteristics (IGBPDIS). 2000. Available online: https://daac.ornl.gov (accessed on 13 February 2018).
45. Food and Agricultural Organization, Digital Soil Map of the World and Derived Soil Properties, Land Water Digital Media Ser. 1 [CR-ROM], Rome. 2003. Available online: http://www.fao.org/ag/agl/lwdms.stm (accessed on 13 February 2018).
46. Xiao, Z.; Liang, S.; Wang, J.; Chen, P.; Yin, X.; Zhang, L.; Song, J. Use of General Regression Neural Networks for Generating the GLASS Leaf Area Index Product from Time Series MODIS Surface Reflectance. *IEEE Trans. Geosci. Remote Sens.* **2013**, *52*, 209–223. [CrossRef]
47. Huang, D.; Knyazikhin, Y.; Wang, W.; Deering, D.W.; Stenberg, P.; Shabanov, N.V.; Tan, B.; Myneni, R.B. Stochastic transport theory for investigating the three-dimensional canopy structure from space measurements. *Remote Sens. Environ.* **2008**, *112*, 35–50. [CrossRef]
48. Dorigo, W.A.; Wagner, W.; Albergel, C.; Albrecht, F.; Balsamo, G.; Brocca, L.; Chung, D.; Ertl, M.; Forkel, M.; Gruber, A.; et al. ESA CCI Soil Moisture for improved Earth system understanding: State-of-the art and future directions. *Remote Sens. Environ.* **2017**, *15*, 185–215. [CrossRef]
49. Gruber, A.; Dorigo, W.A.; Crow, W.; Wagner, W. Triple Collocation-Based Merging of Satellite Soil Moisture Retrievals. *IEEE Trans. Geosci. Remote* **2017**, *55*, 6780–6792. [CrossRef]
50. Liu, Y.Y.; Dorigo, W.A.; Parinussa, R.M.; de Jeu, R.; Wagner, W.; McCabe, M.F.; Evans, J.P.; van Dijk, A.I.J.M. Trend-preserving blending of passive and active microwave soil moisture retrievals. *Remote Sens. Environ.* **2012**, *123*, 280–297. [CrossRef]
51. Dorigo, W.A.; Wagner, W.; Hohensinn, R.; Hahn, S.; Paulik, C.; Drusch, M.; Mecklenburg, S.; van Oevelen, P.; Robock, A.; Jackson, T. The International Soil Moisture Network: A data hosting facility for global in situ soil moisture measurements. *Hydrol. Earth Syst. Sci.* **2011**, *15*, 1675–1698. [CrossRef]

52. Lebel, T.; Cappelaere, B.; Galle, S.; Hanan, N.; Kergoat, L.; Levis, S.; Vieux, B.; Descroix, L.; Gosset, M.; Mougin, E.; et al. AMMA-CATCH studies in the Sahelian region of West-Africa: An overview. *J. Hydrol.* **2009**, *375*, 3–13. [CrossRef]
53. Ardö, J. A 10-Year Dataset of Basic Meteorology and Soil Properties in Central Sudan. *Dataset Pap. Geosci.* **2013**, *2013*, 297973. [CrossRef]
54. Tagesson, T.; Fensholt, R.; Guiro, I.; Rasmussen, M.O.; Huber, S.; Mbow, C.; Garcia, M.; Horion, S.; Sandholt, I.; Holm-Rasmussen, B.; et al. Ecosystem properties of semiarid savanna grassland in West Africa and its relationship with environmental variability. *Glob. Chang. Biol.* **2015**, *21*, 250–264. [CrossRef] [PubMed]
55. Sawada, Y.; Tsutsui, H.; Koike, T. Ground Truth of Passive Microwave Radiative Transfer on Vegetated Land Surfaces. *Remote Sens.* **2017**, *9*, 655. [CrossRef]
56. Duan, Q.; Sorooshian, S.; Gupta, V. Effective and Efficient Global Optimization for Conceptual Rainfall-Runoff Models. *Water Resour. Res.* **1992**, *28*, 1015–1031. [CrossRef]
57. Jackson, R.B.; Canadell, J.G.; Ehleringer, J.R.; Mooney, H.A.; Sala, O.E.; Schulze, E.D. A global analysis of root distributions for terrestrial biomes. *Oecologia* **1996**, *108*, 389–411. [CrossRef] [PubMed]
58. Minamide, M.; Zhang, F. Adaptive Observation Error Inflation for Assimilating All-Sky Satellite Radiance. *Mon. Weather Rev.* **2017**, *145*, 1063–1081. [CrossRef]
59. Jaranilla-Sanchez, P.A.; Wang, L.; Koike, T. Modeling the hydrological responses of the Pampanga River basin, Philippines: A quantitative approach for identifying droughts. *Water Resour. Res.* **2011**, *47*, W03514. [CrossRef]
60. Jonsson, P.; Eklundh, L. TIMESAT—A program for analysing time-series of satellite sensor data. *Comput. Geosci.* **2004**, *30*, 833–845. [CrossRef]
61. Jarlan, L.; Balsamo, G.; Lafont, S.; Beljaars, A.; Calvet, J.-C.; Mougin, E. Analysis of leaf area index in the ECMWF land surface model and impact on latent heat and carbon fluxes: Application to West Africa. *J. Geophys. Res.* **2008**, *113*, D24117. [CrossRef]
62. Liu, Y.; Xiao, J.; Ju, W.; Zhu, G.; Wu, X.; Fan, W.; Li, D.; Zhou, Y. Satellite-derived LAI products exhibit large discrepancies and can lead to substantial uncertainty in simulated carbon and water fluxes. *Remote Sens. Environ.* **2018**, *206*, 174–188. [CrossRef]
63. Crow, W.T.; Chan, S.; Entekhabi, D.; Hsu, A.; Jackson, T.J.; Njoku, E.; O'Neill, P.; Shi, J. An observing system simulation experiment for hydros radiometer-only soil moisture and freeze-thaw products. *IEEE Trans. Geosci. Remote* **2005**, *43*, 1289–1303. [CrossRef]
64. Kobayashi, C.; Endo, H.; Ota, Y.; Kobayashi, S.; Onoda, H.; Harada, Y.; Onogi, K.; Kamahori, H. Preliminary Results of the JRA-55C, an Atmospheric Reanalysis Assimilating Conventional Observations Only. *SOLA* **2014**, *10*, 78–82. [CrossRef]
65. Kobayashi, S.; Ota, Y.; Harada, Y.; Ebita, A.; Moriya, M.; Onoda, H.; Onogi, K.; Kamahori, H.; Kobayashi, C.; Endo, H.; et al. The JRA-55 Reanalysis: General Specifications and Basic Characteristics. *J. Meteorol. Soc. Jpn.* **2015**, *93*, 5–48. [CrossRef]

remote sensing

MDPI

Article

Using Satellite-Derived Vegetation Products to Evaluate LDAS-Monde over the Euro-Mediterranean Area

Delphine Jennifer Leroux, Jean-Christophe Calvet *, Simon Munier and Clément Albergel

Centre National de Recherches Meteorologiques, UMR3589 (CNRS, Météo-France), 31057 Toulouse, France; delphine.leroux@meteo.fr (D.J.L.); simon.munier@meteo.fr (S.M.); clement.albergel@meteo.fr (C.A.)

* Correspondence: jean-christophe.calvet@meteo.fr

Received: 25 June 2018; Accepted: 27 July 2018; Published: 31 July 2018

Abstract: Within a global Land Data Assimilation System (LDAS-Monde), satellite-derived Surface Soil Moisture (SSM) and Leaf Area Index (LAI) products are jointly assimilated with a focus on the Euro-Mediterranean region at 0.5° resolution between 2007 and 2015 to improve the monitoring quality of land surface variables. These products are assimilated in the CO_2 responsive version of ISBA (Interactions between Soil, Biosphere and Atmosphere) land surface model, which is able to represent the vegetation processes including the functional relationship between stomatal aperture and photosynthesis, plant growth and mortality (ISBA-A-gs). This study shows the positive impact on SSM and LAI simulations through assimilating their satellite-derived counterparts into the model. Using independent flux estimates related to vegetation dynamics (evapotranspiration, Sun-Induced Fluorescence (SIF) and Gross Primary Productivity (GPP)), it is also shown that simulated water and CO_2 fluxes are improved with the assimilation. These vegetation products tend to have higher root-mean-square deviations in summer when their values are also at their highest, representing 20–35% of their absolute values. Moreover, the connection between SIF and GPP is investigated, showing a linear relationship depending on the vegetation type with correlation coefficient values larger than 0.8, which is further improved by the assimilation.

Keywords: land surface model; land data assimilation system; accuracy; fluorescence

1. Introduction

Land Surface Models (LSMs) were implemented to simulate energy, mass and momentum fluxes between the land surface and the atmosphere. In LSMs, two vegetation properties control the water and CO_2 fluxes to a large extent: leaf stomatal conductance (gs) and Leaf Area Index (LAI). ISBA (Interaction between Soil, Biosphere and Atmosphere; [1]) is an LSM designed for use in numerical weather prediction and climate models. Unlike most LSMs that use an LAI climatology, the CO_2 responsive version of ISBA (ISBA-A-gs; [2,3]) dynamically computes LAI along with the stomatal conductance and the associated photosynthesis rate. Houghton, J. et al. [4] pointed out the high uncertainty affecting estimates of carbon fluxes and their evolutions at the global scale. Integrating satellite-derived estimates of vegetation properties could help reduce theses uncertainties.

Satellite observations can be used to better constrain LSMs using data assimilation techniques that are able to integrate satellite observations into model simulations [5–15]. It was shown that a model performing better for soil moisture does not necessarily give the best results for plant productivity, which highlights the need to jointly assimilate soil moisture and vegetation observations in order to better constrain the hydrological and the carbon cycle models [16–18]. Moreover, soil moisture plays an essential role in partitioning the incoming water and energy over land, which affects the evapotranspiration, the runoff and all energy fluxes [19]. Joint assimilation of Surface

Soil Moisture (SSM) and LAI satellite products were implemented for ISBA in the SURFEX (SURface EXternalisee [20]) modeling platform over France [12,14], and it was recently extended to the global scale. The Land Data Assimilation System resulting from these efforts is called LDAS-Monde [15].

The LDAS-Monde analyzed variables combine information from the model itself and from the assimilated satellite observations. They are generated for each model time step and model grid point. This makes them highly valuable in terms of spatial and time availability.

The recently available observations of Sun-Induced chlorophyll Fluorescence (SIF) offer a new perspective on vegetation monitoring. A number of authors showed that SIF can be retrieved using observations from the Greenhouse gas Observing Satellite (GOSAT; [21–24]), SCIAMACHYon board the ENVISAT satellite [25] and GOME-2 on board the MetOp-A and MetOp-B satellites [26]. All these studies found a link between SIF and Gross Primary Productivity (GPP), which represents the CO_2 uptake by the vegetation through photosynthesis. Zhang, Y. et al. [27] compared multiple GPP models to GOME-2 SIF data over North America at a coarse spatial resolution ($0.5^\circ \times 0.5^\circ$) and found a good agreement in their spatial distributions and seasonal dynamics. They also showed that, at the biome scale, there was a clear, almost linear relationship between SIF and GPP, highlighting the potential of SIF products to be used as a validation testbed for GPP models. Sun, Y. et al. [28] compared the SIF product from OCO-2 satellite observations to in situ GPP measured from three flux towers and found that the SIF-GPP relationship was consistent across different vegetation types. They found that SIF is related to GPP through empirical orthogonal function analysis, which revealed that the spatio-temporal variations of SIF and GPP were highly consistent (high correlations between the first modes).

LDAS-Monde has already been evaluated over the Euro-Mediterranean area for the 2000–2012 period [15] using a different version of ISBA (diffusive scheme) with (i) agricultural statistics over France; (ii) river discharge observations; (iii) the GLEAM (Global Land surface Evaporation: the Amsterdam Methodology) evaporation product [29] and (iv) the FLUXNET-MTE (FLUXNET network-Multi Tree Ensemble) GPP product [30]. The goal of this paper is to evaluate LDAS-Monde using the evaporation and GPP products for the 2007–2015 period and to assess for the first time the ability of the GOME-2 SIF observations to be an additional independent source of validation of the modeled GPP before and after the joint assimilation of SSM and LAI products in LDAS-Monde. Section 2 is dedicated to the presentation of the model, the data assimilation system and the satellite products. Section 3 shows the impact of the assimilation on the assimilated variables, and the SIF-GPP relationship is investigated. Finally, subgrid variability issues are discussed in Section 4, together with prospects for assimilating new products.

2. Material and Methods

2.1. ISBA Land Surface Model

The ISBA LSM solves the energy and water budgets at the surface level and describes the exchanges between the land surface and the atmosphere on a sub-hourly basis. The modeling platform SURFEX (SURFace Externalisee, [20]) Version 8.0 was used in this study (source code and documentation available at http://www.umr-cnrm.fr/surfex/) with the three-layer version of the soil model in ISBA [31]. For each model grid cell, the soil is partitioned into three layers: the top surface representing the very first centimeters, the root-zone soil layer defined by the vegetation rooting depth (which includes the first layer) and the recharge layer located below the root-zone with a maximum thickness of 1 m. In the model, the propagation of the information from the surface to the deepest layers relies on a force-restore dynamics: the surface and root-zone layers are forced by the atmospheric conditions and restored towards an equilibrium state where the gravity forces match the capillary forces; the drainage from the root-zone soil layer to the recharge layer supplies water and conserves the total water volume.

Vegetation growth and mortality processes were introduced by [2,32] and implemented in the form of a Nitrogen Dilution Process (NIT option, [3]) in order to simulate LAI interactively. Moreover, a refined representation of plant response to soil moisture deficit was implemented by [33,34], with contrasting drought-avoiding and drought-tolerant behaviors. This ISBA configuration is called ISBA-A-gs and can simulate the CO_2 net assimilation and GPP by considering the functional relationship between the photosynthesis rate (A) and the stomatal aperture (gs) based on the biochemical A-gs model proposed by [35]. The vegetation phenology relies on photosynthesis-driven plant growth and mortality, and photosynthesis is related to the mesophyll conductance. More details can be found in [36,37].

2.2. LDAS-Monde

Within the SURFEX Land Data Assimilation System (LDAS, [9,12,14,38,39]), recently extended at the global scale (LDAS-Monde, [15]), satellite-based products of soil wetness and of vegetation can be jointly assimilated using a simplified extended Kalman filter. This method uses finite differences to compute the flow dependency between the observations and the analyzed variables. Equations of the analysis update equations can be found in [14]. In the three-layer soil hydrology version of ISBA, the analyzed variables are the root-zone soil moisture and LAI, each containing 12 values corresponding to the possible 12 plant functional types of the considered pixel (no vegetation, rock bare soils, snow and ice, deciduous forests, coniferous forests, evergreen forests, summer crops, winter crops, irrigated crops, grasslands, tropical grasslands or wetlands, from ECOCLIMAP, which is the vegetation map used in SURFEX, [40]). These two variables are the main interest in the photosynthesis process since the root-zone soil moisture and LAI are the main drivers of the evapotranspiration process. Once a day, the model is stopped, and the analyzed variables are adjusted according to the observations and their relative errors compared to the ones assigned to the open-loop simulations (when the model runs without any assimilation).

2.3. Satellite Observations

2.3.1. Soil Moisture and LAI

The Copernicus Global Land Service (CGLS) distributes an SWI (Soil Water Index) product, which represents the soil wetness taking values from zero (completely dry) to one (saturated). It is calculated with a recursive exponential filter [41] using backscatter observations from the ASCAT C-band radar on board the MetOP satellites, using a change detection technique developed at the Vienna University of Technology [42,43] where the lowest and the highest values of the backscatter observations are assigned respectively to dry (SWI = 0) and saturated (SWI = 1) soils. Moreover, a timescale is associated with an exponential filter representing the depth of the soil profile. In this study, the SWI-001 Version 3.0 product was used, which has a one-day timescale representing the soil wetness of the surface down to a 5-cm depth and is available daily at a 0.1° resolution. Figure 1 shows the average raw SWI-001 product for the whole 2007–2015 period.

In order to assimilate the SWI product, it needs to be rescaled in the model climatology space so that the assimilation does not introduce any bias in the system [44,45]. Based on the soil texture, ISBA defines a wilting point and a field capacity value for each point of the grid, characterizing the dynamical range of the soil moisture. It is thus necessary to rescale the SWI values in the model dynamics, and the approach proposed by [46] was applied here, which makes the average and the variance (the first two statistical moments) of two datasets match through a linear transformation. Draper, C. et al. [6] and Barbu, A. et al. [12] highlighted the importance of allowing seasonal variability in the rescaling process. The distribution matching procedure was applied on a monthly basis using a three-month moving window. Urban areas and frozen pixels were filtered out beforehand.

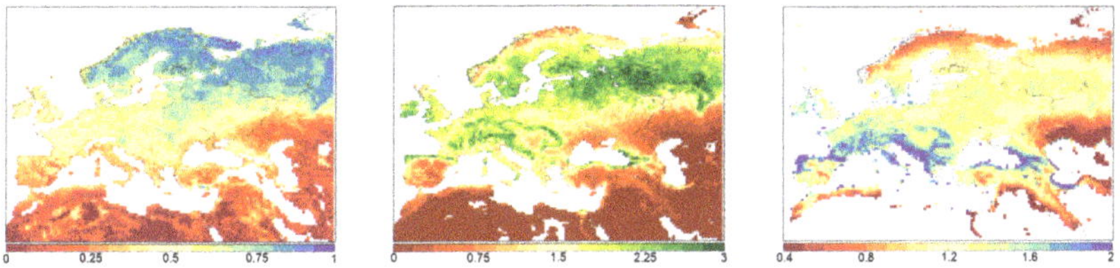

(a) SWI-001 from ASCAT (-) (b) LAI V1 from SPOT-VGT/PROBA-V ($m^2 \cdot m^{-2}$) (c) Evapotranspiration from GLEAM ($kg \cdot m^{-2} \cdot day^{-1}$)

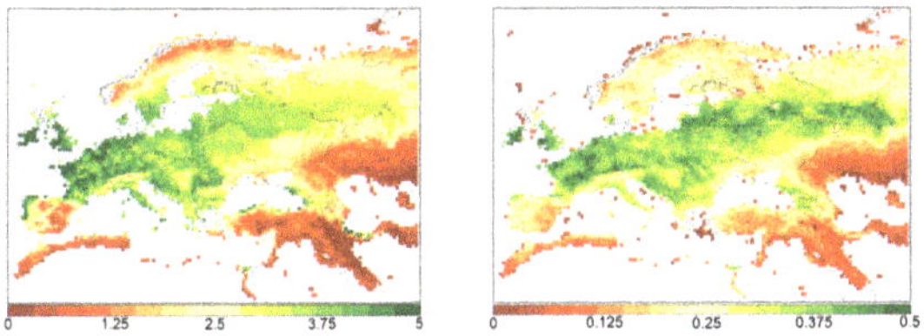

(d) GPP from FLUXNET-MTE ($g(C) \cdot m^{-2} \cdot day^{-1}$) (e) SIF from GOME-2 ($mW \cdot m^{-2} \cdot sr^{-1} \cdot nm^{-1}$)

Figure 1. Averaged satellite-derived products for the whole period from 2007–2015: (**a**) original Soil Water Index (SWI); (**b**) Leaf Area Index (LAI); (**c**) evapotranspiration; (**d**) Gross Primary Production (GPP) and (**e**) Sun-Induced Fluorescence (SIF).

The SWI-001 observations were screened to remove the observations with a quality flag lower than 80%. This threshold value was chosen in order to avoid any persistence effect in the exponential filter (i.e., the same value being automatically prescribed even when observations are missing). It has an impact on the number of available observations, especially at low latitudes, but it was checked that changes in this value have little impact on the scores and the conclusions given in this study. After projection, additional masks for urban regions, steep mountainous terrain and frozen ground indicated by the model simulations, but not detected by ASCAT, have also been applied.

CGLS also distributes an LAI product retrieved from SPOT-VGT and PROBA-V satellite observations using a neural network algorithm [47] trained with two LAI datasets: CYCLOPESV3.1 from the VEGETATION sensor on board SPOT [48] and MODIS Collection 5 on board Terra and Aqua [49]. The GEOV1 LAI product was used in this study. It is available every 10 days as a composite over a 10-day period at a spatial resolution of 1 km × 1 km. Figure 1 shows the average LAI product for the whole 2007–2015 period.

Following [12], the SWI and LAI products are aggregated at the model grid resolution (0.5°) by a simple arithmetic average where and when at least half of the observation grid points are available. Both products have been assimilated at 09:00 UTC as in [12,14,15].

2.3.2. Evapotranspiration

Miralles, D.G. et al. [29] produced global monthly estimates of the land-surface evapotranspiration from multiple satellite-based products from 1980–2016. They used the GLEAM (Global Land surface Evaporation: the Amsterdam Methodology) approach, mainly driven by microwave remote sensing observations while also constrained by satellite-derived soil moisture products. The GLEAM v3.1 product (with several algorithm improvements described in [50]) was used in this study at a spatial resolution of 0.25° × 0.25° and averaged at the model resolution of 0.5°. Figure 1 shows the average evapotranspiration product for the 2007–2015 period considered in this study. Pixels too close to the coasts and pixels where the vegetation coverage was too low (less than 10% of the pixel covered by any kind of vegetation according to ECOCLIMAP) were filtered out.

2.3.3. Gross Primary Production

Jung, M. et al. [30] used machine learning algorithms to convert meteorological parameters to variations of Terrestrial Ecosystem Respiration (TER) and GPP. This dataset is called FLUXNET-MTE (FLUXNET network-Multi Tree Ensemble) and is available monthly at the global scale at 0.5° resolution from 1982–2011. These machine learning algorithms were first trained using FLUXNET [51] in situ TER and GPP fluxes estimated using two flux partitioning methods [52,53]. Figure 1 shows the average GPP product for the whole period 2007–2015. Pixels too close to the coasts and pixels where the vegetation cover was too low (less than 10% of the pixel covered by any kind of vegetation according to ECOCLIMAP) were filtered out.

2.3.4. Fluorescence

The Global Ozone Monitoring Experiment-2 (GOME-2) is an operational scanning spectrometer [54] on board the European Meteorological Satellite (EUMETSAT) Polar System MetOp-A and MetOp-B. GOME-2 has been measuring the Earth's backscattered radiance at wavelengths between 240 and 790 nm. SIF products were derived by [55] from radiance observations at wavelengths between 734 and 758 nm (far-red SIF). The Level-3 v27 SIF product was used in this study, which is a monthly gridded product at 0.5° resolution estimating a daily-averaged SIF. Figure 1 shows the average SIF product for the whole 2007–2015 period. Pixels too close to the coasts and pixels where the vegetation cover was too low (less than 10% of the pixel covered by any kind of vegetation according to ECOCLIMAP) were filtered out.

Thum, T. et al. [56] investigated whether the SIF observations were suitable to assess the performances of the JSBACH (Jena Scheme of Atmosphere Biosphere Coupling in Hamburg) biosphere model [57] at the regional scale of Fenno-Scandinavia and at the site scale with multiple coniferous forests in Finland. Both observations and simulations revealed that SIF can be used to estimate GPP at both site and regional scales. They also concluded that GOME-2-based SIF was a better proxy for GPP (similar slopes of regression for the different sites) than the remotely-sensed FAPAR (Fraction of Absorbed Photosynthetically Active Radiation; different slopes of regression for the different sites). In this study, SIF observations will be compared to simulated GPP.

2.4. Experimental Setup

In this study, the same modeling framework as in [15] was used. It is based on SURFEX Version 8.0, and the same data assimilation method based on the simplified extended Kalman filter was performed. SSM is assimilated on a daily basis, while LAI is assimilated on a 10-day basis. This study differs from [15] because the assimilated SSM was provided by Copernicus Global Land Service instead of the ESA-Climate Change Initiative. Another difference is that we used the three-layer version of ISBA, while [15] used the diffusive representation with ten layers of soil. This study follows the works of [12] and [14], but a wider spatial domain is considered (the Euro-Mediterranean area from 25°N–75.5°N and from 11.75°W–62.5°E) over an extended period of time (from 2007–2015).

In this study, the ISBA-A-gs model was forced with ERA-Interim reanalysis surface atmospheric variables [58]: precipitation, solar radiation (shortwave and longwave), wind speed, surface pressure, air temperature, CO_2 air concentration and air humidity. These forcings are available at the global scale on a 0.5° resolution grid every 3 h from 1989 until the present (with a one-month delay). In order to allow the system to reach equilibrium, all experiments were initialized by performing a 5-year spinup.

3. Results

The impact of the joint assimilation of the SSM and LAI satellite products was evaluated. First, the assimilation performance was assessed by comparing the model open-loop and analysis simulations with the assimilated products (in order to demonstrate that the assimilation systems perform well). In a second step, the consistency between the LDAS analysis and independent vegetation satellite products (evapotranspiration, GPP and fluorescence) was evaluated.

3.1. Impact of the Assimilation on SSM and LAI

In the assimilation process, the available observations were merged with the model simulations, weighted with their errors, as defined in [12,14], to compute an analysis taking into account all this information. It was expected that the analysis would end closer to the observations.

Figure 2 shows the monthly average open-loop simulations, analysis and observations of SSM and LAI over the Euro-Mediterranean area. The SSM analysis was very close to the open-loop since the SSM observations were rescaled before being assimilated. Open-loop and analysis scores for the whole time period and over each pixel were similar, as described in Table 1. These scores were computed using pooled pixel data during one month, so they included both spatial and temporal components.

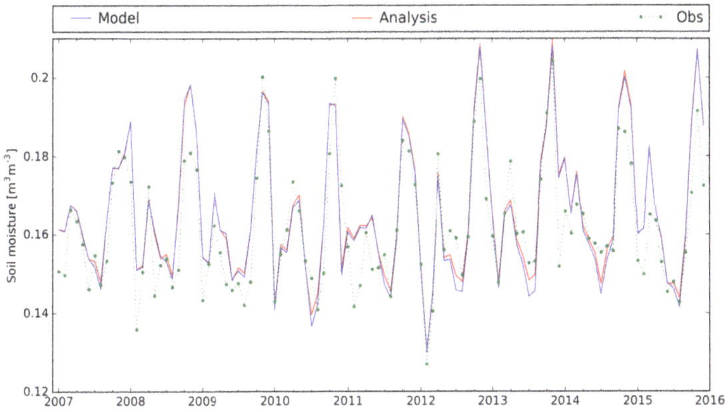

(a) SSM

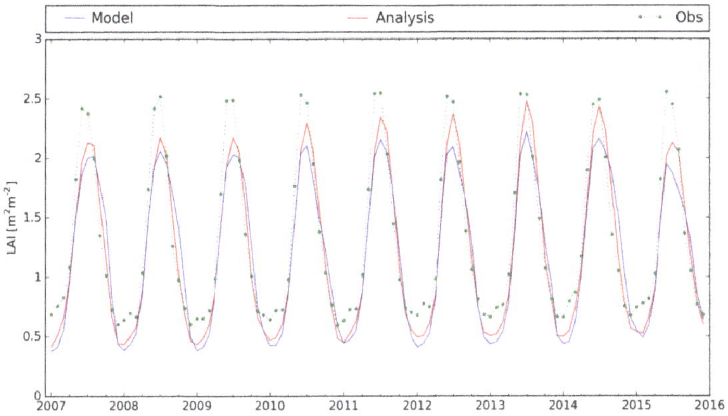

(b) LAI

Figure 2. Monthly time series of the observations (green), open-loop (blue) and analysis (red) simulations from 2007–2015 averaged over the Euro-Mediterranean area for: (**a**) Surface Soil Moisture (SSM) and (**b**) Leaf Area Index (LAI).

Table 1. Statistics between the simulations (open-loop and analysis) and the observations for CGLS SSM, CGLS LAI, GLEAM Evapotranspiration (E), FLUXNET-MTE GPP and between observed GOME-2 SIF and simulated GPP over the Euro-Mediterranean area from 2007–2015: bias, correlation (R), Root-Mean-Square Difference (RMSD), Standard Deviation of Differences (SDD) and the number of observations.

Variable	Exp.	bias	R	RMSD	SDD	No. of obs.
SSM ($m^3 \cdot m^{-3}$)	open-loop	0.002	0.850	0.048	0.048	7,254,829
	analysis	0.004	0.860	0.046	0.046	
LAI ($m^2 \cdot m^{-2}$)	open-loop	−0.114	0.778	0.827	0.819	1,558,568
	analysis	−0.084	0.884	0.594	0.588	
E ($kg.m^{-2} \cdot day^{-1}$)	open-loop	0.015	0.883	0.533	0.533	688,608
	analysis	0.057	0.894	0.525	0.522	
GPP ($kg.m^{-2} \cdot day^{-1}$)	open-loop	−0.630	0.895	1.378	1.225	384,480
	analysis	−0.562	0.916	1.233	1.098	
SIF ($mW \cdot m^{-2} \cdot sr^{-1} \cdot nm^{-1}$) compared to GPP	open-loop	-	0.791	-	-	475,008
	analysis	-	0.813	-	-	

The largest impacts of the assimilation process can be seen on the LAI variable. Figure 2 shows the LAI simulations and observations at the monthly scale. In summer and winter times, the open-loop simulations tended to underestimate LAI (compared to LAI observations), and the analysis simulations were closer to the observations after the assimilation. The statistics from Table 1 indicate a lower RMSD (−28%) and a better correlation R (+14%).

Figure 3 shows seasonal scores for each month of all years from 2007–2015 (RMSD and R). They were computed using all the pixels and dates available for each considered month. These statistics show an improvement for both SSM and LAI analyzed variables for all the months of the year. For SSM, the RMSD was always reduced, with analysis values ranging from about 0.042 $m^3 \cdot m^{-3}$ in March and September to 0.054 $m^3 \cdot m^{-3}$ in May and December. Correlation was slightly improved in winter and spring and more markedly improved in summer and during the autumn, with analysis values ranging from 0.77–0.94. Because of the seasonal rescaling of SSM, the seasonality in the statistics cannot be improved. Regarding LAI, both RMSD and R scores were improved during the year with analysis RMSD and R values ranging from 0.3–0.9 $m^2 \cdot m^{-2}$ and from 0.77–0.91, respectively. The analyzed LAI was closer to the observations than the open-loop simulation throughout the year, in particular during the senescence phase where RMSD dropped from around 0.9 $m^2 \cdot m^{-2}$ (open-loop) to around 0.4–0.6 $m^2 \cdot m^{-2}$ (analysis) in September and October. A seasonality in the scores can clearly be seen with a larger open-loop SSM RMSD in summer and winter associated with a lower correlation in summer and larger open-loop LAI RMSD values in summer associated with a slightly better correlation in summer. After assimilation, the seasonality in the LAI correlation was smoothed out with R values between 0.78 and 0.90.

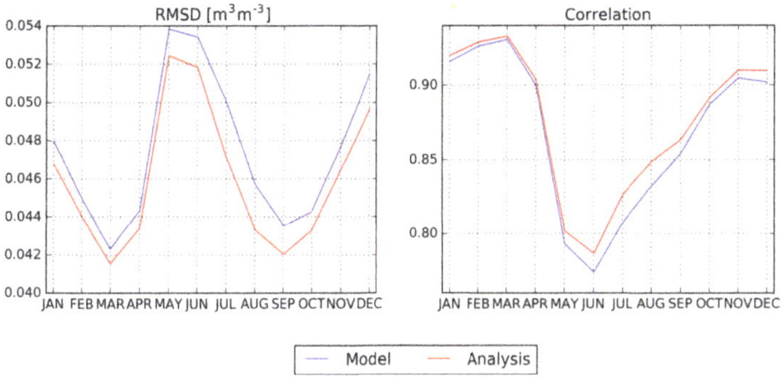

(a) statistics for SSM compared to the CGLS product

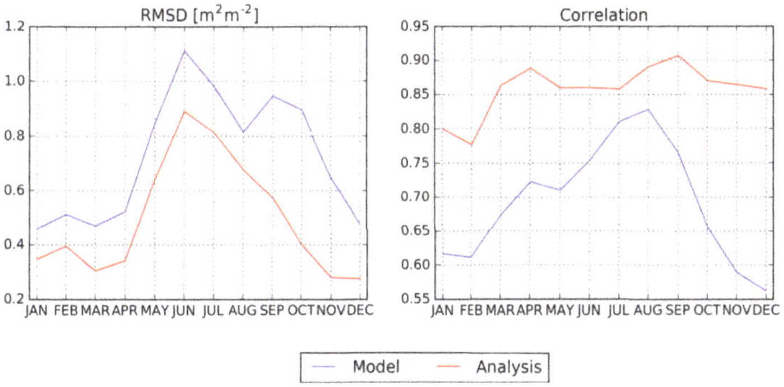

(b) statistics for LAI compared to the CGLS product

Figure 3. Monthly seasonal scores (RMSD and correlation R) of the open-loop (blue) and analysis (red) simulations from 2007–2015 averaged over the Euro-Mediterranean area compared to the observations of: (**a**) SSM and (**b**) LAI. Red shade indicates an improvement from the assimilation process with either a lower error or a higher correlation score.

3.2. *Evaluation Using Satellite-Derived Vegetation Products*

3.2.1. Evapotranspiration and GPP

The consistency of the analysis with the evapotranspiration product from the GLEAM product and the GPP product from the FLUXNET-MTE dataset is evaluated in this section. Since these products are only available at the monthly scale, they were compared to monthly averaged analyzed simulations. The GPP product from the FLUXNET-MTE dataset is only available until 2011, so only the years 2007–2011 are considered in the statistics below.

Figure 4 shows monthly mean time series of evapotranspiration and GPP over the Euro-Mediterranean area of the open-loop and analysis simulations and of the products. The dynamics of the simulated evapotranspiration variable was in line with GLEAM with a slight overestimation in winter and underestimation in summer. The impact of the assimilation was generally small, but is more noticeable during the summer season. The analyzed values were closer to the observations in the summer time. This was confirmed by the seasonal statistics shown in Figure 5, where they tend to be slightly better for the analysis during summer with lower RMSD values and higher correlation scores. Statistics for the whole time period (Table 1) showed that the assimilation of SSM and LAI products slightly improved the consistency of the model with GLEAM.

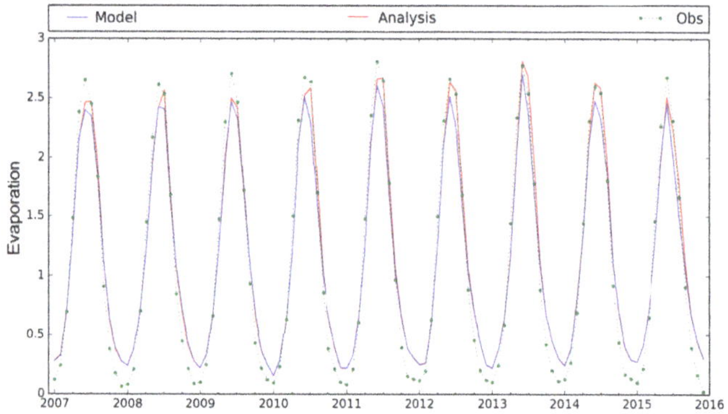

(a) Evapotranspiration (kg·m^{-2}·day^{-1})

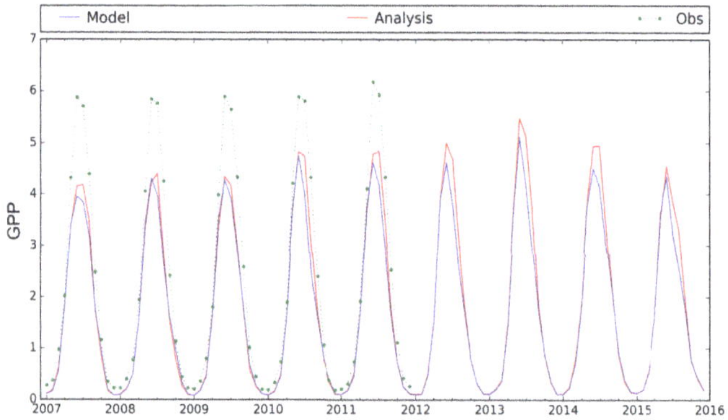

(b) Gross Primary Production (g(C)·m^{-2}·day^{-1})

Figure 4. Monthly time series of the products (green), open-loop (blue) and analysis (red) simulations from 2007 to 2015 averaged over the Euro-Mediterranean area for: (**a**) evapotranspiration; (**b**) gross primary production (observations only until 2011).

Compared to the LDAS analyzed evapotranspiration, the GLEAM product had an RMSD varying from 0.3 kg·m^{-2}·day^{-1} in winter to 0.8 kg·m^{-2}·day^{-1} between May and July. The highest RMSD values mainly occurred over forested areas (not shown).

Regarding the GPP variable, the simulated dynamics was consistent with FLUXNET-MTE, but the analyzed monthly GPP exceeded 5 g(C)·m^{-2}·day^{-1} only once (in 2013), while FLUXNET-MTE seemed to exceed this value for all the years. The assimilation only slightly reduced the mean bias between the simulations and FLUXNET-MTE. However, Figure 5 shows that the assimilation markedly reduced RMSD in summer. The assimilation also significantly improved the correlation scores during the other seasons. The statistics were also slightly improved when looking at the whole time period (Table 1) with a decrease of 10% in the bias and in the Standard Deviation of Differences (SDD) values.

Compared to the LDAS analysis, the FLUXNET-MTE GPP product had a RMSD varying from 0.2 g(C)·m^{-2}·day^{-1} in winter to 2 g(C)·m^{-2}·day^{-1} between June and August. The highest RMSD values mainly occurred over areas covered by crops (not shown).

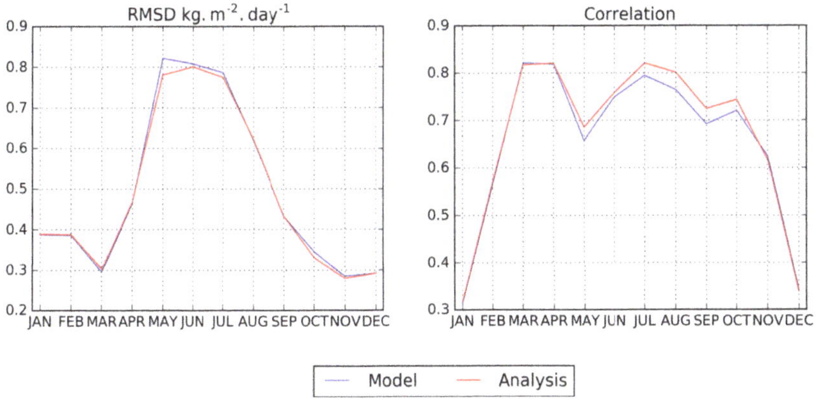

(a) Evapotranspiration

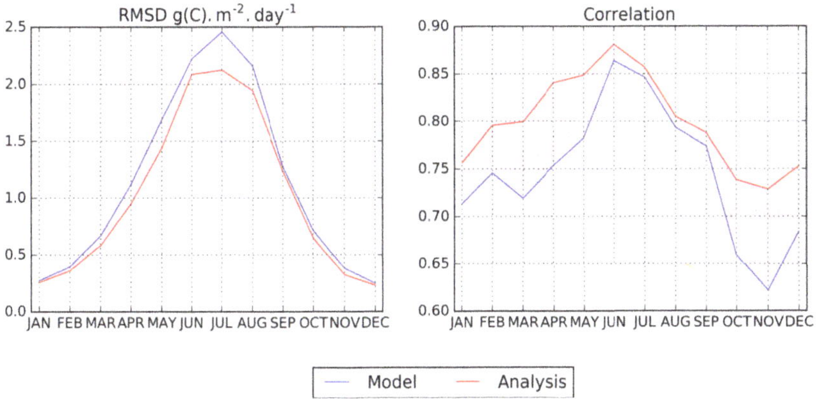

(b) Gross Primary Production

Figure 5. Monthly seasonal scores (RMSD and correlation) of the open-loop (blue) and analysis (red) simulations from 2007–2015 averaged over the Euro-Mediterranean area compared to the products of: (**a**) GLEAM Evapotranspiration (E) and (**b**) FLUXNET-MTE gross primary production (only 2007–2011). Red shade indicates an improvement from the assimilation process with either a lower error or a higher correlation score.

In Figure 6, temporal correlation maps are presented, representative of the time evolution of the variable itself at a specific location. This figure shows the spatial distribution of the correlation for the whole available period (2007–2015 for the evapotranspiration, 2007–2011 for GPP) between the analysis and the observations. The model open-loop simulations already gave very good results in terms of correlation with 65% of the points with a correlation greater than 0.90 for the evapotranspiration and 55% for GPP (not shown). After the joint assimilation of SSM and LAI, 70% and 64% of the points had a correlation higher than 0.90 respectively. As shown on the correlation difference map (right column of Figure 6), these improvements were mainly located on the northern part of the Black Sea and of the Caspian Sea, which are regions mainly covered by crops and grasslands. Some improvements in Western Europe were also visible for the evapotranspiration variable, which is also a region mainly covered by crops and grasslands. Blue areas indicate a degradation of the the correlation after the assimilation at northern latitudes over boreal forests and grasslands. It showed an inconsistency between the assimilated satellite products and the GLEAM evapotranspiration and FLUXNET-MTE GPP products in these regions. Further investigations are needed to explain this inconsistency.

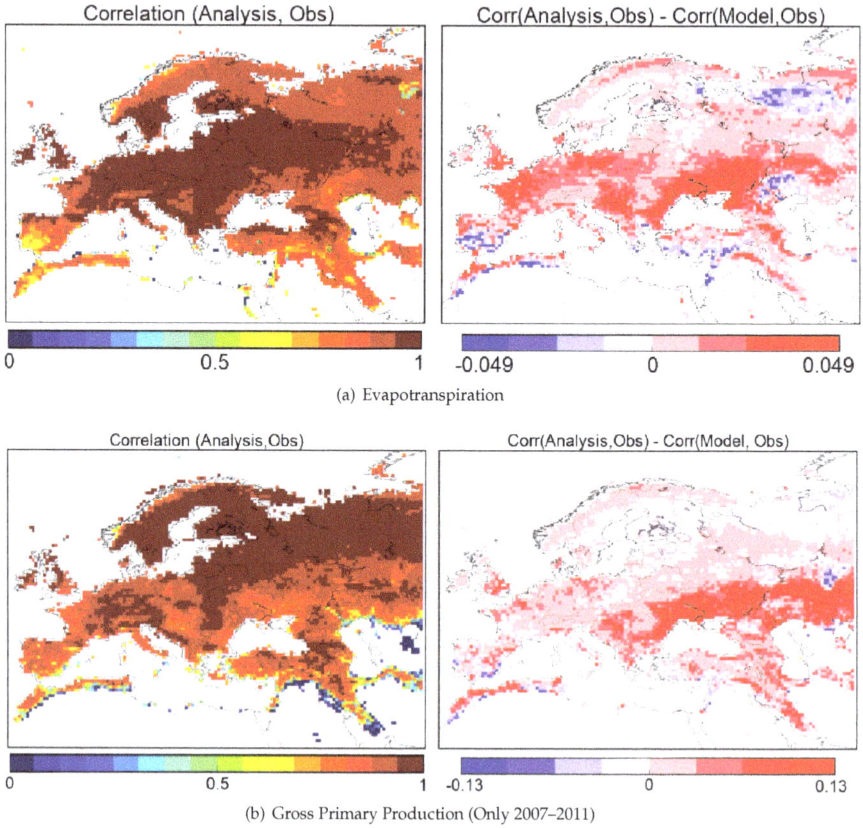

Figure 6. Correlation maps between the analysis and the observations (left column) and changes in correlation triggered by the assimilation (right column, where red means an improvement after assimilation) for the whole 2007–2015 period for: (**a**) E and (**b**) GPP.

3.2.2. Sun-Induced Fluorescence

The modeled GPP values are expressed in g(C)$\cdot$m$^{-2}\cdot$day^{-1}, whereas SIF is an energy flux emitted by the vegetation in units of mW$\cdot$m$^{-2}\cdot$sr$^{-1}\cdot$nm^{-1}. Thus, GPP and SIF cannot be directly compared as they do not represent the same physical quantities. However, their time dynamics and their spatial distributions can be investigated. Figure 7 shows seasonal monthly time series of the modeled GPP (open-loop and analysis) with almost no photosynthesis activity during winter and a peak in the summer time. The SIF observations followed the same time evolution with very low values in winter and higher values in summer. Even if these two quantities are not quantitatively comparable, they followed the same evolution throughout the year.

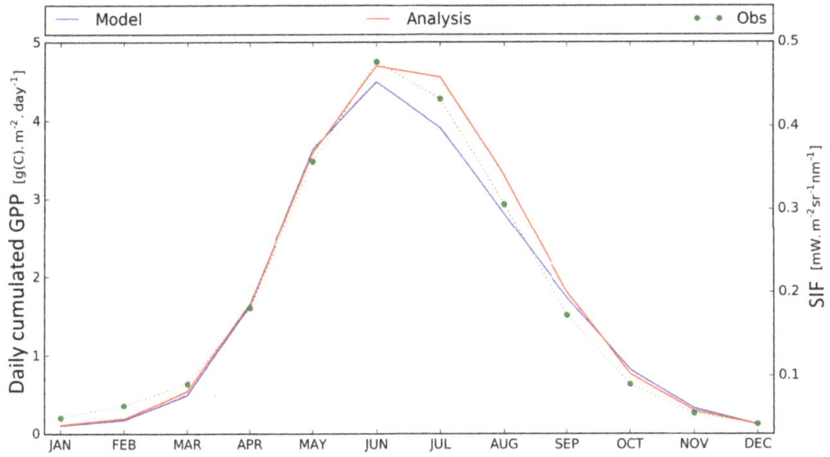

Figure 7. Seasonal monthly time series of the SIF observations (green, right-axis), open-loop (blue, left-axis) and analysis (red, right-axis) GPP simulations from 2007–2015 averaged over the Euro-Mediterranean area.

Figure 8 represents the time correlation between the GPP analysis and the SIF products. The histogram of these correlation values in Figure 9 shows that the assimilation enhanced the consistency between the simulated GPP and SIF products, with more R values larger than 0.8 and fewer R values smaller than 0.8. These improvements were particularly large in cropland areas in Central Europe, the Ukraine and southern Russia, close to the Black Sea and the Caspian Sea. These areas coincide with those in Figure 6 presenting better model R values with GPP. This shows that the joint assimilation of the SSM and LAI products had a positive impact on other vegetation variables.

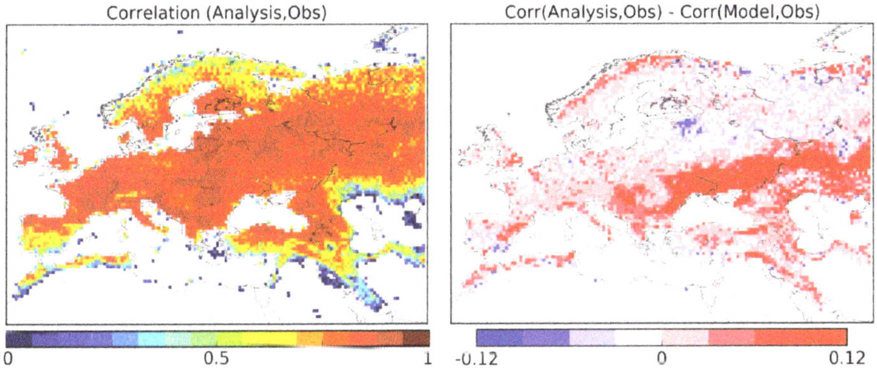

Figure 8. Correlation maps between the SIF observations and the GPP analysis for the whole 2007–2015 period (left column) and the difference map (right column, where red means an improvement after assimilation).

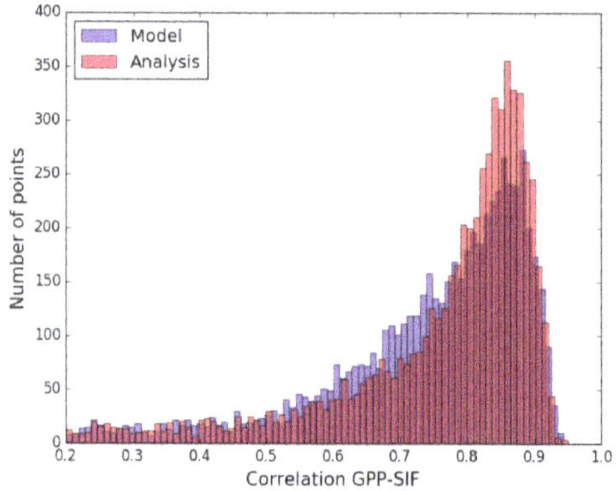

Figure 9. Histogram of correlation values from Figure 8 between the SIF observations and the GPP simulations (open-loop in blue and analysis in red).

As already mentioned in previous studies, such as [24,59] or [60], the relationship between GPP and SIF can vary from one vegetation type to another. For this reason, the SIF-GPP relationship was investigated for each of the four main vegetation types covering the Euro-Mediterranean area. Only pixels with at least 50% of their surface covered by one of these vegetation types were considered. Figure 10 shows the selected pixels: 395 pixels of deciduous forest, 1088 pixels of coniferous forest, 372 pixels of C3 crops and 469 pixels of grassland. Whenever the couple of an observation and simulation is available, a direct comparison can be done as indicated in Figure 10 for the four vegetation types.

Based on these spatial averages, a linear regression was drawn for each vegetation type indicated by the purple lines. The corresponding slope, intercept, correlation coefficient and standard deviation can be found in Table 2. The slope values of the linear regression were different from those found by [24] (last three columns in Table 2) using another GPP model: MPI-BGC (Max Planck Institute-BioGeoChemical, [61,62]). The MPI-BGC GPP product relies on satellite-based estimates of FAPAR and is produced by upscaling in situ measurements of carbon dioxide, water and energy fluxes at the global scale. However, the relative order of the slope values for the different vegetation types is about the same. The largest slope was found for coniferous forests, denoting larger sensitivity of GPP to SIF for this vegetation type. These contrasting slope values indicate that the relationship between GPP and SIF depended on the vegetation type, which is in line with what was already found in [24] even using a different model.

Table 2. Linear regression between averaged monthly values of observed SIF and analyzed GPP over the pixels covered by at least 50% of one vegetation type: deciduous forests, coniferous forests, C3 crops, grasslands. Slope (in g(C).sr.nm.$mW^{-1}\cdot day^{-1}$), intercept (in g(C)$\cdot m^{-2}\cdot day^{-1}$), correlation coefficient (R), Standard Deviation (STD, in g(C)$\cdot m^{-2}\cdot day^{-1}$) and the number of dates are indicated (all statistical scores are significant with a p-value < 0.05). Slope, intercept and correlation values from [24] are also indicated in the right columns for the same vegetation types.

	Leroux et al. (2018) (This Study; Figure 10)					**Adapted from Figure 11 in [24]**		
Veg. Type	**Slope**	**Intercept**	**R**	**STD**	**No. of Dates**	**Slope**	**Intercept**	**R**
deciduous	12.72	−0.10	0.97	0.37	86	7.69	0.23	0.98
coniferous	16.95	−0.74	0.96	0.51	107	12.50	0.00	0.97
C3 crops	11.28	−0.36	0.97	0.28	108	7.69	−0.31	0.99
grasslands	8.97	−0.80	0.80	0.66	108	7.14	−0.36	0.98

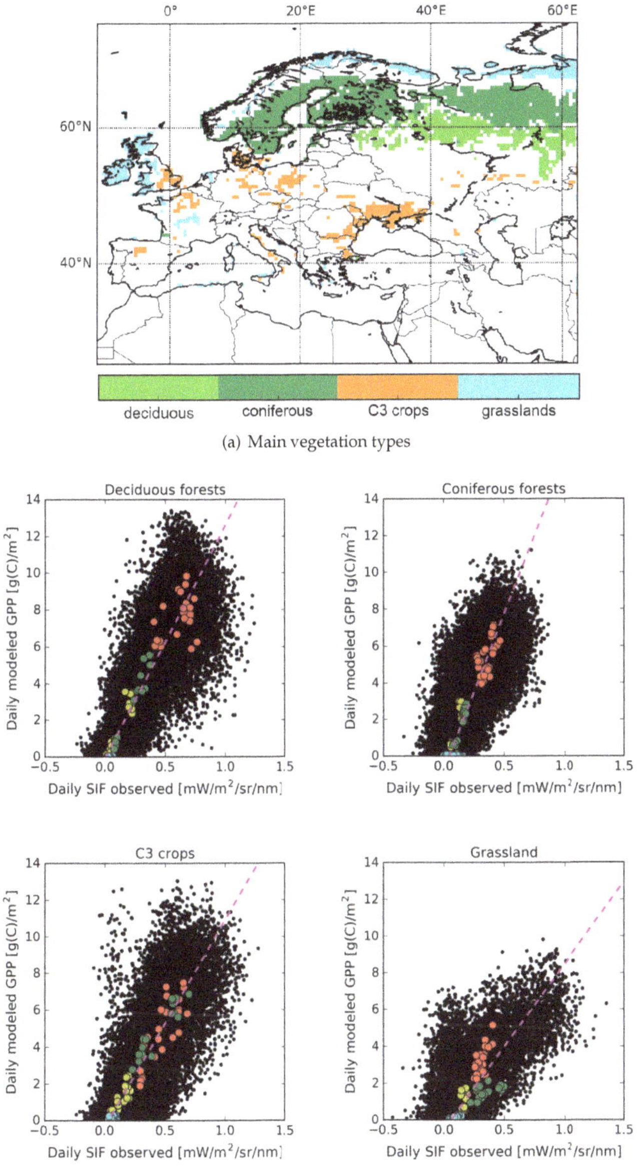

(b) SIF-GPP relationship

Figure 10. Pixels where at least 50% is covered by a single vegetation type (deciduous forests, coniferous forests, C3 crops, grassland, (**a**) used to investigate the relationship between the analyzed GPP and the observed SIF (**b**). Spatial averages are indicated by colored circles depending on the season: summer in red (June-July-August), autumn in yellow (September-October-November), winter in blue (December-January-February) and spring in green (March-April-May). Dashed purple lines represent the regression with the coefficients and statistics shown in Table 2.

4. Discussion

4.1. Investigating the SIF-GPP Relationship

In ISBA, the fluorescence is not simulated directly, but the photosynthesis activity is simulated through the calculation of the GPP, which is driven by plant growth and mortality in the model. Sun, Y. et al. [28] demonstrated that SIF and GPP were driven by the same environmental and biological factors and found that SIF observations from OCO-2 and GPP products from FLUXCOM were highly consistent in time and space. Their study was realized over various types of vegetation near Chicago, Illinois, U.S. At a much larger scale, the results of this study were consistent with [28], as the observed SIF from GOME-2 and the simulated GPP also showed a high correspondence in both time and space. This indicates that the SIF observations could be used as a relevant and independent source of validation for GPP model simulations.

The SCOPEmodel [63] allows the simulations of both SIF and GPP, but according to [64], who used SCOPE within a carbon cycle data assimilation system, SIF and GPP are not sensitive to the same physical parameters (the chlorophyll content and the maximum carboxylation rate, respectively). In this context, it was concluded that the assimilation of SIF observations cannot improve GPP simulations. By simplifying the SCOPE model and by computing the SIF from the GPP directly, [65] showed the potential of SIF to improve photosynthesis simulations. The results of this study showed a very good consistency between SIF and GPP, and the SIF-GPP relationship should be further investigated towards the construction of an observation operator in order to be able to assimilate SIF observations.

Unlike the study of [28], but in line with the studies of [24,59] or [60], the relationship between SIF and GPP was found to be dependent on the vegetation type. When all the pixels were used together to investigate the SIF-GPP relationship, no significant correlation could be found.

The ISBA model was also parameterized differently depending on the vegetation type, which could influence the SIF-GPP relationship in turn. Furthermore, the SIF products and the GPP simulations were at a low resolution (0.5°), which might be too coarse to rule out any heterogeneity within the considered pixels. Moreover, the GOME-2 instrument footprint was 80 × 40 km, then rescaled on a 0.5° grid, so there might have also been some noise introduced by the rescaling process. It was however expected that this would have a small impact on the SIF-GPP relationship.

The Fluorescence EXplorer mission (FLEX, [66]) was selected by the European Space Agency as the eight Earth Explorer mission scheduled for launch in 2022. FLEX is specifically designed for the monitoring of the vegetation fluorescence with an imaging spectrometer covering the spectral range from 500–780 nm. The swath width will be 150 km, and the spatial resolution of the SIF product will be 300 m, allowing further and more precise studies on the SIF-GPP relationship.

4.2. Can SSM and LAI Assimilation Be Improved?

In this study, a single LAI value was assimilated over the whole model grid cell. This means that a single value was used for all the vegetation types present within the considered grid cell. As presented in [12], the Jacobians were calculated individually for each vegetation type, which made the Kalman gain and the increments depend on the vegetation type. Instead of using only one observation of LAI per pixel (especially when the model resolution is 0.5° and the one of the observations is a few hundred meters), the possibility of assimilating different values of LAI for each patch is under investigation. Based on a Kalman filtering technique developed by [67], CGLS LAI data have been disaggregated so that LAI values can be assimilated independently over each individual patch. This has a major impact on the analyzed LAI, but also on the other vegetation variables such as the evapotranspiration and GPP [68].

The assimilation of CGLS SSM (original SWI rescaled in the model dynamic range) presents some caveats, and these data depend on the exponential filter used on multiple observations from the ASCAT radar (backscatter coefficients) at C-band, or 5.3 GHz. In order to be able to assimilate SWI products into the model, a seasonal CDF matching is performed, which projects the observation values into the model space. This process reduces the impact of the SSM assimilation on the analyzed simulations. In order to improve LDAS-Monde analysis, building an observation operator for backscatter coefficients could be a way to overcome the use of the SWI product and directly assimilate the observed backscatter coefficients. Moreover, C-band observations also contain information on vegetation biomass. By building this forward operator, it would be thus possible to consider all the information contained in these observations. SSM products can also be derived from passive microwave satellite observations. In the same way as for the observed backscatter coefficients, building an operator for brightness temperatures (energy that is naturally emitted by the surface of the Earth and dependent on the soil moisture at L-band, or 1.4 GHz) is needed to take full advantage of the available observations.

5. Conclusions

This study investigates the joint assimilation of satellite-derived soil moisture and vegetation observations to yield improved estimates of hydrological and vegetation fields, as well as water and carbon fluxes at the land surface level. SSM and LAI products were assimilated in the CO_2 responsive version of the ISBA land surface model from 2007–2015 over the Euro-Mediterranean area. This joint assimilation in ISBA resulted in improvements in the representation of the vegetation processes.

Independent observations were used to quantify improvements in the simulated water and carbon fluxes, as well as in the modeled vegetation dynamics: LAI from Copernicus Global Land Service, evapotranspiration from the GLEAM dataset, GPP from the FLUXNET-MTE dataset and fluorescence from GOME-2 observations. GPP and evapotranspiration products present a similar temporal evolution with less accuracy during the winter season, but with higher correlation scores. The evapotranspiration product is less accurate over forested areas, whereas the GPP product tends to be less accurate over crop-covered areas.

Fluorescence cannot be directly evaluated using LDAS-Monde, since it is not simulated by the ISBA model, but it is a very good proxy to the photosynthesis process represented by the GPP. Fluorescence observations were compared to the LDAS-Monde-analyzed GPP in the model. This is the first time that fluorescence has been evaluated within the LDAS-Monde framework, and it shows a very good correlation with GPP in the condition that the evaluation is performed independently for each individual vegetation type. At 0.5° resolution, only pixels covered by at least 50% of their surface by one single vegetation type were considered, and a strong and linear relationship has been identified for deciduous forests, coniferous forests, C3 crops and grassland, which could be the translation of a different behavior in the photosynthesis process depending on the vegetation type. This also suggests that LDAS-Monde analysis can be used to study the possible relationship between the fluorescence and the GPP at the continental scale. This constitutes the first step to building an observation operator to assimilate fluorescence observations in view of the preparation of the future satellite mission FLEX, which is scheduled to be launched in 2022.

With more and more available satellite observations, the perspective of assimilating new products such as the surface albedo, radar backscatter coefficients or passive brightness temperatures looks extremely promising and will improve even more the LDAS-Monde analyses and our understanding of the various vegetation processes.

Author Contributions: D.J.L and J.-C.C conceived of and designed the experiments. D.J.L performed the experiments. All the authors analyzed the results. D.J.L wrote the paper.

Funding: The work of Delphine Leroux and Simon Munier was supported by the European Union Seventh Framework Programme (FP7/2007-2013) under Grant Agreement No. 603608, "Global Earth Observation for integrated water resource assessment" (eartH2Observe).

Acknowledgments: The authors acknowledge the Copernicus Global Land service for providing the satellite-derived LAI and soil moisture products.

Conflicts of Interest: The authors declare no conflict of interest.

Abbreviations

The following abbreviations are used in this manuscript:

LDAS	Land Data Assimilation System
SSM	Surface Soil Moisture
LAI	Leaf Area Index
ISBA	Interactions between Soil, Biosphere and Atmosphere
GLEAM	Global Land surface Evaporation: the Amsterdam Methodology
FLUXNET-MTE	Flux Network-Multi-Tree Ensemble
TER	Terrestrial Ecosystem Respiration
GOME	Global Ozone Monitoring Experiment
LSM	Land Surface Model
SURFEX	Surface Externalisée
SIF	Sun-Induced chlorophyll Fluorescence
GOSAT	Greenhouse gas Observing Satellite
GPP	Gross Primary Productivity
GCOS	Global Climate Observing System
CGLS	Copernicus Global Land Service
SWI	Soil Water Index
EUMETSAT	European Meteorological Satellite
JSBACH	Jena Scheme of Atmosphere Biosphere Coupling in Hamburg
RMSD	Root Mean Square Deviation
R	Correlation
STD	STandard Deviation
SDD	Standard Deviation of Differences
FLEX	Fluorescence Explorer mission

References

1. Noilhan, J.; Mahfouf, J.-F. The ISBA land surface parameterisation scheme. *Glob. Planet. Chang.* **1996**, *13*, 145–159. [CrossRef]
2. Calvet, J.-C.; Noilhan, J.; Roujean, J.-L.; Bessemoulin, P.; Cabelguenne, M.; Olioso, A.; Wigneron, J.-P. An interactive vegetation SVAT model tested against data from six contrasting sites. *Agric. For. Meteorol.* **1998**, *92*, 73–95. [CrossRef]
3. Gibelin, A.-L.; Calvet, J.-C.; Roujean, J.-L.; Jarlan, L.; Los, S.O. Ability of the land surface model ISBA-A-gs to simulate leaf area index at the global scale: comparison with satellite products. *J. Geophys. Res.* **2006**, *111*, D18102. [CrossRef]
4. Houghton, J.; Ding, Y.; Griggs, D.; Noguer, M.; van der Linden, P.; Dai, X.; Maskell, K.; Johnson, C. (Eds.) *Climate Change 2001: The Scientific Basis. Contribution of Working Group I to the Third Assessment Report of the Intergovernmental Panel on Climate Change*; Cambridge University Press: New York, NY, USA, 2001.
5. Reichle, R.; Walker, J.; Koster, R.; Houser, P. Extended vs. Ensemble Kalman Filtering for Land Data Assimilation. *J. Hydrometeorol.* **2002**, *3*, 728–740. [CrossRef]
6. Draper, C.; Mahfouf, J.-F.; Calvet, J.-C.; Martin, E.; Wagner, W. Assimilation of ASCAT near-surface soil moisture into the SIM hydrological model over France. *Hydrol. Earth Syst. Sci.* **2011**, *15*, 3829–3841. [CrossRef]
7. Draper, C.; Reichle, R.H.; De Lannoy, G.J.M.; Liu, Q. Assimilation of passive and active microwave soil moisture retrievals. *Geophys. Res. Lett.* **2012**, *39*, L04401. [CrossRef]
8. Dharssi, I.; Bovis, K.J.; Macpherson, B.; Jones, C. Operational assimilation of ASCAT surface soil wetness at the Met Office. *Hydrol. Earth Syst. Sci.* **2011**, *15*, 2729–2746. [CrossRef]
9. Barbu, A.L.; Calvet, J.-C.; Mahfouf, J.-F.; Albergel, C.; Lafont, S. Assimilation of Soil Wetness Index and Leaf Area Index into the ISBA-A-gs land surface model: Grassland case study. *Biogeosciences* **2011**, *8*, 1971–1986. [CrossRef]
10. De Rosnay, P.; Drusch, M.; Vasiljevic, D.; Balsamo, G.; Albergel, C.; Isaksen, L. A simplified Extended Kalman Filter for the global operational soil moisture analysis at ECMWF. *Q. J. R. Meteorol. Soc.* **2013**, *139*, 1199–1213. [CrossRef]

11. De Rosnay, P.; Balsamo, G.; Albergel, C.; Munoz-Sabater, J.; Isaksen, L. Initialisation of land surface variables for Numerical Weather Prediction. *Surv. Geophys.* **2014**, *35*, 607–621. [CrossRef]
12. Barbu, A.L.; Calvet, J.-C.; Mahfouf, J.-F.; Lafont, S. Integrating ASCAT surface soil moisture and GEOV1 leaf area index into the SURFEX modelling platform: A land data assimilation application over France. *Hydrol. Earth Syst. Sci.* **2014**, *18*, 173–192. [CrossRef]
13. Boussetta, S.; Balsamo, G.; Dutra, E.; Beljaars, A.; Albergel, C. Assimilation of surface albedo and vegetation states from satellite observations and their impact on numerical weather prediction. *Remote Sens. Environ.* **2015**, *163*, 111–126. [CrossRef]
14. Fairbairn, D.; Barbu, A.L.; Napoly, A.; Albergel, C.; Mahfouf, J.-F.; Calvet, J.-C. The effect of satellite-derived surface soil moisture and leaf area index land data assimilation on streamflow simulations over France. *Hydrol. Earth Syst. Sci.* **2017**, *21*, 2015–2033. [CrossRef]
15. Albergel, C.; Munier, S.; Leroux, D.J.; Dewaele, H.; Fairbairn, D.; Barbu, A.L.; Gelati, E.; Dorigo, W.; Faroux, S.; Meurey, C.; et al. Sequential assimilation of satellite-derived vegetation and soil moisture products using SURFEX v8.0: LDAS-Monde assessment over the Euro-Mediterranean area. *Geosci. Model Dev.* **2017**, *10*, 3889–3912. [CrossRef]
16. Wang, L.; D'Odorico, P.; Evans, J.P.; Eldridge, D.; McCabe, M.F.; Caylor, K.K.; King, E.G. Dryland ecohydrology and climate change: Critical issues and technical advances. *Hydrol. Earth Syst. Sci.* **2012**, *16*, 2585-2603. [CrossRef]
17. Kaminski, T.; Knorr, W.; Schurmann, G.; Scholze, M.; Rayner, P.J.; Zaehle, S.; Blessing, S.; Dorigo, W.; Gayler, V.; Giering, R.; et al. The BETHY/JSBACH Carbon Cycle Data Assimilation System: Experiences and challenges. *J. Geophys. Res. Biogeosci.* **2013**, *118*, 1414–1426. [CrossRef]
18. Traore, A.K.; Ciais, P.; Vuichard, N.; Poulter, B.; Viovy, N.; Guimberteau, M.; Jung, M.; Myneni, R.; Fisher, J.B. Evaluation of the ORCHIDEE ecosystem model over Africa against 25 years of satellite-based water and carbon measurements. *J. Geophys. Res. Biogeosci.* **2014**, *119*, 2014JG002638. [CrossRef]
19. Mohr, K.I.; Famiglietti, J.S.; Boone, A.; Starks, P.J. Modeling soil moisture and surface flux variability with an untuned land surface scheme: A case study from the Southern Great Plains 1997 Hydrology Experiment. *J. Hydrometeorol.* **2000**, *1*, 154–169. [CrossRef]
20. Masson, V.; Le Moigne, P.; Martin, E.; Faroux, S.; Alias, A.; Alkama, R.; Belamari, S.; Barbu, A.; Boone, A.; Bouyssel, F.; et al. The SURFEXv7.2 land and ocean surface platform for coupled or offline simulation of earth surface variables and fluxes. *Geosci. Model Dev.* **2013**, *6*, 929–960. [CrossRef]
21. Frankenberg, C.; Fisher, J.B.; Worden, J.; Badgley, G.; Saatchi, S.S.; Lee, J.-E.; Toon, G.C.; Butz, A.; Jung, M.; Kuze, A.; et al. New global observations of the terrestrial carbon cycle from GOSAT: Patterns of plant fluorescence with gross primary productivity. *Geophys. Res. Lett.* **2011**, *38*, L17706. [CrossRef]
22. Frankenberg, C.; O'Dell, C.; Guanter, L.; McDuffie, J. Remote sensing of near-infrared chlorophyll fluorescence from space in scattering atmospheres: Implications for its retrieval and interferences with atmospheric CO_2 retrievals. *Atmos. Meas. Tech.* **2012**, *5*, 2081–2094. [CrossRef]
23. Joiner, J.; Yoshida, Y.; Vasilkov, A.P.; Yoshida, Y.; Corp, L.A.; Middleton, E.M. First observations of global and seasonal terrestrial chlorophyll fluorescence from space. *Biogeosciences* **2011**, *8*, 637–651. [CrossRef]
24. Guanter, L.; Frankenberg, C.; Dudhia, A.; Lewis, P.E.; Gomez-Dans, J.; Kuze, A.; Suto, H.; Grainger, R.G. Retrieval and global assessment of terrestrial chlorophyll fluorescence from GOSAT space measurements. *Remote Sens. Environ.* **2012**, *121*, 236–251. [CrossRef]
25. Joiner, J.; Yoshida, Y.; Vasilkov, A.P.; Middleton, E.M.; Campbell, P.K.E.; Yoshida, Y.; Kuze, A.; Corp, L.A. Filling-in of near-infrared solar lines by terrestrial fluorescence and other geophysical effects: simulations and space-based observations from SCIAMACHY and GOSAT. *Atmos. Meas. Tech.* **2012**, *5*, 809–829. [CrossRef]
26. Joiner, J.; Guanter, L.; Lindstrot, R.; Voigt, M.; Vasilkov, A.P.; Middleton, E.M.; Huemmrich, K.F.; Yoshida, Y.; Frankenberg, C. Global monitoring of terrestrial chlorophyll fluorescence from moderate-spectral-resolution near-infrared satellite measurements: methodology, simulations, and application to GOME-2. *Atmos. Meas. Tech.* **2013**, *6*, 2803–2823. [CrossRef]
27. Zhang, Y.; Xiao, X.; Jin, C.; Dong, J.; Zhou, S.; Wagle, P.; Joiner, J.; Guanter, L.; Zhang, Y.; Zhang, G.; et al. Consistency between sun-induced chlorophyll fluorescence and gross primary production of vegetation in North America. *Remote Sens. Environ.* **2016**, *183*, 154–169. [CrossRef]

28. Sun, Y.; Frankenberg, C.; Wood, J.D.; Schimel, D.S.; Jung, M.; Guanter, L.; Drewry, D.T.; Verma, M.; Porcar-Castell, A.; Griffis, T.J.; et al. OCO-2 advances photosynthesis observation from space via solar-induced chlorophyll fluorescence. *Science* **2017**, *358*, 189. [CrossRef] [PubMed]
29. Miralles, D.G.; Holmes, T.R.H.; de Jeu, R.A.M.; Gash, J.H.; Meesters, A.G.C.A.; Dolman, A.J. Global land-surface evaporation estimated from satellite-based observations. *Hydrol. Earth Syst. Sci.* **2011**, *15*, 453–469. [CrossRef]
30. Jung, M.; Reichstein, M.; Bondeau, A. Towards global empirical upscaling of FLUXNET eddy covariance observations: validation of a model tree ensemble approach using a biosphere model. *Biogeosciences* **2009**, *6*, 2001–2013.
31. Boone, A.; Calvet, J.-C.; Noilhan, J. Inclusion of a third soil layer in a land surface scheme using the force-restore method. *J. Appl. Meteorol.* **1999**, *38*, 1611–1630. [CrossRef]
32. Calvet, J.-C.; Soussana, J.-F. Modelling CO_2-enrichment effects using an interactive vegetation SVAT scheme. *Agric. For. Meteorol.* **2001**, *108*, 129–152. [CrossRef]
33. Calvet, J.-C. Investigating soil and atmospheric plant water stress using physiological and micrometeorological data. *Agric. For. Meteorol.* **2000**, *103*, 229–247. [CrossRef]
34. Calvet, J.-C.; Rivalland, V.; Picon-Cochard, C.; Guehl, J.-M. Modelling forest transmiration and CO_2 fluxes-response to soil moisture stress. *Agric. For. Meteorol.* **2004**, *124*, 143–156. [CrossRef]
35. Jacobs, C.M.J.; van den Hurk, B.J.J.M.; de Bruin, H.A.R. Stomatal behaviour and photosynthetic rate of unstressed grapevines in semi-arid conditions. *Agric. For. Meteorol.* **1996**, *80*, 111–134. [CrossRef]
36. Lafont, S.; Zhao, Y.; Calvet, J.-C.; Peylin, P.; Ciais, P.; Maignan, F.; Weiss, M. Modelling LAI, surface water and carbon fluxes at high-resolution over France: Comparison of ISBA-A-gs and ORCHIDEE. *Biogeosciences* **2012**, *9*, 439–456. [CrossRef]
37. Szczypta, C.; Calvet, J.-C.; Maignan, F.; Dorigo, W.; Baret, F.; Ciais, P. Suitability of modelled and remotely sensed essential climate variables for monitoring Euro-Mediterranean droughts. *Geosci. Model Dev.* **2014**, *7*, 931–946. [CrossRef]
38. Mahfouf, J.-F.; Bergaoui, K.; Draper, C.; Bouyssel, F.; Taillefer, F.; Taseva, L. A comparison of two off-line soil analysis schemes for assimilation of screen level observations. *J. Geophys. Res.* **2009**, *114*, D08105. [CrossRef]
39. Albergel, C.; Calvet, J.-C.; Mahfouf, J.-F.; Rüdiger, C.; Barbu, A.L.; Lafont, S.; Roujean, J.-L.; Walker, J.P.; Crapeau, M.; Wigneron, J.-P. Monitoring of water and carbon fluxes using a land data assimilation system: A case study for southwestern France. *Hydrol. Earth Syst. Sci.* **2010**, *14*, 1109–1124. [CrossRef]
40. Faroux, S.; Kaptue Tchuente, A.T.; Roujean, J.-L.; Masson, V.; Martin, E.; Le Moigne, P. ECOCLIMAP-II/Europe: A twofold database of ecosystems and surface parameters at 1 km resolution based on satellite information for use in land surface, meteorological and climate models. *Geosci. Model Dev.* **2013**, *6*, 563–582. [CrossRef]
41. Albergel, C.; Rüdiger, C.; Pellarin, T.; Calvet, J.-C.; Fritz, N.; Froissard, F.; Suquia, D.; Petitpa, A.; Piguet, B.; Martin, E. From near-surface to root-zone soil moisture using an exponential filter: An assessment of the method based on in-situ observations and model simulations. *Hydrol. Earth Syst. Sci.* **2008**, *12*, 1323–1337. [CrossRef]
42. Wagner, W.; Lemoine, G.; Rott, H. A method for estimating soil moisture from ERS scatterometer and soil data. *Remote Sens. Environ.* **1999**, *70*, 191–207. [CrossRef]
43. Bartalis, Z.; Wagner, W.; Naeimi, V.; Hasenauer, S.; Scipal, K.; Bonekamp, H.; Figa, J.; Anderson, C. Initial soil moisture retrievals from the METOP-A Advanced Scatterometer (ASCAT). *Geophys. Res. Lett.* **2007**, *34*, L20401. [CrossRef]
44. Reichle, R.H.; Koster, D. Bias reduction in short records of satellite soil moisture. *Geophys. Res. Lett.* **2004**, *31*, L19501. [CrossRef]
45. Drusch, M.; Wood, E.F.; Gao, H. Observations operators for the direct assimilation of TRMM microwave imager retrieved soil moisture. *Geophys. Res. Lett.* **2005**, *32*, L15403. [CrossRef]
46. Scipal, K.; Drusch, M.; Wagner, W. Assimilation of a ERS scatterometer derived soil moisture index in the ECMWF numerical weather prediction system. *Adv. Water Resour.* **2008**, *31*, 1101–1112. [CrossRef]
47. Baret, F.; Weiss, M.; Lacaze, R.; Camacho, F.; Makhmara, H.; Pacholcyzk, P.; Smets, B. GEOV1: LAI and FAPAR essential climate variables and FCOVER global time series capitalizing over existing products. Part1: Principles of development and production. *Remote Sens. Environ.* **2013**, *137*, 299–309. [CrossRef]

48. Baret, F.; Hagolle, O.; Geiger, B.; Bicheron, P.; Miras, B.; Huc, M.; Berthelot, B.; Nino, F.; Weiss, M.; Samain, O.; et al. LAI, fAPAR and fCover CYCLOPES global products derived from VEGETATION. Part 1: Principles of the algorithm. *Remote Sens. Environ.* **2007**, *110*, 275–286. [CrossRef]
49. Yang, W.; Shabanov, N.V.; Huang, D.; Wang, W.; Dickinson, R.E.; Nemani, R.R.; Knyazikhin, Y.; Myneni, R.B. Analysis of leaf area index products from combination of MODIS Terra and Aqua data. *Remote Sens. Environ.* **2006**, *104*, 297–312. [CrossRef]
50. Martens, B.; Miralles, D.G.; Lievens, H.; van der Schalie, R.; de Jeu, R.A.M.; Fernandez-Prieto, D.; Beck, H.E.; Dorigo, W.A.; Verhoest, N.E.C. GLEAM v3: Satellite-based land evaporation and root-zone soil moisture. *Geosci. Model Dev.* **2017**, *10*, 1903–1925. [CrossRef]
51. Baldocchi, D.T. Breathing of the terrestrial biosphere: Lessons learned from a global network of carbon dioxide flux measurement systems. *Aust. J. Bot.* **2008**, *56*, 1–26. [CrossRef]
52. Reichstein, M.; Falge, E.; Baldocchi, D.; Papale, D.; Aubinet, M.; Berbigier, P.; Bernhofer, C.; Buchmann, N.; Gilmanov, T.; Granier, A.; et al. On the separation of net ecosystem exchange into assimilation and ecosystem respiration: review and improved algorithm. *Glob. Chang. Biol.* **2005**, *11*, 1424–1439. [CrossRef]
53. Lasslop, G.; Reichstein, M.; Papale, D.; Richardson, A.D.; Arneth, A.; Barr, A.; Stoy, P.; Wohlfahrt, G. Separation of net ecosystem exchange into assimilation and respiration using a light response curve approach: Critical issues and global evaluation. *Glob. Chang. Biol.* **2010**, *16*, 187–208. [CrossRef]
54. Munro, R.; Eisinger, M.; Anderson, C.; Callies, J.; Corpaccioli, E.; Lang, R.; Lefebvre, A.; Livschitz, Y.; Perez Albinana, A. GOME-2 on MetOp: From in-orbit verification to routine operations. In Proceedings of the EUMETSAT Meteorological Satellite Conference, Helsinki, Finland, 12–16 June 2006.
55. Joiner, J.; Yoshida, Y.; Guanter, L.; Middleton, E.M. New methods for the retrieval of chlorophyll red fluorescence from hyperspectral satellite instruments: Simulations and application to GOME-2 and SCIAMACHY. *Atmos. Meas. Tech.* **2016**, *9*, 3939–3967. [CrossRef]
56. Thum, T.; Zaehle, S.; Kohler, P.; Aalto, T.; Aurela, M.; Guanter, L.; Kolari, P.; Laurila, T.; Lohila, A.; Magnani, F.; et al. Modelling sun-induced fluorescence and photosynthesis with a land surface model at local and regional scales in northern Europe. *Biogeosciences* **2017**, *14*, 1969–1987. [CrossRef]
57. Reick, C.H.; Raddatz, T.; Brovkin, V.; Gayler, V. Representation of natural and anthropogenic land cover change in MPI-ESM. *J. Adv. Model. Earth Syst.* **2013**, *5*, 459–482. [CrossRef]
58. Dee, D.P.; Uppala, S.M.; Simmons, A.J.; Berrisford, P.; Poli, P.; Kobayashi, S.; Andrae, U.; Balmaseda, M.A.; Balsamo, G.; Bauer, P.; et al. The ERA-Interim reanalysis: Configuration and performance of the data assimilation system. *Q. J. R. Meteorol. Soc.* **2011**, *137*, 553–597. [CrossRef]
59. Guanter, L.; Zhang, Y.; Jung, M.; Joiner, J.; Voigt, M.; Berry, J.A.; Frankenberg, C.; Huete, A.R.; Zarco-Tejada, P.; Lee, J.-E.; et al. Global and time-resolved monitoring of crop photosynthesis with chlorophyll fluorescence. *Proc. Natl. Acad. Sci. USA* **2014**, *111*, E1327–E1333. [CrossRef] [PubMed]
60. Duveiller, G.; Cescatti, A. Spatially downscaling sun-induced chlorophyll fluorescence leads to an improved temporal correlation with gross primary productivity. *Remote Sens. Environ.* **2016**, *182*, 72–89. [CrossRef]
61. Jung, M.; Reichstein, M.; Margolis, H.A.; Cescatti, A.; Richardson, A.D.; Altaf Arain, M.; Arneth, A.; Bernhofer, C.; Bonal, D.; Chen, J.; et al. Global patterns of land-atmosphere fluxes of carbon dioxide, latent heat, and sensible heat from eddy covariance, satellite, and meteorological observations. *J. Geophys. Res. Biogeosci.* **2011**, *116*, G00J07. [CrossRef]
62. Jung, M.; Reichstein, M.; Margolis, H.A.; Cescatti, A.; Richardson, A.D.; Altaf Arain, M.; Arneth, A.; Bernhofer, C.; Bonal, D.; Chen, J.; et al. Corrections to Global patterns of land-atmosphere fluxes of carbon dioxide, latent heat, and sensible heat from eddy covariance, satellite, and meteorological observations. *J. Geophys. Res. Biogeosci.* **2012**, *117*, G04011. [CrossRef]
63. Van der Tol, C.; Verhoef, W.; Timmermans, J.; Verhoef, A.; Su, Z. An integrated model of soil-canopy spectral radiances, photosynthesis, fluorescence, temperature and energy balance. *Biogeosciences* **2009**, *6*, 3109–3129. [CrossRef]
64. Koffi, E.N.; Rayner, P.J.; Norton, A.J.; Frankenberg, C.; Scholze, M. Investigating the usefulness of satellite-derived fluorescence data in inferring gross primary productivity within the carbon cycle data assimilation system. *Biogeosciences* **2015**, *12*, 4067–4084. [CrossRef]
65. Lee, J.-E.; Berry, J.A.; van der Tol, C.; Yang, X.; Guanter, L.; Damm, A.; Baker, I.; Frankenberg, C. Simulations of chlorophyll fluorescence incorporated into the Community Land Model version 4. *Glob. Chang. Biol.* **2015**, *21*, 3469–3477. [CrossRef] [PubMed]

66. Drusch, M.; Moreno, J.; Del Bello, U.; Franco, R.; Goulas, Y.; Huth, A.; Kraft, S.; Middleton, E.M.; Miglietta, F.; Mohammed, G.; et al. The Fluorescence EXplorer Mission Concept—ESA's Earth Explorer 8. *IEEE Trans. Geosci. Remote Sens.* **2017**, *55*, 1273–1284. [CrossRef]
67. Carrer, D.; Meurey, C.; Ceamanos, X.; Roujean, J.-L.; Calvet, J.-C.; Liu, S. Dynamic mapping of snow-free vegetation and bare soil albedos at global 1 km scale from 10 year analysis of MODIS satellite products. *Remote Sens. Environ.* **2014**, *140*, 420–432. [CrossRef]
68. Munier, S.; Carrer, D.; Planque, C.; Camacho, F.; Albergel, C.; Calvet, J.C. Satellite Leaf Area Index: Global scale analysis of the tendencies per vegetation type over the last 17 years. *Remote Sens.* **2018**, *10*, 424. [CrossRef]

remote sensing

MDPI

Article

LDAS-Monde Sequential Assimilation of Satellite Derived Observations Applied to the Contiguous US: An ERA-5 Driven Reanalysis of the Land Surface Variables

Clement Albergel [1,*], Simon Munier [1], Aymeric Bocher [1], Bertrand Bonan [1], Yongjun Zheng [1], Clara Draper [2], Delphine J. Leroux [1] and Jean-Christophe Calvet [1]

1 CNRM UMR 3589, Météo-France/CNRS, 31057 Toulouse, France; simon.munier@meteo.fr (S.M.); aymeric.bocher@polytechnique.edu (A.B.); bertrand.bonan@meteo.fr (B.B.); yongjun.zheng@meteo.fr (Y.Z.); delphine.leroux@meteo.fr (D.J.L.); jean-christophe.calvet@meteo.fr (J.-C.C.)

2 CIRES/NOAA Earth System Research Laboratory, Boulder, CO 80309, USA; clara.draper@noaa.gov

* Correspondence: clement.albergel@meteo.fr

Received: 5 September 2018; Accepted: 10 October 2018; Published: 12 October 2018

Abstract: Land data assimilation system (LDAS)-Monde, an offline land data assimilation system with global capacity, is applied over the CONtiguous US (CONUS) domain to enhance monitoring accuracy for water and energy states and fluxes. LDAS-Monde ingests satellite-derived surface soil moisture (SSM) and leaf area index (LAI) estimates to constrain the interactions between soil, biosphere, and atmosphere (ISBA) land surface model (LSM) coupled with the CNRM (Centre National de Recherches Météorologiques) version of the total runoff integrating pathways (CTRIP) continental hydrological system (ISBA-CTRIP). LDAS-Monde is forced by the ERA-5 atmospheric reanalysis from the European Center for Medium Range Weather Forecast (ECMWF) from 2010 to 2016 leading to a seven-year, quarter degree spatial resolution offline reanalysis of land surface variables (LSVs) over CONUS. The impact of assimilating LAI and SSM into LDAS-Monde is assessed over North America, by comparison to satellite-driven model estimates of land evapotranspiration from the Global Land Evaporation Amsterdam Model (GLEAM) project, and upscaled ground-based observations of gross primary productivity from the FLUXCOM project. Taking advantage of the relatively dense data networks over CONUS, we have also evaluated the impact of the assimilation against in situ measurements of soil moisture from the USCRN (US Climate Reference Network), together with river discharges from the United States Geological Survey (USGS) and the Global Runoff Data Centre (GRDC). Those data sets highlight the added value of assimilating satellite derived observations compared with an open-loop simulation (i.e., no assimilation). It is shown that LDAS-Monde has the ability not only to monitor land surface variables but also to forecast them, by providing improved initial conditions, which impacts persist through time. LDAS-Monde reanalysis also has the potential to be used to monitor extreme events like agricultural drought. Finally, limitations related to LDAS-Monde and current satellite-derived observations are exposed as well as several insights on how to use alternative datasets to analyze soil moisture and vegetation state.

Keywords: land surface modeling; data assimilation; remote sensing

1. Introduction

One of the major scientific challenges in relation to the adaptation to climate change is observing and simulating the response of land biophysical variables to extreme events, making land surface models (LSMs) constrained by high-quality gridded atmospheric variables and coupled with river-routing models key tools to address these challenges [1,2]. The modelling of terrestrial variables can be improved through the dynamical integration of observations. Remote sensing observations are particularly useful in this context because of their global coverage and higher spatial resolution (10 km and below). The current fleet of Earth observation missions holds an unprecedented potential to quantify land surface variables (LSVs) [3] and many satellite-derived products relevant to the hydrological and vegetation cycles are already available at high spatial resolutions. However, satellite remote sensing observations exhibit spatial and temporal gaps and not all key LSVs can be observed. LSMs are able to provide LSV estimates at all times and locations using physically-based equations, but as remotely sensed observations, they are affected by uncertainties (e.g., parametrization representation, atmospheric forcing, initialisation). Through a weighted combination of both, LSVs can be better estimated than by either source of information alone [4]; data assimilation techniques enable one to spatially and temporally integrate observed information into LSMs in a consistent way to unobserved locations, time steps, and variables.

In the past recent years, several land data assimilation system (LDAS) have emerged at different spatial scales: "site level", like the data assimilation system for LSMs using CLM4.5 (Community Land Model 4.5, [5]); regional, like the coupled land vegetation LDAS (CLVLDAS, [6,7]) and the famine early warning systems network (FEWSNET) LDAS (FLDAS, [8]); continental, like the North American LDAS (NLDAS, [9,10]) and the National Climate Assessment LDAS (NCA-LDAS [11]); as well as at global scale, like the global land data assimilation (GLDAS, [12]) and, more recently, LDAS-Monde [13]. LDAS-Monde has been developed to constrain the CO_2-responsive version of the ISBA (interactions between soil, biosphere, and atmosphere) LSM [14–17] using satellite derived observations within the open-source SURFEX modelling platform (SURFace Externalisee, [18]) of Meteo-France. LDAS-Monde has been implemented in a monitoring chain of terrestrial water and carbon fluxes. Unlike most of the above mentioned LDAS, LDAS-Monde is able to jointly and sequentially assimilate vegetation products such as leaf area index (LAI) together with surface soil moisture (SSM) observations [13,19–21]. Ref. [13] tested LDAS-Monde over Europe and the Mediterranean basin for the 2000–2012 period. A long term, global scale, multi-sensor satellite-derived surface soil moisture dataset (ESA CCI SSM, [22–25]) along with satellite derived LAI (GEOV1, http://land.copernicus.eu/global/ last access, June 2018), were jointly assimilated. LDAS-Monde was forced by WFDEI (WATCH-forcing-data-ERA-interim) observations based atmospheric forcing dataset [26,27] at half degree spatial resolution. Analysis impact was successfully carried out using (i) agricultural statistics over France; (ii) river discharge observations; (iii) satellite-derived estimates of land evapotranspiration from the global land evaporation amsterdam model (GLEAM) project; and (iv) spatially gridded observations-based estimates of up-scaled gross primary production and evapotranspiration from the FLUXNET network [13].

In this study, LDAS-Monde is applied and tested in a data-rich area: the CONtiguous US (CONUS, defined here as longitudes from 130.0°W to 60.0°W, latitudes from 20.0°N to 55.0°N, as shown in Figure 1). LDAS-Monde is forced by the latest ERA-5 atmospheric reanalysis from the European Center for Medium Range Weather Forecast (ECMWF) from 2010 to 2016 leading to a seven-year, quarter degree spatial resolution offline reanalysis of the land surface variables (LSVs). Ref. [28] assessed ERA-5 ability to force the ISBA LSM by comparison to satellite-derived products and in situ observations covering a substantial part of the land surface storage and fluxes. They found that using ERA-5 in place of its predecessor, ERA-Interim, led to significant improvements in the representation of the LSVs linked to the terrestrial water cycle (surface soil moisture, river discharges, snow depth, and turbulent atmospheric fluxes), but did not improve the LSVs linked to the vegetation cycle (evapotranspiration, carbon uptake, and LAI). In that respect, the assimilation of LAI through ERA-5 driven reanalysis from LDAS-Monde is expected to bring clear improvements [13]. In this study, the impact of LDAS-Monde analysis with respect to an open-loop (i.e., model run without assimilation) is assessed using satellite-driven model estimates of land evapotranspiration from the Global Land Evaporation Amsterdam Model (GLEAM) project and upscaled ground-based observations of gross primary productivity from the FLUXCOM project, together with river discharges from the United

States Geophysical Survey (USGS) and the Global Runoff Data Centre (GRDC). Over CONUS, in situ measurements of soil moisture from the USCRN network (US Climate Reference Network) are also used in the evaluation. Section 2 describes the different components of LDAS-Monde as well as the evaluation data sets and strategy. Section 3 provides a set of statistical diagnostics to assess and evaluate the impact of the assimilation. Finally, Section 4 provides perspectives and future research directions.

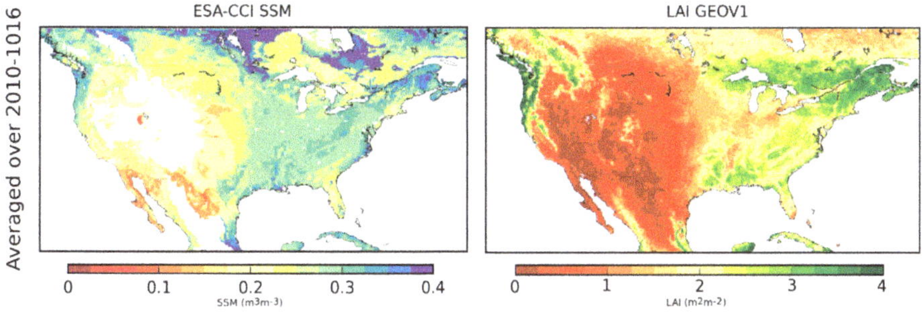

(**a**) (**b**)

Figure 1. Averaged (**a**) surface soil moisture from the Climate Change Initiative project of ESA (for pixels with less than 15% of urban areas and with an elevation of less than 1500 m above sea level); (**b**) GEOV1 leaf area index from the Copernicus Global Land Service project (for pixels covered by more than 90% of vegetation) from 2010 to 2016. ESA CCI SSM—European Space Agency and Climate Change Initiative surface soil moisture; LAI—leaf area index.

2. Data and Methods

2.1. LDAS-Monde System Components

LDAS-Monde allows sequential assimilation of satellite derived land observations at a global scale. The assimilation is performed into the open-access SURFEX modelling platform of Météo-France (SURFace Externalisée, [18]). It produces offline re-analyses of LSVs using (i) an LSM along with data assimilation techniques, (ii) observations, and (iii) atmospheric forcing. Those components of LDAS-Monde are briefly described below.

2.1.1. The SURFEX Modelling Platform

LDAS-Monde uses the CO_2-responsive version of ISBA embedded within the SURFEX platform. The most recent version of SURFEX (version 8.1) is used in this study with the "NIT" plant biomass monitoring option for ISBA. In this configuration, ISBA simulates leaf-scale physiological processes and plant growth [15–17]. The dynamic evolution of the vegetation biomass and LAI variables is driven by photosynthesis in response to atmospheric and climate conditions. Photosynthesis enables vegetation growth resulting from the CO_2 uptake. During the growing phase, enhanced photosynthesis corresponds to a CO_2 uptake, which results in vegetation growth from the LAI minimum threshold (prescribed as 1 m^2 m^{-2} for coniferous forest or 0.3 m^2 m^{-2} for other vegetation types). Transfers of water and heat through the soil rely on a multilayer diffusion scheme [29,30]. The ISBA parameters are defined for 12 generic land surface patches. They include nine plant functional types (needle leaf trees, evergreen broadleaf trees, deciduous broadleaf trees, C3 crops, C4 crops, C4 irrigated crops, herbaceous, tropical herbaceous, and wetlands) as well as bare soil, rocks, and permanent snow and ice surfaces.

This version of ISBA is coupled to the CTRIP river routing model through OASIS-MCT [31] in order to simulate streamflows of the main rivers [32–35]. Besides, a single-source energy budget of a soil/vegetation composite is computed. SURFEX also involves data assimilation techniques to analyse LSVs from the ISBA LSM.

This study makes use of the simplified version of an extended Kalman filter (SEKF), as already used and described in [13,19,21,36]. The SEKF uses finite differences from perturbed simulations to estimate the linear tangent model linking the model state control variables to the observed variables. Satellite derived surface soil moisture (SSM) and leaf area index (LAI) are simultaneously assimilated to update eight model state control variables (i.e., control variables); LAI and soil moisture from seven layers of soil, from 1 cm to 100 cm. Assimilating SSM and LAI within LDAS-Monde results in updates of the LSM variables in different ways. Model variables corresponding to the observations are first updated through the Kalman gain computed by the SEKF. Secondly, control variables are updated through their sensitivity to the observed variables. For example, the assimilation of LAI impacts LAI itself, but also soil moisture from the seven layers present in the state vector and the assimilation of SSM impacts LAI. Finally, other variables are indirectly modified by the analysis through biophysical processes and feedbacks in the model by updates of the control variables.

2.1.2. ESA CCI Surface Soil Moisture and CGLS Leaf Area Index

In this study the European Space Agency and Climate Change Initiative (ESA CCI) SSM-combined version of the product (v4.1) is assimilated into LDAS-Monde (http://www.esa-soilmoisture-cci.org, last access June 2018). The CCI merges SSM observations from seven different microwave radiometers (SMMR, SSM/I, TMI, ASMR-E, WindSat, AMSR2, SMOS) and four different scatterometers (ERS-1 and 2 AMI, and MetOp-A and B ASCAT) into a single combined data set covering the time period from November 1978 to December 2016. Data are expressed in volumetric (m^3 m^{-3}) units and quality flags are provided (i.e., snow coverage or temperature below 0° and dense vegetation). For a more comprehensive overview of the product, see [24,25]. Topographic relief is known to negatively affect satellite remote sensing retrievals of SSM [37], hence the time series for pixels whose average altitude exceeds 1500 m above sea level were not accounted for. Data on pixels with urban land cover fractions larger than 15% were discarded too, to limit the effects of artificial surfaces. These thresholds were set according to [13,20,38], who processed satellite-based SSM retrievals for data assimilation experiments with the ISBA LSM. Data are available almost every day with a spatial resolution of 0.25° $\times$ 0.25°. To assimilate SSM data, it is important to rescale the observations such that they are consistent with the model climatology [39,40]. Hence, similarly to previous studies, the ESA CCI SSM product has been transformed into the model-equivalent SSM to address possible misspecification of physiographic parameters, such as the wilting point and the field capacity. The linear rescaling approach described in [41] (using the first two moments of the cumulative distribution function, CDF) was used. It consists of a linear rescaling enabling a correction of the differences in the mean and variance of the distribution. It has been applied at a seasonal scale (i.e., for each specific month) following [13].

The GEOV1 LAI is also assimilated. It is produced by the European Copernicus Global Land Service project (http://land.copernicus.eu/global/, last access June 2018). Ref. [42] proposed an evaluation of this product in the context of Numerical Weather Prediction (NWP). LAI observations are retrieved from the SPOT-VGT (from 1999 to 2014) and then from PROBA-V (from 2014 to present) satellite data according to the methodology proposed by [43]. The 1 km spatial resolution observations are interpolated by an arithmetic average to the 0.25° $\times$ 0.25° model grid points, if at least 50% of the observation grid points are observed (i.e., half the maximum amount). LAI observations have a temporal frequency of 10 days at best (e.g., in presence of clouds, no observations are available). Both assimilated datasets are illustrated by Figure 1, averaged over 2010–2016.

2.1.3. ERA-5 Atmospheric Reanalysis

ERA-5 [44] is the fifth generation of European re-analyses produced by the ECMWF and a key element of the EU-funded Copernicus Climate Change Service (C3S). ERA-5 important changes relative to ERA-interim former ECMWF's atmospheric reanalysis include (i) a higher spatial and temporal resolution as well as (ii) a more recent version of ECMWF Earth system model physics and data assimilation system (corresponding to ECMWF's cycle CY41R2, https://www.ecmwf.int/en/forecasts/documentation-and-support/changes-ecmwf-model/ifs-documentation, last access June 2018). It makes it possible to use modern parameterizations of Earth processes compared with older versions used in ERA-interim. For instance, in addition to being applied to satellite observations, a variational bias scheme is also applied to aircraft and surface ozone and pressure data. ERA-5 also benefits from reprocessed data sets that were not ready yet during the production of ERA-interim. Two other important features of ERA-5 are the more frequent model output and improved model spatial resolution, going from six-hourly output in ERA-interim to hourly output analysis in ERA-5, and from 79 km (horizontal dimension) and 60 levels (vertical dimension) to 31 km and 137 levels in ERA-5. Finally, ERA-5 also provides an estimate of uncertainty through the use of a 10-member ensemble of data assimilations (EDA) at a coarser resolution (63 km horizontal resolution) and three-hourly frequency. ERA-5 is foreseen to replace ERA-interim re-analysis. All ERA-5 atmospheric variables were interpolated at $0.25° \times 0.25°$ spatial resolution. A bilinear interpolation from the native reanalysis grid to the regular grid was made.

2.2. Evaluation Datasets and Methods

The LDAS-Monde analysis impact was assessed with respect to the open-loop model run (i.e., no assimilation). The system was spun-up by running year 2010 twenty times. Table 1 presents the set up of the different experiments used in this study, the open-loop and the analysis, as well as two additional model runs: (i) Ini_model, a 12-month model run starting on 1 January 2016 (initialised by the model, that is, the openloop with no data assimilation, simulation run from 2010 to 2015); and (ii) Ini_analysis, a 12-month model run initialised by initial conditions from the analysis on 1 January 2016. The two above-mentioned assimilated datasets (ESA CCI SSM and LAI GEOV1) were used as a way to check to what extent the assimilation system was able to produce analyses closer to these two datasets that were assimilated than the open-loop. Then, two independent spatially distributed datasets, namely evapotranspiration from the GLEAM project [45,46] and gross primary production (GPP) from the FLUXCOM project [47,48], were used in the evaluation process. Ground based measurements of soil moisture from the USCRN (US Climate Reference Network, [49]) were also used, along with river discharge observations from the United States Geophysical Survey (USGS) and the Global Runoff Data Centre (GRDC).

The ability of LDAS-Monde to represent SSM, LAI, evapotranspiration, and GPP was assessed using the correlation coefficient (R) and root mean square difference (RMSD). These metrics were applied at a seasonal scale (i.e., for each month) over 2010–2016. For ground-based measurements of SSM, R was calculated for both absolute and anomaly time series in order to remove the strong impact from the SSM seasonal cycle on this specific metric (see e.g., [13,28]). Ground measurements at a depth of 5 cm were compared with soil moisture of the third layer of soil (between 4 and 10 cm depth) from both the model and the analysis for months from April to September over the 2010–2016 time period to avoid frozen conditions. Only stations with significant R values for the two experiments (with p-value < 0.05) were kept for the evaluation.

Table 1. Set up of the experiments used in this study. ISBA—interactions between soil, biosphere, and atmosphere; DA—data assimilation; SEKF—simplified version of an extended Kalman filter; ESA CCI SSM—European Space Agency and Climate Change Initiative surface soil moisture; LAI—leaf area index; CTRIP—Centre National de Recherches Météorologiques version of the total runoff integrating pathways.

Experiments (Time Period)	Model	Domain & Spatial Resolution	Atmospheric Forcing	DA Method	Assimilated Observations	Model Equivalents of the Observations	Control Variables	Additional Options
Model or Open-loop (2010–2016)	ISBA Multi-layer soil model CO_2-responsive version (Interactive vegetation)	CONtiguous US (CONUS), 0.25° × 0.25°	ERA-5	N/A	N/A	N/A	N/A	Coupling with CTRIP (0.5°)
Analysis (2010–2016)	ISBA Multi-layer soil model CO_2-responsive version (Interactive vegetation)	CONtiguous US (CONUS), 0.25° × 0.25°	ERA-5	SEKF	SSM (ESA CCI) LAI (GEOV1)	Rescaled WG2 (Second layer of soil (1–4 cm)) LAI	Layers of soil 2 to 8 (WG2 to WG8, 1–100 cm) LAI	Coupling with CTRIP (0.5°)
Ini_Model (2016)	ISBA Multi-layer soil model CO_2-responsive version (Interactive vegetation)	CONtiguous US (CONUS), 0.25° × 0.25°	ERA-5	12-month model run starting on 1 January 2016 (initialised by the model simulation, i.e., Open-loop, run from 2010 to 2015)				Coupling with CTRIP (0.5°)
Ini_Analysis (2016)	ISBA Multi-layer soil model CO_2-responsive version (Interactive vegetation)	CONtiguous US (CONUS), 0.25° × 0.25°	ERA-5	12-month model run starting on 1 January 2016 (initialised by the analysis run from 2010 to 2015)				Coupling with CTRIP (0.5°)

In order to provide an easier measurement of the added value of the analysis, statistics were also normalized with respect to the model. The so-called normalized information contribution index (NIC as in [28,50]) was applied to the correlation coefficient (Equation (1), for both volumetric and anomaly time-series) and to RMSD (Equation (2)) to quantify the improvement or degradation from the analysis with respect to the model.

$$NIC_{R} = \frac{R_{(Analysis)} - R_{(Model)}}{1 - R_{(Model)}} \times 100 \quad (1)$$

$$N_{RMSD} = \frac{RMSD_{(Analyse)} - RMSD_{(Model)}}{RMSD_{(Model)}} \times 100 \quad (2)$$

NIC scores were then classified according to three categories: (i) negative impact from the analysis with respect to the model with values smaller than −3%, (ii) positive impact from the analysis with respect to the model with values greater than +3%, and (iii) neutral impact from the analysis with respect to the model with values between −3% and 3%.

Over the 2010–2016 time period, river discharge from the analysis and model runs were compared with daily streamflow data from USGS and GRDC. Data were selected for sub-basins with rather large drainage areas (10,000 km^2 or greater) because of the low resolution of CTRIP ($0.5^\circ \times 0.5^\circ$) and with a long observation time series (48 months or more). As commonly found in the literature, observed and simulated river discharge (Q) data are expressed in $m^3\ s^{-1}$. However, given that the observed drainage areas may differ from the simulated ones, specific discharge in mm d^{-1} (the ratio of Q to the drainage area) was used in this study, similarly to [13,28]. Stations with drainage areas differing by more than 20% from the simulated ones were discarded. Impact on Q was evaluated using the Kling–Gupta Efficiency (KGE, [51]) score:

$$KGE = 1 - \sqrt{RE_{\sigma}^2 + RE_{\mu}^2 + (1 - R)^2} \quad (3)$$

with RE_μ and RE_σ representing the relative error of simulated or analysed mean and standard deviation (Equations (4) and (5)), respectively; R representing the correlation coefficient between the observed discharges and either the modelled or analysed river discharges.

$$RE_{\mu} = \frac{Q_{\mu}}{Q_{(obs.)_{\mu}}} - 1 \quad (4)$$

$$RE_{\sigma} = \frac{Q_{\sigma}}{Q_{(obs.)_{\sigma}}} - 1 \quad (5)$$

KGE represents the Euclidean distance from the ideal point in the [RE_μ, RE_σ, R] score space. RE_μ, RE_σ, and R constitute a set of mathematically independent metrics quantifying the fit of simulated/analysed discharge time series. At best, Re_μ and RE_σ are equal to 0 and R is equal to 1 (leading to a perfect KGE value of 1), indicating that simulated or analysed time series are identical to the measured one. NIC (Equation (1)) was applied to KGE (Equation (6)) as well, only for stations with KGE values greater than 0. Finally, RE_μ and RE_σ metrics were normalised, following Equation (7), as well as Equation (8) to appreciate the added value from the analysis with respect to the model.

$$NIC_{KGE} = \frac{KGE_{(Analysis)} - KGE_{(Model)}}{1 - KGE_{(Model)}} \times 100 \quad (6)$$

$$N_{RE_{\mu}} = \frac{100 * RE_{\mu_{(Analysis)}} - RE_{\mu_{(Model)}}}{RE_{\mu_{(Model)}}} \quad (7)$$

$$N_{RE_{\sigma}} = \frac{100 * RE_{\sigma_{(Analysis)}} - RE_{\sigma_{(Model)}}}{RE_{\sigma_{(Model)}}} \quad (8)$$

3. Results

3.1. Analysis Impact on Assimilated Variables

Being the model equivalents of the assimilated observations, LAI and soil moisture from the second layer of soil are expected to be the two variables most affected by the assimilation. Figure 2 presents a 10-day time series of LAI averaged over the whole domain for the 2010–2016 time period. From Figure 2, one can see that the open-loop simulation tends to overestimate the observed LAI in winter periods and that the senescence phase of vegetation is too late over the autumn when compared with the observations. In that respect, the assimilation is efficiently correcting the model; however, analysed LAI does not reach LAI maximal values of the observations. Figure 3 shows maps of LAI for the model (Figure 3a), the observations (Figure 3b), and the analysis (Figure 3c) averaged over 2010–2016. It is clearly visible that the model overestimates LAI in the eastern part of the domain. Also, some geographical patterns visible in the observations (e.g., the Mississippi area in Figure 3b) are not represented in the model (Figure 3a). After assimilation, the analysis presents reduced LAI values in the eastern part of the domain and the above-mentioned geographical patterns are visible too (Figure 3c). This shows the ability of the assimilation to integrate geographical information into the model. Figure 3 also presents seasonal scores between the model and the observations and between the analysis and the observations for RMSD and R over the 2010–2016 time period. The analysis leads to a better fit between the model forecasts and the subsequent assimilated observations for both metrics. On average for the whole period, RMSD values drop from 1.10 $m^2\ m^{-2}$ (model vs. observations) to 0.65 $m^2\ m^{-2}$ (analysis vs. observations), while R values increase from 0.69 (model vs. observations) to 0.88 (analysis vs. observations). Figure 4 presents the same information content for soil moisture. As ESA CCI SSM was rescaled in order to match the first two moments of the modelled SSM cumulative distribution function, the impact is marginal and differences are hardly visible from Figure 4a–c. From Figure 4d,e, however, one can appreciate the added value of the analysis: RMSD values drop from 0.046 $m^3\ m^{-3}$ (model vs. observations) to 0.044 $m^3\ m^{-3}$ (analysis vs. observations), while R values increase from 0.85 (model vs. observations) to 0.87 (analysis vs. observations). It is worth mentioning that the good level of scores (prior as well as after assimilation) are linked to the rescaling of the ESA CCI SSM data to the model climatology. Finally, Figure 5 shows maps of analysis increments for 4 (out of 8) control variables averaged over the whole 2010–2016 time period (LAI, second, fourth, and sixth layers of soil from left to right, respectively). It can be noticed that the magnitude of increments is decreasing with depth. It can also be noticed that over almost the whole domain, the analysis tends to add water in the soil near the surface (positive increments), while it dries layers where the roots are mainly located (from layer 4 to 6, negative increments).

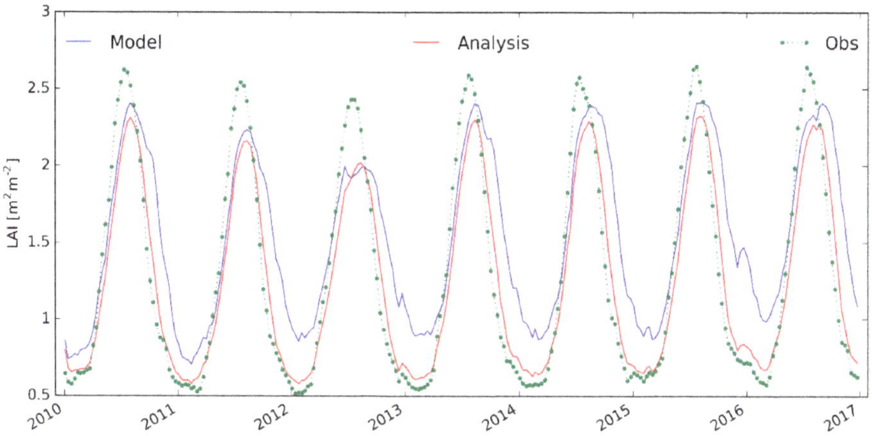

Figure 2. Leaf area index time series from the model (blue line), the observations (green dots and dashed line), and the analysis (red line) averaged over the whole domain from 2010 to 2017.

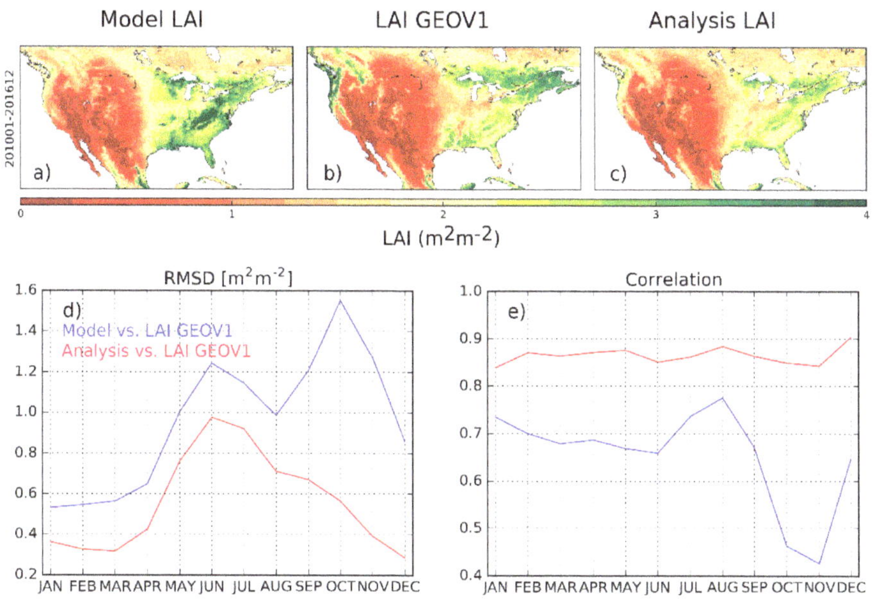

Figure 3. Top row; leaf area index from (**a**) the model, (**b**) the observations, and (**c**) the analysis averaged over the 2010–2016 time period. Bottom row: seasonal (**d**) root mean square difference (RMSD) and (**e**) correlation values between leaf area index (LAI) from the model (in blue), the analysis (in red) and GEOV1 LAI estimates from the Copernicus Global Land Service project from 2010 to 2016.

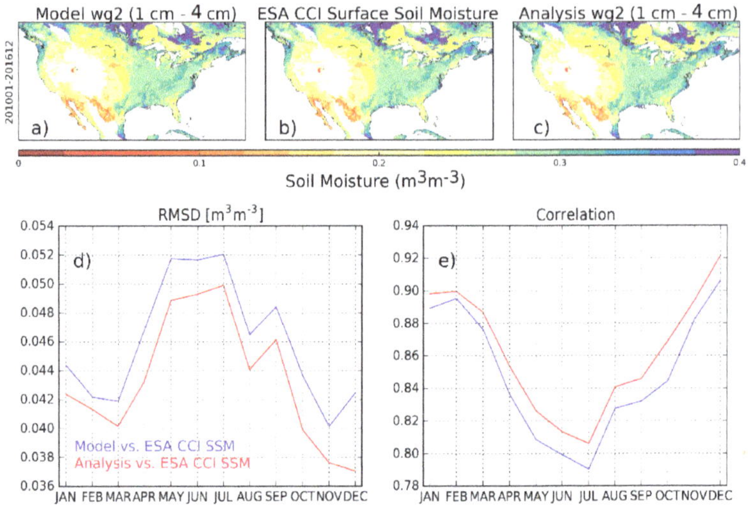

Figure 4. (**a**–**e**) same as Figure 3 for soil moisture.

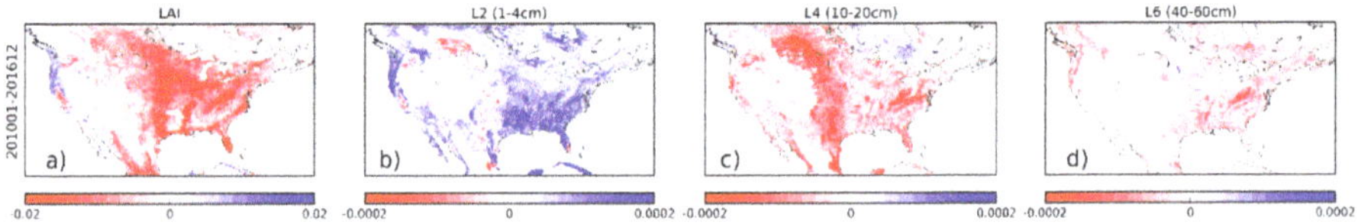

Figure 5. Analysis increments averaged over the 2010–2016 time period for (**a**) LAI in $m^2\ m^{-2}$, (**b**) second, (**c**) fourth, and (**d**) sixth layer of soil moisture in $m^3\ m^{-3}$.

3.2. Evaluation Using Independent Datasets

3.2.1. Evapotranspiration and GPP

Table 2 presents the statistical scores from the evaluation of both the open-loop and the analysis with respect to evapotranspiration and GPP averaged over 2010–2016, as well as the mean of the evaluation data set over the considered area. On average, R increases from 0.80 to 0.81 when comparing evapotranspiration from the model and from the analysis, respectively, to the independent estimates. Average RMSD decreases from 0.89 kg $m^{-2}\ d^{-1}$ to 0.85 kg $m^{-2}\ d^{-1}$. When compared with GPP estimates, averaged correlations rise from 0.74 to 0.78 and RMSD drops from 2.20 g(C) $m^{-2}\ d^{-1}$ to 1.91 g(C) $m^{-2}\ d^{-1}$ when considering the model or the analysis, respectively. Figure 6 presents spatial maps of N_{RMSD} (Figure 6a,c) and NIC_R (Figure 6b,d) resulting from the comparison with evapotranspiration (Figure 6, top row) and GPP (Figure 6, bottom row) of their modelled and analysed equivalent. For N_{RMSD} (Figure 6a,c), blue colours represent an improvement from the analysis regarding RMSDs (i.e., the latter better represents either evapotranspiration or GPP than the model), while for NIC_R (Figure 6b,d), red colours depict an improvement from the analysis. Figure 6 shows that both evapotranspiration and GPP are improved almost everywhere in terms of correlation and RMSD, and that the impact of the assimilation is stronger for GPP than for evapotranspiration. At the seasonal scale (not shown), the assimilation leads to a positive impact all year long in the representation of GPP in terms of both RMSD and R values. Impact from the assimilation on evapotranspiration is smaller (as seen in Table 2), while RMSD values are slightly improved all year long, R values are slightly improved from April to October and slightly degraded from November to March.

Table 2. Statistical scores from the evaluation of both the open-loop and the analysis with respect to evapotranspiration and gross primary production (GPP) averaged over 2010–2016. RMSD—root mean square difference.

	Mean of the Evaluation Data Set	Experiments	RMSD	R
Evapotranspiration	1.46 $kg/m^2/d$	Open-loop	0.87 $kg/m^2/d$	0.80
		Analysis	0.85 $kg/m^2/d$	0.81
Gross Primary Production	1.76 $g(C)/m^2/d$	Open-loop	2.20 $g(C)/m^2/d$	0.74
		Analysis	1.91 $g(C)/m^2/d$	0.78

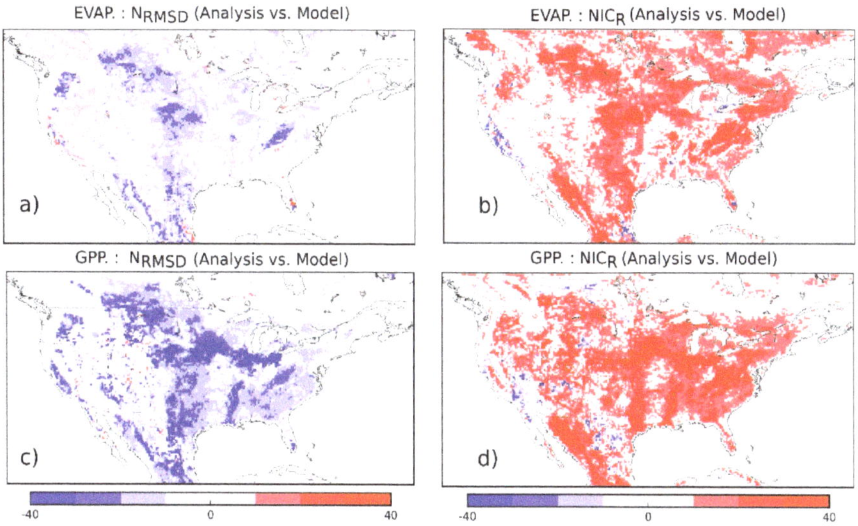

Figure 6. Top row: (**a**) normalized root mean square difference (RMSD) (blue colours indicate an improvement) and (**b**) normalized information contribution (NIC) applied on correlations values (red colours indicate an improvement) for evapotranspiration from the analysis with respect to the model. Bottom row: same as top row for gross primary production. Units are percentages.

Geographical patterns, as seen on the middle of Figure 6a,c (transition from relative wet to dry areas) areas of the Midwest, are also visible in the soil moisture increments maps of Figure 5b,c, where stronger increments occur. It is difficult to state whether or not those strong corrections reflect atmospheric forcing errors, for example, precipitation, in this area as to date, very few studies have evaluated ERA5 precipitation errors. Ref. [52] have evaluated a large amount of precipitation products, including ERA5, using the high-resolution (4 km) stage-IV gauge-radar precipitation data set as a reference over CONUS using several metrics for 2008–2017. Figure 1d of [52], illustrates the Kling–Gupta efficiency (KGE) scores between ERA5 precipitation and stage-IV reference. If lower values seem to be observed in the above-mentioned transition area, it is not clear enough to incriminate precipitation errors. However, this geographical pattern corresponds to a specific soil moisture regime, the Ustic regime, where moisture is present in the soil, but limited, at times in which conditions are suitable for plant growth (as visible on the following USDA map: https://www.nrcs.usda.gov/Internet/FSE_MEDIA/nrcs142p2_050436.jpg, last access October 2018). This area has specific soil properties including clay, with high swelling potential [53]. Such soil properties are likely to be misrepresented in the model, possibly leading to stronger increments in the analysis.

3.2.2. Soil Moisture

The statistical scores for surface soil moisture from the model and the analysis (third layer of soil between 4 and 10 cm depth) over 2010–2016 when compared with ground measurements from the USCRN network (at 5 cm depth) are presented in Table 3. Median R values on volumetric soil moisture time-series (anomaly time series), along with their 95% confidence interval of the median, derived from 10,000-sample bootstrapping are as follows: 0.72 ± 0.02 (0.60 ± 0.02) and 0.74 ± 0.02 (0.60 ± 0.02), while median ubRMSDs are 0.049 ± 0.004 and 0.048 ± 0.004 for the model and the analysis, respectively. Figure 7a,b illustrate correlation values on volumetric and anomaly time-series between the model and the observations, respectively, for each station. Figure 7c,d represent the added value of the analysis expressed through the NIC index (Equation (1)) applied for correlations (NIC_R) values on volumetric and anomaly time-series; large blue circles represent a positive impact from the analysis at NIC_R greater that +3 (i.e., R values are better when the analysis is used than when the model is used), large red circles a degradation from the analysis at NIC_R smaller than −3, while diamond symbols represent a rather neutral impact at NIC_R in between [−3; +3]. While 46% (81%) of the pool of stations present a rather neutral impact for R values on volumetric (anomaly) time series, stations more impacted by the analysis tend to be positively impacted at 46% (18%), to be compared with 8% (1%) of negative impacts. Although differences between the model run and the analysis are rather small, these results underline the added value of the analysis with respect to the model run.

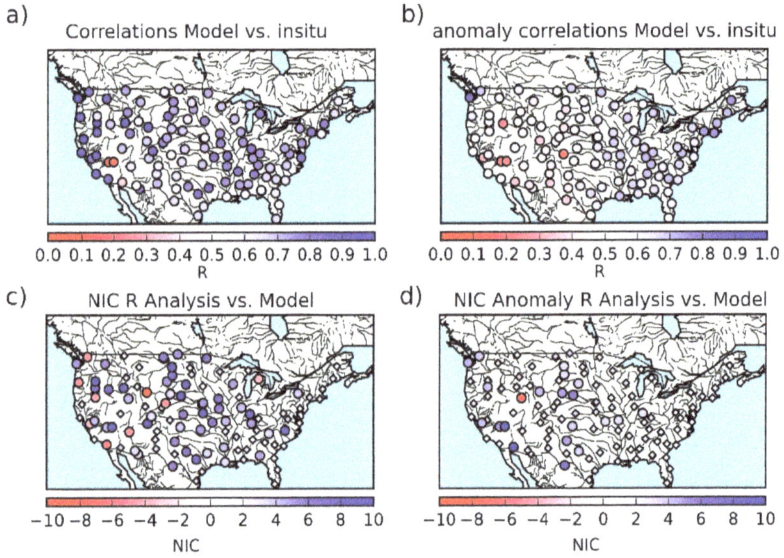

Figure 7. Maps of correlation (R) on volumetric time-series (**a**) and anomaly time-series (**b**) between in situ measurements at 5 cm depth from the US Climate Reference Network (USCRN) network and soil moisture from the model (third layer of soil between 4 cm and 10 cm) from 2010 to 2016. NIC applied on R (anomaly R) values (**c**,**d**); analysis with respect to the model. NIC scores are classified according to three categories: (i) negative impact from the analysis with respect to the model with values smaller than −3% (red circles), (ii) positive impact from the analysis with respect to the model with values greater than +3% (blue circles), and (iii) neutral impact from the analysis with respect to the model with values between −3% and 3% (diamonds).

Table 3. Analysis impact evaluation against in situ measurements of soil moisture from the US Climate Reference Network (USCRN) network. In situ measurements at a depth of 5 cm are used to evaluate soil moisture from the third layer of soil (4–10 cm) from either the model or analysis experiment over 2010–2016. The normalized information contribution (NIC) is applied to the correlation (anomaly correlations) values. NIC scores are classified according to three categories: (i) negative impact from the analysis with respect to the model with values smaller than −3%, (ii) positive impact from the analysis with respect to the model with values greater than +3%, and (iii) neutral impact from the analysis with respect to the model with values between −3% and 3%.

110 (110) Stations with Significant R (Anomaly R)	Median R (Anomaly R)	Median ubRMSD	Positive Impact: >+3	←3 Negative Impact: <−3	Neutral Impact [−3; +3]
Model	0.72 ± 0.02 * (0.60 ± 0.02 *)	0.049 ± 0.004 *	N/A	N/A	N/A
Analysis	0.74 ± 0.02 * (0.60 ± 0.02 *)	0.048 ± 0.004 *	46% (18%)	8% (1%)	46% (81%)

* 95% confidence interval of the median derived from a 10,000 samples bootstrapping.

3.2.3. Streamflow

A subset of 258 out of 531 gauging stations was selected for the evaluation according to the criteria described in the methodology section, with KGE scores within the [0, 1] interval. Figure 8 presents the performance of analysed streamflow with respect to the one from the model run for this pool of stations, with a focus on the eastern part of the domain. NIC_{KGE} values are presented following the same classification as NIC_R applied to soil moisture. Scores are presented in Table 4. Looking at NIC_{KGE}, 62% of the pool of stations (258 stations) present a rather neutral impact (at NIC_{KGE} between [−3; 3]) and 26% of the stations present a positive impact (at NIC_{KGE} > +3), while only 12% of stations have a negative impact (at NIC_{KGE} < −3). NICR, $N_{RE\sigma}$, and $N_{RE\mu}$ follow the same classification (with even a smaller percentage of stations being negatively affected by the analysis; 1%); when the analysis is impacting streamflow representation, it tends to be a positive impact.

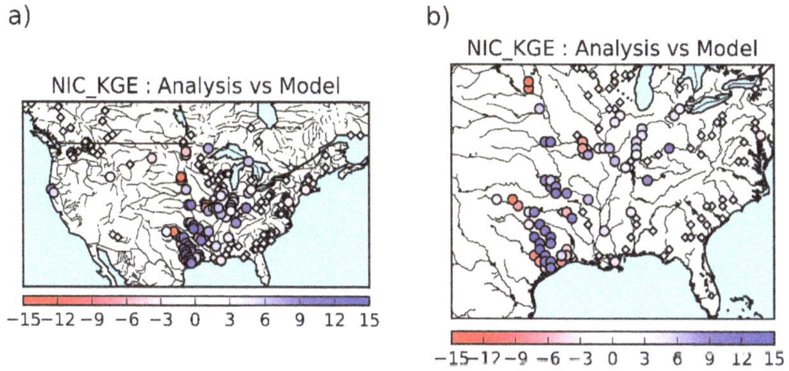

Figure 8. Normalized information contribution scores based on Kling–Gupta efficiency (KGE) scores (NIC_{KGE}) (**a**) analysis with respect to the model, (**b**) zoom over the eastern part of the domain. Small diamonds represent stations for which NIC_{KGE} are between [−3; +3]. NIC_{KGE} greater than 3 (blue large circles) suggest an improvement from the analysis over the model, values smaller than −3 (red large circles) suggest a degradation. For sake of clarity, a factor of 100 has been applied to NIC.

Table 4. Analysis impact evaluation against daily streamflow over 2010–2016. The impact from the analysis with respect to the model is assessed through the normalized information contribution (NIC) applied to the Kling–Gupta efficiency (KGE) score, as well as using normalized relative error of simulated or analysed mean (RE_{μ}) and standard deviation (RE_{σ}). Scores are classified according to three categories: (i) negative impact from the analysis with respect to the model with values smaller than −3, (ii) positive impact from the analysis with respect to the model with values greater than +3, and (iii) neutral impact from the analysis with respect to the model with values between −3 and 3.

258 out of 531 Stations with KGE Greater than 0	Positive Impact: >+3	Negative Impact: <−3	Neutral Impact [−3; +3]
NIC_{KGE}	26%	12%	62%
$N_{RE\sigma}$	22%	1%	77%
$N_{RE\mu}$	34%	1%	65%

4. Potential Applications, Discussions, and Perspectives

4.1. Could LDAS-Monde be Used to Monitor Agricultural Droughts?

The previous section has highlighted the LDAS-Monde ability to enhance the monitoring accuracy for land surface variables. It should then be possible to use it to better represent extreme events like agricultural droughts. Figure 9 represents monthly LAI anomalies averaged over the U.S. corn belt (simplified as a box from 110°W to 70°W and 30°N to 50°N) with respect to 2010–2016 means from the model, the analysis, and the observations. As shown by Figure 9, for the second part of the year 2012, LAI observations exhibit a strong negative anomaly at this domain scale. While it is also visible in the model, the latter clearly overestimates the intensity of the observed anomaly. The analysed LAI anomaly is closer to the observed one than the model. This extreme drought event is known as the August 2012 U.S. corn belt drought. The U.S. Department of Agriculture (USDA, www.nass.usda.gov, last access June 2018) estimated that corn yield (per acre of planted crop) was 26% below the expectation that they had at the beginning of the 2012 growing season. The 2012 corn yield deficit and the implied climatic impact was classified as a *'historic event'* [54]. As visible on Figure 9, spring 2012 presents a positive anomaly for vegetation. Ref. [55] defined spring 2012 as the earliest false spring in the North American record (i.e., a period of weather in late winter or early spring allowing vegetation to be prematurely brought out of dormancy). This false spring has contributed to an earlier dry out of the soil. Figure 10 presents maps of the LAI anomaly for this specific month for the model, observations, and analysis from left to right, respectively. Compared with the observations (Figure 10b), the area affected by the anomaly in the model (Figure 10a) is too large and too intense, while the analysis (Figure 10c) better matches the observed pattern in both space and intensity. This impact is valid when compared with most of the severe droughts events that occurred over CONUS (data from the National Oceanic and Atmospheric Administration (NOAA) state of the climate website, last access April 2018 https://www.ncdc.noaa.gov/sotc/drought/201803, not shown). Hence, LDAS-Monde provides a better tool than the model alone to monitor extreme events like agricultural droughts.

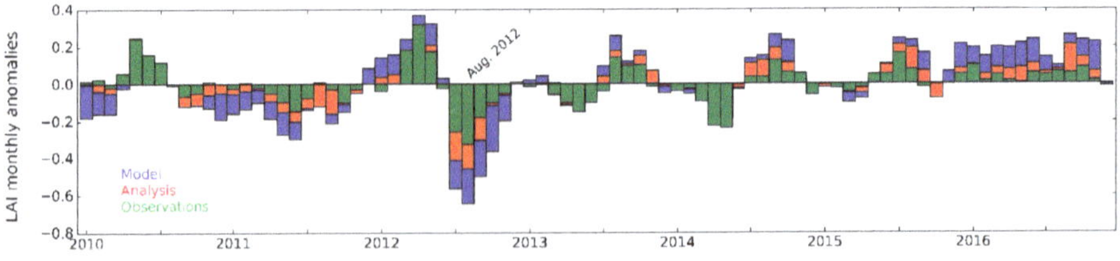

Figure 9. Leaf area index (LAI) monthly anomalies over the 2010–2016 time period for the model (blue bars), the analysis (red bars), and the CGLS GEOV1 observations (in green) over the corn belt drought defined as a box from 110°W to 70°W and 30°N to 50°N.

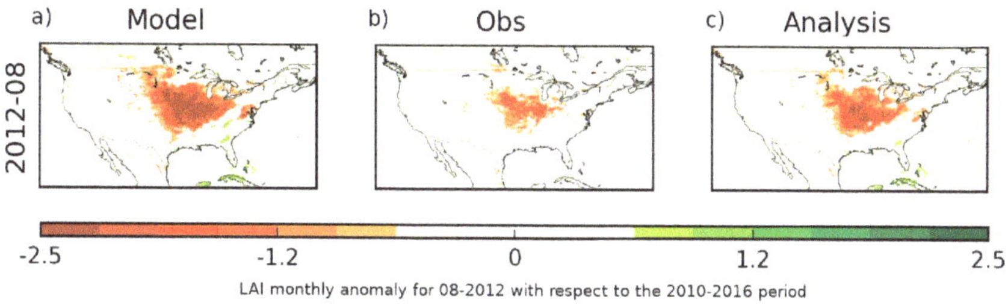

Figure 10. (**a**) Monthly anomaly of leaf area index for August 2012 with respect to the 2010–2016 period, (**b**) same as (**a**) for observed leaf area index, and (**c**) same as (**a**) for analysed leaf area index.

It is also worth mentioning that if LDAS-Monde brings a clear improvement in the representation of LAI, reducing the overestimation duration as well as the minimal values, as its model counterpart, it fails capturing the observed LAI peak intensity. Ref. [13] has evaluated the model sensitivity of the observation for Europe over 2000–2012 reflected in the SEKF Jacobians. The Jacobians depend on the model physics and their examination provides useful insight that explains the data assimilation system performances [21,56]. Ref. [13] suggests a seasonal dependency of the model sensitivity to the observed LAI. High sensitivity is found in autumn. Smaller model sensitivity at the time of the year where the peak LAI occurs (late spring) prevails the analysis to match the observations correctly.

As highlighted by [57], who have evaluated the capacity of several LSMs (including ISBA) to accurately simulate observed energy and water fluxes during droughts, there is a need to re-examine existing model components in LSMs to improve simulations of soil hydrological processes and water–plant interactions. It appears from Figure 2 that although the analysis is able to correct the overestimated LAI values in winter, the minimum LAI thresholds used in ISBA has to be revisited. Recently, the satellite derived LAI data have been disaggregated following a Kalman filtering technique developed by [58]. This enables the LAI signal for each vegetation type to be separated within the pixel, which provides a dynamic vegetation-dependent estimate of the assimilated LAI within the pixel [59]. This new dataset will make it possible to modify the minimum LAI thresholds accordingly.

4.2. Could LDAS-Monde Provide Accurate Initial Conditions for Vegetation Forecasts?

In the context of NWP, assimilation of satellite observations in atmospheric models has the capacity to mitigate model deficiencies, leading to better estimates of system states. This has been the main driver of the improvement of both weather forecast skill and lead time [60]. Data assimilation is able to produce similar benefits for LSVs forecasting. Seeking to foster a link with applications, LDAS-Monde could not only be used to monitor the LSVs, but could also be integrated into a forecasting system (at different time scales) assuming that it can provide better initial conditions than a model run and that its impact lasts in time. Many applications could benefit from a better representation of the LSVs, from NWP [61], to early warning systems of, for example, agricultural drought and yield forecasts. As a first step towards such early warning systems, Figure 11 shows a comparison between LAI from the two last simulations presented in Table 1: Ini_Model and Ini_Analysis. Figure 11a,b shows monthly RMSD (R) values for the year 2016 for LAI. A strong impact is visible not only from the beginning of the two simulations, but also a few months later (up to April). The four maps of Figure 11c show RMSD differences between Ini_analysis and Ini_model from January to April. All maps are dominated by negative values (in blue), suggesting that Ini_analysis presents a better match with the observed LAI and that analysis effects last in time. Southeastern part of the domain is mainly affected, these areas are dominated by broadleaves forest as well as C3 crops. Large differences between the model and the analysis during winter time (as shown on Figure 2) suggest that the minimum LAI threshold used in the model for these vegetation types has to be revisited. Such results are strongly linked to the time of the year when the simulation is initialised by the analysis, that is, the greater the prior difference between the model and the analysis experiments will be, the stronger the impact. As for LAI, and according to Figure 2a, marked impact would be expected from July to March. It is also very promising that the impact of LAI initialisation lasts in time for several weeks or even months. Those results are in line with findings from [5] who have assimilated biomass and LAI observations at "site-level" in CLM4.5. They found that monthly forecasts (and even longer forecasts range) were improved by data assimilation compared with forecasts without data assimilation.

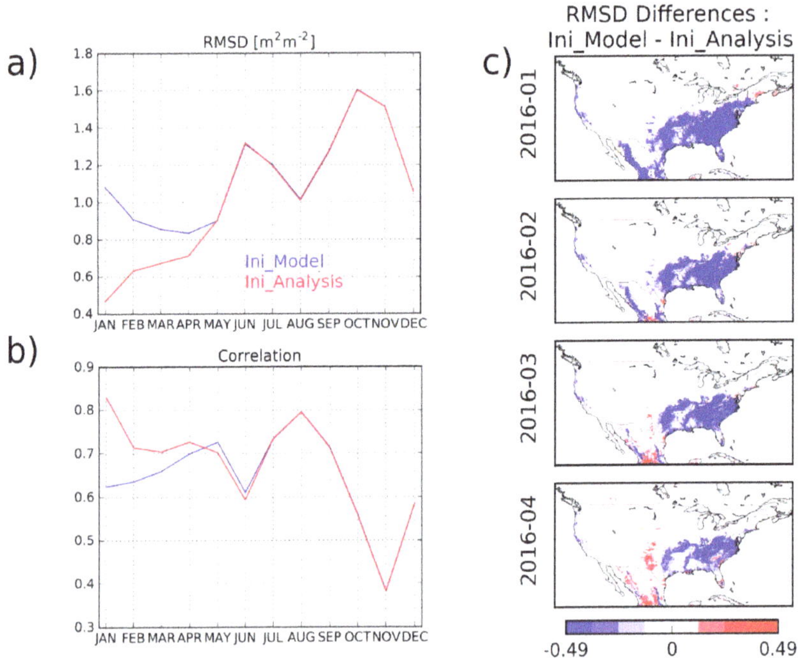

Figure 11. Seasonal (**a**) root mean square differences (RMSD) and (**b**) correlation values between observed leaf area index (LAI) and (in blue) a 12-month model run, (in red) a 12-month model run initialised by analysed conditions from LDAS-Monde. (**c**) RMSD differences values between a 12-month model run and a 12-month model run initialised by analysed conditions from LDAS-Monde.

Although the ISBA LSM does not directly represent grain yield (GY), it is assumed that the regional-scale simulations of above-ground biomass from a generic LSM can provide the inter-annual variability as a proxy for GY [62,63]. Refs. [13,64] have also found that the LDAS-Monde analysed above-ground biomass inter-annual variability was in better agreement with that of GY than its open-loop counterpart. These studies have been performed over France using straw cereal GY values from the Agreste French agricultural statistics portal (http://agreste.agriculture.gouv.fr). If more evaluations are required to assess LDAS-Monde capacity to represent GY, it paves the way towards potential productivity and yield forecasting system.

4.3. Which Alternative Data to Better Constrain LDAS-Monde?

LDAS-Monde re-analyses presented above were repeated, assimilating only SSM or LAI, the results suggested that most of the skill came from the assimilation of LAI (not shown). Assimilating LAI permits to analyze not only LAI itself, but also the root zone soil moisture. While assimilating SSM does mainly affect the first layers of soil (layer 2, 1 cm to 4 cm and layer 3, 4 cm to 10 cm), assimilating LAI has an impact on deeper layers (up to 60 cm) and is more efficient to analyse the root zone soil moisture too. This has also been suggested by [13], when analysing the ISBA LSM sensitivity to the assimilated observations through the SEKF Jacobians.

However, the LAI product used in this study is available every 10 days at best, making it less efficient to constrain the ISBA LSM, particularly in areas of the world affected by clouds for long periods of time (e.g., areas affected by the monsoon regime). Another caveat is the use of a single LAI value for all vegetation types that are represented in SURFEX. As detailed in [20], the innovation is computed from the difference between the observed LAI and the modelled LAI aggregated over all the vegetation types. Then, the Kalman gain is calculated for each individual vegetation type. The analysis increment is added to the background for each vegetation type, producing a vegetation-dependent analysis update. The vegetation dependence is introduced in the Kalman gain via the Jacobian elements. As mentioned already, the possibility of having LAI estimates for each type of vegetation is under investigation [59] and has the capacity of overcoming the above-mentioned weakness. An extension of this work is under development to assimilate in LDAS-Monde the disaggregated LAI product for each vegetation type independently, with very promising preliminary results [65].

Microwave remote sensing over land has mainly focused on soil moisture retrieval [66,67] and vegetation was mostly considered during the retrieval of surface soil moisture as a by-pass product affecting the signal penetration to the surface [68,69]. The attenuation of the signal (i.e., when passing through the vegetation) depends on the vegetation optical depth (VOD). VOD describes the attenuation of radiation due to scattering and absorption within the vegetation layer, which is caused by the water contained in the vegetation. It is function of the frequency of the microwave sensor, the water content of the plant (trunk, branches, leaves), as well as the biomass (e.g., [70–73]. VOD can be retrieved from microwave data, for example, from the L-band soil moisture and ocean salinity (SMOS) mission [74] or the C-band advanced scatterometer (ASCAT) mission on-board the meteorological operational satellite A (MetOp-A) [75,76]. VOD can be related to LAI (e.g., [77–80]. Figure 12 presents a map of temporal correlation coefficient values between modelled LAI and microwave-derived VOD from radar backscatter measurements of ASCAT [75,76] (Figure 12a), as well as their distribution (Figure 12b) for 2010–2016. High correlations values are found in large parts of the domain, with a median value of 0.57. The northern part of the domain shows R values greater than 0.7, while smaller R values (and even negative R values) are found in the southern part of the domain. Over dry soils, sub-surface scattering from the microwave signal potentially affects the VOD estimates (Wolfgang Wagner, TU WIEN, personal communication, April 2018). The same VOD dataset has a higher median R value with the observed LAI, that is, 0.88. Consequently it better correlates with the analysis (median R values of 0.61) than with the model. If a strong statistical relationship between C-band VOD and LAI can be obtained through the use of, for example, machine learning techniques (like neural network techniques, [81], it could enable obtaining a surrogate of LAI based on C-band VOD that would have the advantage of having higher temporal frequency than the current LAI product (low frequency microwave observations are not affected much by clouds and are not affected by solar elevation). Such a product could then be assimilated into LDAS-Monde to better constrain the system. Looking at such a relationship for data assimilation purposes is currently under study at CNRM.

Also, retrieved soil moisture is assimilated in LDAS-Monde, from active radar backscatter (σ_o) observations. Retrieval methods usually make use of land surface parameters and auxiliary information (like vegetation, texture, and temperature), possibly being inconsistent with specific model simulations (which also include these parameters, but potentially from different sources). Also, if retrievals and model simulations rely on similar types of auxiliary information, their errors may be cross-correlated, potentially degrading the system performance [82]. This leads to an increasing tendency towards the direct assimilation of σ_o observations (and of passive radiometer brightness temperature, T_b, as well) [83–87]. CNRM is also investigating the direct assimilation of σ_o. It requires the implementation of a forward model for σ_o in the ISBA LSM. Ref. [85] used the Water Cloud Model [88] to relate surface soil moisture from the Global Land Evaporation Amsterdam Model (GLEAM, [45,89]) to σ_o for data assimilation purposes. Within LDAS-Monde, both surface soil moisture and leaf area index could be related to the radar backscatter.

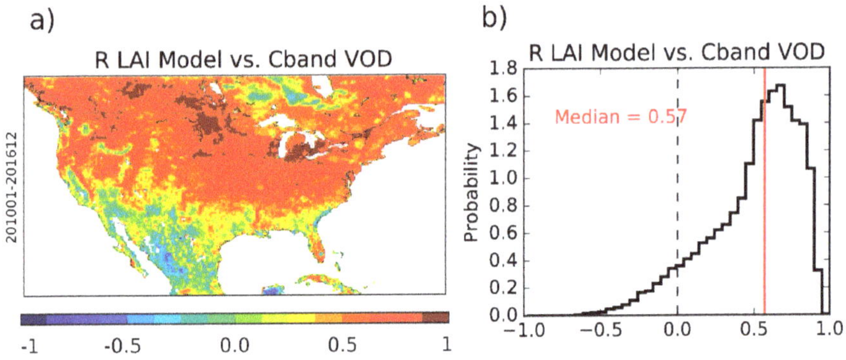

Figure 12. (**a**) Correlation coefficient values between modelled LAI and C-band vegetation optical depth (VOD) over 2010–2016, (**b**) probability distribution of the correlation coefficient values over the same period.

5. Conclusions

In this study, LDAS-Monde sequential assimilation of satellite derived surface soil moisture and leaf area index, forced by ERA-5 latest atmospheric re-analysis, was applied to the CONtiguous US domain. Ref. [28] have highlighted the added value of using the ERA-5 atmospheric reanalysis to force the ISBA Land Surface Model over the CONtiguous US for the 2010–2016 period. They found that the use of ERA-5 instead of ERA-interim leads to significant improvements in the representation of the land surface variables linked to the terrestrial water cycle (e.g., surface soil moisture, river discharges, snow depth, and turbulent fluxes), but to a rather neutral impact on land surface variables linked to the vegetation cycle (e.g., evapotranspiration, carbon uptake, and leaf area index). Assimilating satellite derived observations linked to vegetation (LAI in this application) through LDAS-Monde forced by ERA-5 not only leads to a clear improvement in the representation of the vegetation cycle in ISBA, but brings further improvement on the representation of the terrestrial water cycle. The results have highlighted the stronger impact of LAI observations assimilation with respect to soil moisture assimilation. Other vegetation-related observations such as vegetation optical depth could be used, under specific circumstances, as a surrogate of LAI limiting the negative impact of the rather low temporal frequency of the LAI product. LDAS-Monde is a powerful tool to track the evolution of land surface variables and to monitor extreme events such as agricultural drought. Since LDAS-Monde analysis is more accurate than a simple model run, it can be used to initialise a forecast experiment of the land surface variables. Preliminary results suggest that its impact on forecast experiments, in particular with respect to vegetation, is positive and lasts in time. It opens the way towards applications from monitoring to forecasting land surface states. For that purpose, LDAS-Monde will be forced by other, higher spatial resolution, ECMWF atmospheric products like the high resolution forecast (HRES, current spatial resolution of ~9 km), which also gives daily forecasts up to 10 days ahead and/or the ensemble forecast (ENS, current spatial resolution of ~18 km), giving daily forecasts up to 15 days (46 days twice a week).

Author Contributions: C.A. and S.M. conceived and designed the experiments; C.A. performed the experiments; all the authors analysed the results, C.A. wrote the paper.

Funding: This research received no external funding.

Acknowledgments: The authors would like to thank the Copernicus Global Land Service for providing the satellite-derived LAI products and the Vienna University of Technology (Vienna, Austria) for the Vegetation Optical Depth datasets. Emanuel Dutra (Instituto Dom Luiz, IDL, Faculty of Sciences, University of Lisbon, Portugal) is thanked for his help processing the ERA-5 data.

Conflicts of Interest: The authors declare no conflict of interest.

Code Availability: LDAS-Monde is a part of the ISBA land surface model and is available as open source via the surface modelling platform called SURFEX. SURFEX can be downloaded freely at http://www.umr-cnrm.fr/surfex/ using a CECILL-C Licence (a French equivalent to the L-GPL licence; http://www.cecill.info/licences/Licence_CeCILL-C_V1-en.txt). It is updated at a relatively low frequency (every 3 to 6 months). If more frequent updates are needed, or if what is required is not in Open-SURFEX (DrHOOK, FA/LFI formats, GAUSSIAN grid), you are invited to follow the procedure to get an SVN account and access real-time modifications of the code (see the instructions at the first link). The developments presented in this study stemmed on SURFEX version 8.0 and are now part of the version 8.1 (revision number 4621).

Data Availability: The ERA-5 datasets are distributed by ECMWF (http://apps.ecmwf.int/datasets/, ECMWF, last access: June 2018). The ECOCLIMAP dataset is distributed by CNRM (https://opensource.umr-cnrm.fr/projects/ecoclimap, CNRM, 2013). The SURFEX model code is distributed by CNRM (http://www.umr-cnrm.fr/surfex/, CNRM, 2016). The satellite-derived LAI GEOV1 observations are freely accessible from the Copernicus Global Land Service (http://land.copernicus.eu/global/; last access: June 2018). The ESA CCI surface soil moisture dataset is distributed by ESA (http://www.esa-soilmoisture-cci.org/, last access: June 2018, [25]). The satellite-driven model estimates of land evapotranspiration are freely accessible at http://www.gleam.eu (last access: June 2018; [46]). The upscaled estimates of gross primary production are freely accessible at https://www.bgc-jenna.mpg.de/geodb/projects/Home.php (last access: June 2018; [48]). In situ measurements of soil moisture are freely available at https://www.ncdc.noaa.gov/crn (last access: June 2018; [49]). In situ measurements of streamflow are freely available at https://nwis.waterdata.usgs.gov/nwis (last access: June 2018, USGS).

References

1. Dirmeyer, P.A.; Gao, X.; Zhao, M.; Guo, Z.; Oki, T.; Hanasaki, N. The Second Global Soil Wetness Project (GSWP-2): Multi-model analysis and implications for our perception of the land surface. *Bull. Am. Meteorol. Soc.* **2006**, *87*, 1381–1397. [CrossRef]
2. Schellekens, J.; Dutra, E.; Martínez-de la Torre, A.; Balsamo, G.; van Dijk, A.; Sperna Weiland, F.; Minvielle, M.; Calvet, J.-C.; Decharme, B.; Eisner, S.; et al. A global water resources ensemble of hydrological models: The eartH2Observe Tier-1 dataset. *Earth Syst. Sci. Data* **2017**, *9*, 389–413. [CrossRef]
3. Lettenmaier, D.P.; Alsdorf, D.; Dozier, J.; Huffman, G.J.; Pan, M.; Wood, E.F. Inroads of remote sensing into hydrologic science during the WRR era. *Water Resour. Res.* **2015**, *51*, 7309–7342. [CrossRef]
4. Reichle, R.H.; Koster, R.D.; Liu, P.; Mahanama, S.P.; Njoku, E.G.; Owe, M. Comparison and assimilation of global soil moisture retrievals from the Advanced Microwave Scanning Radiometer for the Earth Observing System (AMSR-E) and the Scanning Multichannel Microwave Radiometer (SMMR). *J. Geophys. Res.* **2007**, *112*, D09108. [CrossRef]
5. Fox, A.M.; Hoar, T.J.; Anderson, J.L.; Arellano, A.F.; Smith, W.K.; Litvak, M.E.; MacBean, N.; Schimel, D.S.; Moore, D.J.P. Evaluation of a Data Assimilation System for Land Surface Models using CLM4.5. *J. Adv. Model. Earth Syst.* **2018**. [CrossRef]
6. Sawada, Y.; Koike, T. Simultaneous estimation of both hydrological and ecological parameters in an ecohydrological model by assimilating microwave signal. *J. Geophys. Res. Atmos.* **2014**, *119*. [CrossRef]
7. Sawada, Y.; Koike, T.; Walker, J.P. A land data assimilation system for simultaneous simulation of soil moisture and vegetation dynamics. *J. Geophys. Res. Atmos.* **2015**, *120*. [CrossRef]
8. McNally, A.; Arsenault, K.; Kumar, S.; Shukla, S.; Peterson, P.; Wang, S.; Funk, C.; Peters-Lidard, C.D.; Verdin, J.P. A land data assimilation system for sub-Saharan Africa food and water security applications. *Sci. Data* **2017**, *4*, 170012. [CrossRef] [PubMed]
9. Mitchell, K.E.; Lohmann, D.; Houser, P.R.; Wood, E.F.; Schaake, J.C.; Robock, A.; Cosgrove, B.A.; Sheffield, J.; Duan, Q.; Luo, L.; et al. The multi-institution North American Land Data Assimilation System (NLDAS): Utilizing multiple GCIP products and partners in a continental distributed hydrological modeling system. *J. Geophys. Res.* **2004**, *109*, D07S90. [CrossRef]

10. Xia, Y.; Mitchell, K.; Ek, M.; Sheffield, J.; Cosgrove, B.; Wood, E.; Luo, L.; Alonge, C.; Wei, H.; Meng, J.; et al. Continental-scale water and energy flux analysis and validation for the North American Land Data Assimilation System project phase 2 (NLDAS-2): 1. Intercomparison and application of model products. *J. Geophys. Res.* **2012**, *117*, D03109. [CrossRef]
11. Kumar, S.V.; Jasinski, M.; Mocko, D.; Rodell, M.; Borak, J.; Li, B.; Kato Beaudoing, H.I.R.O.K.O.; Peters-Lidard, C.D. NCA-LDAS land analysis: Development and performance of a multisensor, multivariate land data assimilation system for the National Climate Assessment. *J. Hydrometeorol.* **2018**. [CrossRef]
12. Rodell, M.; Houser, P.R.; Jambor, U.E.; Gottschalck, J.; Mitchell, K.; Meng, C.J.; Arsenault, K.; Cosgrove, B.; Radakovich, J.; Bosilovich, M.; et al. The Global Land Data Assimilation System. *Bull. Am. Meteorol. Soc.* **2004**, *85*, 381–394. [CrossRef]
13. Albergel, C.; Munier, S.; Leroux, D.J.; Dewaele, H.; Fairbairn, D.; Barbu, A.L.; Gelati, E.; Dorigo, W.; Faroux, S.; Meurey, C.; et al. Sequential assimilation of satellite-derived vegetation and soil moisture products using SURFEX_v8.0: LDAS-Monde assessment over the Euro-Mediterranean area. *Geosci. Model Dev.* **2017**, *10*, 3889–3912. [CrossRef]
14. Noilhan, J.; Mahfouf, J.-F. The ISBA land surface parameterisation scheme. *Glob. Planet. Chang.* **1996**, *13*, 145–159. [CrossRef]
15. Calvet, J.-C.; Noilhan, J.; Roujean, J.-L.; Bessemoulin, P.; Cabelguenne, M.; Olioso, A.; Wigneron, J.-P. An interactive vegetation SVAT model tested against data from six 780 contrasting sites. *Agric. For. Meteorol.* **1998**, *92*, 73–95. [CrossRef]
16. Calvet, J.-C.; Rivalland, V.; Picon-Cochard, C.; Guehl, J.-M. Modelling forest transpiration and CO_2 fluxes—Response to soil moisture stress. *Agric. For. Meteorol.* **2004**, *124*, 143–156. [CrossRef]
17. Gibelin, A.-L.; Calvet, J.-C.; Roujean, J.-L.; Jarlan, L.; Los, S.O. Ability of the land surface model ISBA-A-gs to simulate leaf area index at global scale: Comparison with satellite products. *J. Geophys. Res.* **2006**, *111*, 1–16. [CrossRef]
18. Masson, V.; Le Moigne, P.; Martin, E.; Faroux, S.; Alias, A.; Alkama, R.; Belamari, S.; Barbu, A.; Boone, A.; Bouyssel, F.; et al. The SURFEXv7.2 land and ocean surface platform for coupled or offline simulation of earth surface variables and fluxes. *Geosci. Model Dev.* **2013**, *6*, 929–960. [CrossRef]
19. Barbu, A.L.; Calvet, J.-C.; Mahfouf, J.-F.; Albergel, C.; Lafont, S. Assimilation of Soil Wetness Index and Leaf Area Index into the ISBA-A-gs land surface model: Grassland case study. *Biogeosciences* **2011**, *8*, 1971–1986. [CrossRef]
20. Barbu, A.L.; Calvet, J.C.; Mahfouf, J.F.; Lafont, S. Integrating ASCAT surface soil moisture and GEOV1 leaf area index into the SURFEX modelling platform: A land data assimilation application over France. *Hydrol. Earth Syst. Sci.* **2014**, *18*, 173–192. [CrossRef]
21. Fairbairn, D.; Barbu, A.L.; Napoly, A.; Albergel, C.; Mahfouf, J.-F.; Calvet, J.-C. The effect of satellite-derived surface soil moisture and leaf area index land data assimilation on streamflow simulations over France. *Hydrol. Earth Syst. Sci.* **2017**, *21*, 2015–2033. [CrossRef]
22. Liu, Y.Y.; Parinussa, R.M.; Dorigo, W.A.; De Jeu, R.A.; Wagner, W.; Van Dijk, A.I.; McCabe, M.F.; Evans, J.P. Developing an improved soil moisture dataset by blending passive and active microwave satellite-based retrievals. *Hydrol. Earth Syst. Sci.* **2011**, *15*, 425–436. [CrossRef]
23. Liu, Y.Y.; Dorigo, W.A.; Parinussa, R.M.; de Jeu, R.A.; Wagner, W.; McCabe, M.F.; Evans, J.P.; Van Dijk, A.I. Trend-preserving blending of passive and active microwave soil moisture retrievals. *Remote Sens. Environ.* **2012**, *123*, 280–297. [CrossRef]
24. Dorigo, W.A.; Gruber, A.; De Jeu, R.A.; Wagner, W.; Stacke, T.; Loew, A.; Albergel, C.; Brocca, L.; Chung, D.; Parinussa, R.M.; et al. Evaluation of the ESA CCI soil moisture product using ground-based observations. *Remote Sens. Environ.* **2015**. [CrossRef]
25. Dorigo, W.; Wagner, W.; Albergel, C.; Albrecht, F.; Balsamo, G.; Brocca, L.; Chung, D.; Ertl, M.; Forkel, M.; Gruber, A.; et al. ESA CCI Soil Moisture for improved Earth system understanding: State-of-the art and future directions. *Remote Sens. Environ.* **2017**. [CrossRef]
26. Weedon, G.P.; Gomes, S.; Viterbo, P.; Shuttleworth, W.J.; Blyth, E.; Österle, H.; Adam, J.C.; Bellouin, N.; Boucher, O.; Best, M. Creation of the WATCH forcing data and its use to assess global and regional reference crop evaporation over land during the twentieth century. *J. Hydrometeorol.* **2011**, *12*, 823–848. [CrossRef]

27. Weedon, G.P.; Balsamo, G.; Bellouin, N.; Gomes, S.; Best, M.J.; Viterbo, P. The WFDEI meteorological forcing data set: WATCH Forcing data methodology applied to ERA- interim reanalysis data. *Water Resour. Res.* **2014**, *50*, 7505–7514. [CrossRef]
28. Albergel, C.; Dutra, E.; Munier, S.; Calvet, J.-C.; Munoz-Sabater, J.; de Rosnay, P.; Balsamo, G. ERA-5 and ERA-Interim driven ISBA land surface model simulations: Which one performs better? *Hydrol. Earth Syst. Sci.* **2018**, *22*, 3515–3532. [CrossRef]
29. Boone, A.; Masson, V.; Meyers, T.; Noilhan, J. The influence of the inclusion of soil freezing on simulations by a soil vegetation–atmosphere transfer scheme. *J. Appl. Meteorol.* **2000**, *39*, 1544–1569. [CrossRef]
30. Decharme, B.; Martin, E.; Faroux, S. Reconciling soil thermal and hydrological lower boundary conditions in land surface models. *J. Geophys. Res.-Atmos.* **2013**, *118*, 7819–7834. [CrossRef]
31. Voldoire, A.; Decharme, B.; Pianezze, J.; Lebeaupin Brossier, C.; Sevault, F.; Seyfried, L.; Garnier, V.; Bielli, S.; Valcke, S.; Alias, A.; et al. SURFEX v8.0 interface with OASIS3-MCT to couple atmosphere with hydrology, ocean, waves and sea-ice models, from coastal to global scales. *Geosci. Model Dev.* **2017**, *10*, 4207–4227. [CrossRef]
32. Decharme, B.; Alkama, R.; Douville, H.; Becker, M.; Cazenave, A. Global evaluation of the ISBA-TRIP continental hydrologic system, Part 2: Uncertainties in river routing simulation related to flow velocity and groundwater storage. *J. Hydrometeorol.* **2010**, *11*, 601–617. [CrossRef]
33. Decharme, B.; Alkama, R.; Papa, F.; Faroux, S.; Douville, H.; Prigent, C. Global offline evaluation of the ISBA-TRIP flood model. *Clim. Dynam.* **2012**, *38*, 1389–1412. [CrossRef]
34. Vergnes, J.-P.; Decharme, B. A simple groundwater scheme in the TRIP river routing model: Global off-line evaluation against GRACE terrestrial water storage estimates and observed river discharges. *Hydrol. Earth Syst. Sci.* **2012**, *16*, 3889–3908. [CrossRef]
35. Vergnes, J.-P.; Decharme, B.; Habets, F. Introduction of groundwater capillary rises using subgrid spatial variability of topography into the ISBA land surface model. *J. Geophys. Res.-Atmos.* **2014**, *119*, 11065–11086. [CrossRef]
36. Leroux, D.J.; Calvet, J.-C.; Munier, S.; Albergel, C. Using Satellite-Derived Vegetation Products to Evaluate LDAS-Monde over the Euro-Mediterranean Area. *Remote Sens.* **2018**, *10*, 1199. [CrossRef]
37. Mätzler, C.; Standley, A. Relief effects for passive microwave remote sensing. *Int. J. Remote Sens.* **2000**, *21*, 2403–2412. [CrossRef]
38. Draper, C.; Mahfouf, J.-F.; Calvet, J.-C.; Martin, E.; Wagner, W. Assimilation of ASCAT near-surface soil moisture into the SIM hydrological model over France. *Hydrol. Earth Syst. Sci.* **2011**, *15*, 3829–3841. [CrossRef]
39. Reichle, R.H.; Koster, D. Bias reduction in short records of satellite soil moisture. *Geophys. Res. Lett.* **2004**, *31*, L19501. [CrossRef]
40. Drusch, M.; Wood, E.F.; Gao, H. Observations operators for the direct assimilation of TRMM microwave imager retrieved soil moisture. *Geophys. Res. Lett.* **2005**, *32*, L15403. [CrossRef]
41. Scipal, K.; Drusch, M.; Wagner, W. Assimilation of a ERS scatterometer derived soil moisture index in the ECMWF numerical weather prediction system. *Adv. Water Resour.* **2008**, *31*, 1101–1112. [CrossRef]
42. Boussetta, S.; Balsamo, G.; Dutra, E.; Beljaars, A.; Albergel, C. Assimilation of surface albedo and vegetation states from satellite observations and their impact on numerical weather prediction. *Remote Sens. Environ.* **2015**, *163*, 111–126. [CrossRef]
43. Baret, F.; Weiss, M.; Lacaze, R.; Camacho, F.; Makhmared, H.; Pacholczyk, P.; Smetse, B. GEOV1: LAI and FAPAR essential climate variables and FCOVER global time series capitalizing over existing products, Part 1: Principles of development and production. *Remote Sens. Environ.* **2013**, *137*, 299–309. [CrossRef]
44. Hersbach, H.; Dee, D. *ERA-5 Reanalysis is in Production*; ECMWF Newsletter, Number 147; ECMWF: Reading, UK, 2016; p. 7.
45. Miralles, D.G.; Holmes, T.R.H.; De Jeu, R.A.M.; Gash, J.H.; Meesters, A.G.C.A.; Dolman, A.J. Global land-surface evaporation estimated from satellite-based observations. *Hydrol. Earth Syst. Sci.* **2011**, *15*, 453–469. [CrossRef]
46. Martens, B.; Miralles, D.G.; Lievens, H.; van der Schalie, R.; de Jeu, R.A.M.; Fernández-Prieto, D.; Beck, H.E.; Dorigo, W.A.; Verhoest, N.E.C. GLEAM v3: Satellite-based land evaporation and root-zone soil moisture. *Geosci. Model Dev.* **2017**, *10*, 1903–1925. [CrossRef]

47. Tramontana, G.; Jung, M.; Schwalm, C.R.; Ichii, K.; Camps-Valls, G.; Ráduly, B.; Reichstein, M.; Arain, M.A.; Cescatti, A.; Kiely, G.; et al. Predicting carbon dioxide and energy fluxes across global FLUXNET sites with regression algorithms. *Biogeosciences* **2016**, *13*, 4291–4313. [CrossRef]
48. Jung, M.; Reichstein, M.; Schwalm, C.R.; Huntingford, C.; Sitch, S.; Ahlström, A.; Arneth, A.; Camps-Valls, G.; Ciais, P.; Friedlingstein, P.; et al. Compensatory water effects link yearly global land CO_2 sink changes to temperature. *Nature* **2017**, *541*, 516–520. [CrossRef] [PubMed]
49. Bell, J.E.; Palecki, M.A.; Collins, W.G.; Lawrimore, J.H.; Leeper, R.D.; Hall, M.E.; Kochendorfer, J.; Meyers, T.P.; Wilson, T.; Baker, B.; et al. U.S. Climate Reference Network soil moisture and temperature observatons. *J. Hydrometeorol.* **2013**, *14*, 977–988. [CrossRef]
50. Kumar, S.; Reichle, R.H.; Koster, R.D.; Crow, W.T.; Peters-Lidard, C. Role of Subsurface Physics in the Assimilation of Surface Soil Moisture Observations. *J. Hydrometeor.* **2009**, *10*, 1534–1547. [CrossRef]
51. Gupta, H.V.; Kling, H.; Yilmaz, K.K.; Martinez, G.F. Decomposition of the mean squared error and NSE performance criteria: Implications for improving hydrological modelling. *J. Hydrol.* **2009**, *377*, 80–91. [CrossRef]
52. Beck, H.E.; Pan, M.; Roy, T.; Weedon, G.P.; Pappenberger, F.; van Dijk, A.I.J.M.; Huffman, G.J.; Adler, R.F.; Wood, E.F. Daily evaluation of 26 precipitation datasets using Stage-IV gauge-radar data for the CONUS, Hydrol. *Earth Syst. Sci. Discuss.* **2018**. [CrossRef]
53. Olive, W.W.; Chleborad, A.F.; Frahme, C.W.; Schlocker, J.; Schneider, R.R.; Schuster, R.L. *Swelling Clays Map of the Conterminous United States*; USGS: Reston, VA, USA, 1989.
54. Hoerling, M.; Eischeid, J.; Kumar, A.; Leung, R.; Mariotti, A.; Mo, K.; Schubert, S.; Seager, R. Causes and Predictability of the 2012 Great Plains Drought. *Bull. Am. Meteorol. Soc.* **2014**, *95*, 269–282. [CrossRef]
55. Ault, T.R.; Henebry, G.M.; De Beurs, K.M.; Schwartz, M.D.; Betancourt, J.L.; Moore, D. The False Spring of 2012, Earliest in North American Record. *EOS* **2013**, *94*, 181–182. [CrossRef]
56. Rüdiger, C.; Albergel, C.; Mahfouf, J.-F.; Calvet, J.-C.; Walker, J.P. Evaluation of Jacobians for Leaf Area Index data assimilation with an extended Kalman filter. *J. Geophys. Res.* **2010**, *115*, D09111. [CrossRef]
57. Ukkola, A.M.; De Kauwe, M.G.; Pitman, A.J.; Best, M.J.; Abramowitz, G.; Haverd, V.; Decker, M.; Haughton, N. Land surface models systematically overestimate the intensity, duration and magnitude of seasonal-scale evaporative droughts. *Environ. Res. Lett.* **2016**, *11*, 104012. [CrossRef]
58. Carrer, D.; Meurey, C.; Ceamanos, X.; Roujean, J.-L.; Calvet, J.-C.; Liu, S. Dynamic mapping of snow-free vegetation and bare soil albedos at global 1 km scale from 10 year analysis of MODIS satellite products. *Remote Sens. Environ.* **2014**, *140*, 420–432. [CrossRef]
59. Munier, S.; Carrer, D.; Planque, C.; Camacho, F.; Albergel, C.; Calvet, J.-C. Satellite Leaf Area Index: Global Scale Analysis of the Tendencies Per Vegetation Type Over the Last 17 Years. *Remote Sens.* **2018**, *10*, 424. [CrossRef]
60. Bauer, P.; Thorpe, A.; Brunet, G. The quiet revolution of numerical weather prediction. *Nature* **2015**, *525*, 47–55. [CrossRef] [PubMed]
61. De Rosnay, P.; Drusch, M.; Vasiljevic, D.; Balsamo, G.; Albergel, C.; Isaksen, L. A simplified extended Kalman filter for the global operational soil moisture analysis at ECMWF. *Q. J. R. Meteorol. Soc.* **2013**, *139*, 1199–1213. [CrossRef]
62. Calvet, J.-C.; Lafont, S.; Cloppet, E.; Souverain, F.; Badeau, V.; Le Bas, C. Use of agricultural statistics to verify the interannual variability in land surface models: A case study over France with ISBA-A-gs. *Geosci. Model Dev.* **2012**, *5*, 37–54. [CrossRef]
63. Canal, N.; Calvet, J.-C.; Decharme, B.; Carrer, D.; Lafont, S.; Pigeon, G. Evaluation of root water uptake in the ISBA-A-gs land surface model using agricultural yield statistics over France. *Hydrol. Earth Syst. Sci.* **2014**, *18*, 4979–4999. [CrossRef]
64. Dewaele, H.; Munier, S.; Albergel, C.; Planque, C.; Laanaia, N.; Carrer, D.; Calvet, J.-C. Parameter optimisation for a better representation of drought by LSMs: Inverse modelling vs. sequential data assimilation. *Hydrol. Earth Syst. Sci.* **2017**, *21*, 4861–4878. [CrossRef]
65. Munier, S.; Leroux, D.; Albergel, C.; Carrer, D.; Calvet, J.C. Hydrological impacts of the assimilation of satellite-derived disaggregated Leaf Area Index into the SURFEX modelling platform. *Hydrol. Earth Syst. Sci. Discuss.* **2018**, in preparation.

66. Entekhabi, D.; Nakamura, H.; Njoku, E.G. Solving the inverse problem for soil moisture and tem-perature profiles by the sequential assimilation of multifrequency remotely sensed observations. *IEEE Trans. Geosci. Remote Sens.* **1994**, *32*, 438–448. [CrossRef]
67. Reichle, R.H.; Entekhabi, D.; McLaughlin, D.B. Downscaling of radio brightness measurements for soil moisture estimation: A four-dimensional variational data assimilation approach. *Water Resour. Res.* **2001**, *37*, 2353–2364. [CrossRef]
68. Kurum, M.; Lang, R.H.; O'Neill, P.E.; Joseph, A.T.; Jackson, T.J.; Cosh, M. A first-order radiative transfer model for microwave radiometry of forest canopies at L-band. *IEEE Trans. Geosci. Remote Sens.* **2011**, *49*, 3167–3179. [CrossRef]
69. Kurum, M.; O'Neill, P.E.; Lang, R.H.; Joseph, A.T.; Cosh, M.H.; Jackson, T.J. Effective tree scattering and opacity at L-band. *Remote Sens. Environ.* **2012**, *118*, 1–9. [CrossRef]
70. Meesters, A.G.C.A.; Jeu, R.A.M.D.; Owe, M. Analytical derivation of the vegetation optical depth from the microwave polarization difference index. *IEEE Geosci. Remote Sens. Lett.* **2005**, *2*, 121–123. [CrossRef]
71. Liu, Y.Y.; Jeu, R.D.; McCabe, M.F.; Evans, J.P.; van Dijk, A.I.J.M. Global lon-term passive microwave satellite-based retrievals of vegetation optical depth. *Geophys. Res. Lett.* **2011**, *38*. [CrossRef]
72. Konings, A.G.; Gentine, P. Global variations in ecosystem-scale isohydricity. *Glob. Chang. Biol.* **2016**, *23*, 891–905. [CrossRef] [PubMed]
73. Tian, F.; Brandt, M.; Liu, Y.Y.; Verger, A.; Tagesson, T.; Diouf, A.A.; Rasmussen, K.; Mbow, C.; Wang, Y.; Fensholt, R. Remote sensing of vegetation dynamics in drylands: Evaluating vegetation optical depth (VOD) using AVHRR NDVI and in situ green biomass data over West African Sahel. *Remote Sens. Environ.* **2016**, *177*, 265–276. [CrossRef]
74. Fernandez-Moran, R.; Wigneron, J.-P.; De Lannoy, G.; Lopez-Baeza, E.; Parrens, M.; Mialon, A.; Mahmoodi, A.; Al-Yaari, A.; Bircher, S.; Al Bitar, A.; et al. A new calibration of the effective scattering albedo and soil roughness parameters in the SMOS SM retrieval algorithm. *Int. J. Appl. Earth Obs. Geoinf.* **2017**, *62*, 27–38. [CrossRef]
75. Vreugdenhil, M.; Dorigo, W.A.; Wagner, W.; de Jeu, R.A.M.; Hahn, D.; Marle, M.J.E. Analyzing the vegetation parameterization in the TU-Wien ASCAT soil moisture retrieval. *IEEE Trans. Geosci. Remote Sens.* **2016**, *54*, 3513–3531. [CrossRef]
76. Vreugdenhil, M.; Hahn, S.; Melzer, T.; Bauer-Marschallinger, B.; Reimer, C.; Dorigo, W.A.; Wagner, W. Assessing vegetation dynamics over Mainland Australia with Metop ASCAT. *IEEE J. Sel. Top. Appl. Earth Obs. Remote Sens.* **2016**, *10*, 2240–2248. [CrossRef]
77. Zribi, M.; Chahbi, A.; Shabou, M.; Lili-Chabaane, Z.; Duchemin, B.; Baghdadi, N.; Amri, R.; Chehbouni, A. Soil surface moisture estimation over a semi-arid region using ENVISAT ASAR radar data for soil evaporation evaluation. *Hydrol. Earth Syst. Sci. Discuss.* **2011**, *15*, 345–358. [CrossRef]
78. Kim, Y.; Jackson, T.; Bindlish, R.; Lee, H.; Hong, S. Radar Vegetation Index for Estimating the Vegetation Water Content of Rice and Soybean. *IEEE Geosci. Remote Sens. Lett.* **2012**, *9*, 564–568. [CrossRef]
79. Sawada, Y.; Tsutsui, H.; Koike, T.; Rasmy, M.; Seto, R.; Fuji, H. A Field Verification of an Algorithm for Retrieving Vegetation Water Content from Passive Microwave Observations. *IEEE Trans. Geosci. Remote Sens.* **2016**, *54*, 2082–2095. [CrossRef]
80. Momen, M.; Wood, J.D.; Novick, K.A.; Pangle, R.; Pockman, W.T.; McDowell, N.G.; Konings, A.G. Interacting effects of leaf water potential and biomass on vegetation optical depth. *J. Geophys. Res. Biogeosci.* **2017**, *122*, 3031–3046. [CrossRef]
81. Rodríguez-Fernández, N.J.; Muñoz Sabater, J.; Richaume, P.; de Rosnay, P.; Kerr, Y.H.; Albergel, C.; Drusch, M.; Mecklenburg, S. SMOS near-real-time soil moisture product: Processor overview and first validation results. *Hydrol. Earth Syst. Sci.* **2017**, *21*, 5201–5216. [CrossRef]
82. De Lannoy, G.J.M.; Reichle, R.H.; Pauwels, V.R.N. Global calibration of the GEOS-5 L-band microwave radiative transfer model over nonfrozen land using SMOS observations. *J. Hydrometeorol.* **2013**, *14*, 765–785. [CrossRef]
83. De Lannoy, G.J.M.; Reichle, R.H. Global assimilation of multiangle and multipolarization SMOS brightness temperature observations into the GEOS-5 catchment Land Surface Model for soil moisture estimation. *J. Hydrometeorol.* **2016**, *17*, 669–691. [CrossRef]

84. Han, X.; Hendricks Franssen, H.-J.; Montzka, C.; Vereecken, H. Soil moisture and soil properties estimation in the community land model with synthetic brightness temperature observations. *Water Resour. Res.* **2014**, *50*, 6081–6105. [CrossRef]
85. Lievens, H.; Al Bitar, A.; Verhoest, N.E.C.; Cabot, F.; De Lannoy, G.J.M.; Drusch, M.; Dumedah, G.; Hendricks Franssen, H.J.; Kerr, Y.H.; Kumar Tomer, S.; et al. Optimization of a radiative transfer forward operator for simulating SMOS brightness temperatures over the Upper Mississippi Basin. *J. Hydrometeorol.* **2015**, *16*, 1109–1134. [CrossRef]
86. Lievens, H.; Martens, B.; Verhoest, N.E.C.; Hahn, S.; Reichle, R.H.; Miralles, D.G. Assimilation of global radar backscatter and radiometer brightness temperature observations to improve soil moisture and land evaporation estimates. *Remote Sens. Environ.* **2016**, *189*, 194–210. [CrossRef]
87. Zhao, L.; Yang, Z.-L.; Hoar, T.J. Global Soil Moisture Estimation by Assimilating AMSR-E Brightness Temperatures in a Coupled CLM4–RTM–DART System. *J. Hydrometeorol.* **2016**, *17*, 2431–2454. [CrossRef]
88. Attema, E.; Ulaby, F. Vegetation modeled as a water cloud. *Radio Sci.* **1978**, *13*, 357–364. [CrossRef]
89. Miralles, D.G.; De Jeu, R.A.M.; Gash, J.H.; Holmes, T.R.H.; Dolman, A.J. Magnitude and variability of land evaporation and its components at the global scale. *Hydrol. Earth Syst. Sci.* **2011**, *15*, 967–981. [CrossRef]

MDPI
St. Alban-Anlage 66
4052 Basel
Switzerland
Tel. +41 61 683 77 34
Fax +41 61 302 89 18
www.mdpi.com

Remote Sensing Editorial Office
E-mail: remotesensing@mdpi.com
www.mdpi.com/journal/remotesensing

www.ingramcontent.com/pod-product-compliance
Lightning Source LLC
Chambersburg PA
CBHW051837210326
41597CB00033B/5686
* 9 7 8 3 0 3 9 2 1 6 4 0 6 *